Lecture Notes in Electrical Engineering 1036

The book series *Lecture Notes in Electrical Engineering* (LNEE) publishes the latest developments in Electrical Engineering—quickly, informally and in high quality. While original research reported in proceedings and monographs has traditionally formed the core of LNEE, we also encourage authors to submit books devoted to supporting student education and professional training in the various fields and applications areas of electrical engineering. The series cover classical and emerging topics concerning:

- Communication Engineering, Information Theory and Networks
- Electronics Engineering and Microelectronics
- Signal, Image and Speech Processing
- Wireless and Mobile Communication
- Circuits and Systems
- Energy Systems, Power Electronics and Electrical Machines
- Electro-optical Engineering
- Instrumentation Engineering
- Avionics Engineering
- Control Systems
- Internet-of-Things and Cybersecurity
- Biomedical Devices, MEMS and NEMS

For general information about this book series, comments or suggestions, please contact leontina.dicecco@springer.com.

To submit a proposal or request further information, please contact the Publishing Editor in your country:

China

Jasmine Dou, Editor (jasmine.dou@springer.com)

India, Japan, Rest of Asia

Swati Meherishi, Editorial Director (Swati.Meherishi@springer.com)

Southeast Asia, Australia, New Zealand

Ramesh Nath Premnath, Editor (ramesh.premnath@springernature.com)

USA, Canada

Michael Luby, Senior Editor (michael.luby@springer.com)

All other Countries

Leontina Di Cecco, Senior Editor (leontina.dicecco@springer.com)

**** This series is indexed by EI Compendex and Scopus databases. ****

Riccardo Berta · Alessandro De Gloria
Editors

Applications in Electronics Pervading Industry, Environment and Society

APPLEPIES 2022

Springer

Editors
Riccardo Berta
DITEN
University of Genoa
Genoa, Italy

Alessandro De Gloria
DITEN
University of Genoa
Genoa, Italy

ISSN 1876-1100 ISSN 1876-1119 (electronic)
Lecture Notes in Electrical Engineering
ISBN 978-3-031-30332-6 ISBN 978-3-031-30333-3 (eBook)
https://doi.org/10.1007/978-3-031-30333-3

This Springer imprint is published by the registered company Springer Nature Switzerland AG
The registered company address is: Gewerbestrasse 11, 6330 Cham, Switzerland

Preface

The 2022 edition of the conference on Applications in Electronics Pervading Industry, Environment and Society was held in Genova, Italy, on September 26–27, 2022, at the Polytechnic School (Villa Giustiniani-Cambiaso). During the two days, 140 registered participants, from 35 different entities (28 universities and 7 industries), discussed electronic applications in several domains, demonstrating how electronics has become pervasive and ever more embedded in everyday objects and processes.

The conference had the technical and financial support of University of Genoa, Italian Association for Electronics (SIE) and Giakova.

After a strict blind-review selection process, 25 full lectures and 27 short presentations have been accepted in six sessions focused on circuits and electronic systems and their relevant applications in the following fields: Internet of Things (S1), Hardware Acceleration (S2), Machine Learning (S3), and Transport, Energy, Security, and Health (S4).

There were also two scientific keynotes, given by Luca Benini (ETH Zurich) and Enrico Sangiorgi (University of Bologna); two industrial keynotes, by Michele Chiabrera (Inventvm Semiconductors) and Carlo Cavazzoni (Leonardo Spa); and a special industrial session by STMicroelectronics.

The contribution collected in this book, together with the special talks and the industrial events, proves that nowadays electronic systems have reached a pervasive level of penetration on all aspects of everyday life and industry, supporting all kind of domains, like mobility, security, energy, health, etc.

To fully exploit the capabilities in terms of computing, storage, and networking, a multidisciplinary approach is needed to support the design, prototyping, and testing of all new products and services.

During the conference, authors from academia and industry show the design of different electronics-enabled systems characterized by innovation, high performance, real-time operations, and budget compliance (in terms of time, cost, device size, weight, power consumption, etc.).

All these challenging aspects call for the importance of the role of ApplePies and similar conferences as a place where idea of cutting-the-edge technological solutions coming from a variety of application domains can be discussed and shared, enabling a fruitfully cross-fertilization.

The APPLEPIES 2022 conference has reached in its tenth edition, becoming a reference point for the growing research community in the field of electronics systems applications.

Riccardo Berta
General Chair

Alessandro De Gloria
Honorary Chair

Contents

Internet of Things

Enabling Predictive Maintenance on Electric Motors Through a Self-sustainable Wireless Sensor Node

Andrea Bentivogli[2(✉)], Tommaso Polonelli[1], Michele Magno[1], and Guido Comai[2]

[1] ETH Zürich, D-ITET, Gloriastrasse 35, 8092 Zürich, Switzerland
{tommaso.polonelli,michele.magno}@pbl.ee.ethz.ch
[2] MediConIngegneria s.r.l., via E. Mattei, 20, 40054 Budrio, Italy
{andrea.bentivogli,guido.comai}@mediconingegneria.it
https://www.mediconingegneria.it

Abstract. Periodic maintenance and unpredictable equipment failure of industrial machinery are expensive elements in a company's balance and potentially hazardous for human operators. Periodic inspections at predefined intervals are commonly applied to limit unplanned production downtime and safety concerns. The latest advancements in smart sensor technology enables online equipment monitoring that can directly anticipate the deterioration and incoming breakages on operating machines, reducing maintenance costs. This paper presents a *deploy and forget* sensor node for predictive maintenance on industrial electric motors, which targets three-phase asynchronous motors, supporting the data collection from multiple sensors, such as vibrations, environmental noise, temperature, and the external magnetic field. The sensor node features ultra-low-power design, achieving self-sustainability by exploiting a 4×4 cm thermal electric generator with a ΔT of 20 °C for at least 72 s. Moreover, it features short-long wireless data transfer over WiFi and the NB-IoT protocol. Results report the energy harvesting efficiency and the circuit power consumption from a real-world tests.

Keywords: predictive maintenance · energy harvesting · low power · IoT · NB-IoT

1 Introduction

Academic and industry researchers are developing new technologies for automated production, typically named Industry 4.0 [1]. Other than wireless sensor networks, machine learning, IoT devices [2] and distributed heterogeneous measurements, a new industrial trend has been arising: condition monitoring and predictive maintenance [3]. In Industry 4.0, predictive maintenance is the leading technology to increase the production yield of future factories [3,4]. Indeed, it

A. Bentivogli—March 10, 2023. This work was partially supported by Simotop s.p.a.. Moreover, the authors are grateful to Ing. Alessandro Zanna for the technical support.

R. Berta and A. De Gloria (Eds.): ApplePies 2022, LNEE 1036, pp. 3–8, 2023.
https://doi.org/10.1007/978-3-031-30333-3_1

is a key element to enable intelligent production chains, in which the equipment automatically signals failures [3,4], or predict incoming malfunctions [1].

The growing request for *deploy and forget* [5] wireless sensors is pushing researchers to design optimized low-power and long-lasting sensor nodes that are flexible, low cost, and easily scalable [6]. In the industry, electric motors are exploited on a broad scale and ubiquitous in any sector [7], from electric cars to production lines. Thus, a flexible and self-sustaining wireless sensor is increasingly required, featuring long-range or low-power wireless connectivity and a virtually infinite lifetime thanks to local energy sources and an optimized ultra-low-power electronic design [5].

This paper proposes a self-sustaining device for long-term monitoring of industrial three-phase electric motors. It enables predictive maintenance through a heterogeneous set of low-power Micro-Electro-Mechanical Systems (MEMS), i.e., one three axial accelerometer, two Hall effect sensors, a microphone, and a temperature sensor. The energy harvester relies on a ThermoElectric Generator (TEG) capable of supplying up to 77.3 mW with a ΔT of 25.4 °C, enough to support a measurement every 72 s. Moreover, the device supports long-range connectivity directly to the cloud through an NB-IoT module or to a local gateway by the WiFi protocol. A backup non-rechargeable battery supports wireless connectivity in all possible operating conditions. Power consumption in different operative modes, together with the TEG efficiency and the wireless current envelope, are reported in this work as a result of real-world measurements and in-field assessments.

2 Self-sustainable Wireless Sensor Node

The proposed design consists of four main blocks, the smart power management, the digital ICs with the microcontroller, sensors, and the dual band wireless interface. Every sub-block is designed to support the system functionally at the minimum energy cost.

The source of energy relies on a battery and a 4×4 cm TEG, which are a non-rechargeable lithium element of 2.6 Ah at 3.7 V and an SP1848-27145 Peltier. The latter works between $-60 - 125$ °C and starts to generate an active power from a temperature difference <20 °C. The TEG is tightened to the motor heatsink (hot side) supported by an aluminum wedge, optimized to maximize the contact surface with the motor frame; thus, decreasing the series thermal resistance. A passive cooler is strengthened with thermal insulators to the TEG and the wedge on the cold side. The assembly is in Fig. 1. The passive heatsink is mounted on the sensor node cold side, which is actively cooled by the motor blower thanks to a perpendicular alignment between the airflow and the aluminum slats.

The sensor block is composed of four different elements listed below. The vibrometer IIS3DWB from STMicroelectronics is selected for accelerometer analysis due to its high sampling rate (26.7 kHz) and low noise density of $75\mu g\sqrt{Hz}^{-1}$. Audio analysis is based on the SPH6611LR5H MEMS microphone

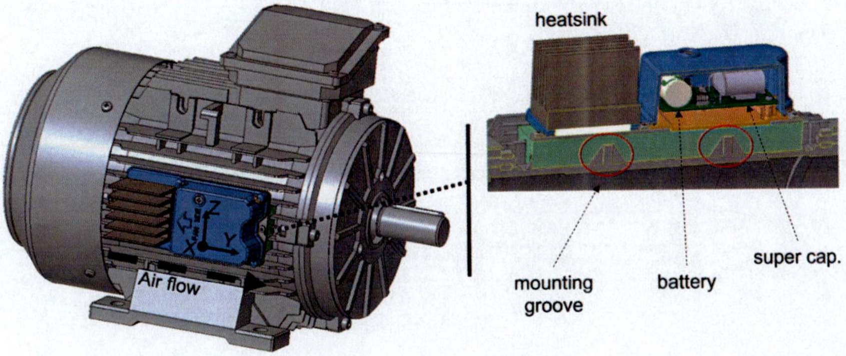

Fig. 1. Overview of the final prototype. Wedge details show the thermal and rigid coupling between the sensor node and the motor frame. Measurement exes and the installation orientation are related to reported results.

from Knowles, which features active power consumption of 435 μW. The microphone is filtered and amplified respectively in the range 5 Hz–100 kHz and with a 21× gain. Regarding the magnetic field, we selected a single axis ratiometric linear Hall effect sensor, the DRV5055, from Texas Instruments. It provides a sensitivity of $10\,\mathrm{mVG}^{-1}$ with a noise density of $2.15\,\mathrm{mG}\sqrt{\mathrm{Hz}}^{-}1$ and a maximum range of ±210 G. The sensor output is amplified and filtered by a low-noise operational amplifier with a gain of 70. The bandwidth is 20 kHz and the current consumption is 5.6 mA at 3 V. The X Y directions are measured using two DRV5055, see Fig. 1.

The digital core of the sensor relies on an STM32L4, a low-power ARM Cortex-M4f MCU family from ST Microelectronics. It is always connected to a power source, in the range of 1.8–3.6 V, which is automatically managed by a μW ORing that selects the best power source between the battery and the supercapacitor. It is driven directly by a BQ25505 from Texas Instruments through the active push-pull driver for primary and secondary battery management.

Sensors and the external memory are powered through a power switch, activated by the MCU only during the data acquisition. Also, wireless transceivers are directly driven by the STM32L4, with an intermediate buck-boost converted to stabilize the supply voltage at 3.8 V with an input range between 1.8–3.6 V and an average conversion efficiency of 85%. The NB-IoT protocol is supported by the ME310G1 module from Telit, while the WiFi link is enable by the ATWINC1500 from Microchip.

Table 1 proposes the measured generated power on a 4 kW electric motor from Simotop. At the nominal power, the TEG generates 48.3 mW, while in heavy load conditions, it reaches 77.3 mW, with a frame temperature of almost 60°C. The charging relies on the BQ25505, which features an efficiency of >80% and the possibility to support a super-capacitor.

Table 1. 4 kW electric motor characterization for TEG energy harvesting. Powered at 380 Vrms 50 Hz. Room temperature: 23 °C

Pot [kW]	T. frame [°C]	T. heatsink [°C]	ΔT	TEG [mW]
1.900	45	30.0	15.0	23.7
2.255	47.5	29.5	18.0	42.1
3.956	50.0	30.0	20.0	48.3
4.552	54.6	30.5	24.1	66.4
5.456	55.3	30.6	24.7	73.5
6.555	56.9	31.5	25.4	77.3

In the worst case, the energy harvester needs to sustain 290 mJ for the sensor acquisition and to transmit approximately 19 kB, it requires 18 J and 7 J, respectively in the worst and best wireless link condition. However, extracting the most relevant features directly on-board leads to energy transmission savings. For example, we extract the highest 15 local maxima; moreover, peaks with an intra-distance below a predefined threshold are grouped. Then, the transmission energy is reduced respectively to 3 J and 1.2 J. On the other hand, due to the increased on-board computational complexity, the digital part is estimated to ~500mJ. Hence requiring a total energy budget of 3.5 J in the worst case.

Digital and analogical circuits can operate in the voltage supply range between 3 V and 3.6 V, a voltage drop in which the rechargeable energy storage has to accumulate the estimated 3.5 J plus an energy buffer used for connections and re-transmission occurred during the wireless transmission. Then, we selected a commercial super-capacitor, SCMT22F505MRBA0, from Kyocera AVX, which features a capacitance of 5 F and a nominal voltage of 5.5 V. However, the maximum nominal voltage is never reached to prolong the component lifetime up to 50×. With this configuration, it is possible to safely connect sensors and the microcontroller directly to energy sources without needing a DC/DC converter. Moreover, considering that the MCU can operate lower to 1.8 V, the super-capacitor includes an extra energy buffer for on-board data analysis.

Taking into consideration Table 1 and the system energy requirement, the TEG energy harvester can sustain a complete acquisition every 72 s on an electric motor operating at the nominal power, 4 kW. Whereas to start from a non-operative state, power supply below 1.8 V, the energy harvester needs 11min to fully recharge the ~32 J energy storage. The minimum operating ΔT reported in Table 1 is 15 °C, below this temperature the SP1848-27145 generates a voltage below 0.7 V, where the BQ25505 cannot operate. Thus, in conditions where the monitored electric motor is active at least for 72 s within every measurement interval, the non-rechargeable battery is not used.

2.1 Power Consumption Measurement

The power consumption was characterized with an NPower Profiler Kit II from Nordic Semiconductor. In sleep, the leakage current on the battery is 2.2 µA when the super-capacitor is charged; otherwise, the current scales up to 9.3 µA

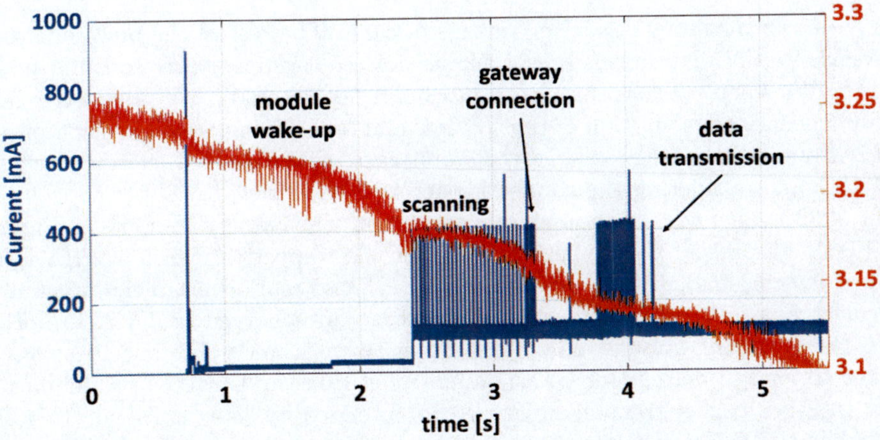

Fig. 2. WiFi current profile including the connection and the transmission of 4 kB. The super-capacitor voltage is overlapped to show the voltage envelope vs. time.

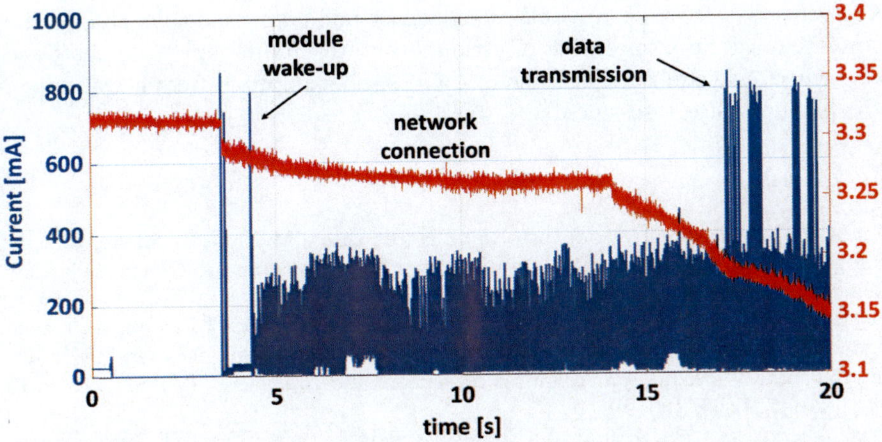

Fig. 3. NB-IoT current profile including the connection and the transmission of 4 kB. The super-capacitor voltage is overlapped to show the voltage envelope vs. time.

mainly due to the super-capacitor leakage of 8 µA. The rest of the powered system needs 22 µW. During the sensor acquisition, the system power consumption is 114 mW, where the sensor part contributes for 53 mW. We measured the dynamic energy consumption of the NB-IoT and WiFi transmission, with a payload between 4 kB and 90 kB. In the worst case, the measured energy reaches 6.2 J without considering the connection interval. However, the super-capacitor buffer of 32 J can cover this boundary case without the battery support. In Fig. 2 and Fig. 3, we present the dynamic current profile respectively with a WiFi and NB-IoT data upload, with a payload of 4 kB. These plots overlap the dynamic current compared to the super-capacitor voltage, pre-charged at 3.3 V. In both cases, the operating voltage allows the circuit to work under ICs specifications, leaving an energy margin for transmission repetitions and further communications. In Fig. 2 packet transmissions are characterized by short spikes up to 400 mA, while the RX modes requires a constant 100 mA. For the NB-IoT, Fig. 3, more than 10 s are required to connect the module to the internet, while the data exchange is characterized by a current peak of almost 800 mA.

3 Conclusion

This paper presented a *deploy-and-forget* sensor node designed to enable predictive maintenance on a operating electric motor. Results demonstrate the possibility to check the equipment status every 72 s, allowing high frequency analysis and real-time failure assessments.

References

1. Zonta, T., da Costa, C.A., da Rosa Righi, R., de Lima, M.J., da Trindade, E.S., Li, G.P.: Predictive maintenance in the industry 4.0: A systematic literature review. Comput. Indust. Eng., 106889 (2020)
2. Polonelli, T., Brunelli, D., Girolami, A., Demmi, G.N., Benini, L.: A multi-protocol system for configurable data streaming on iot healthcare devices. In: IEEE 8th International Workshop on Advances in Sensors and Interfaces (IWASI), vol. 2019, pp. 112–117. IEEE (2019)
3. Pech, M., Vrchota, J., Bednář, J.: Predictive maintenance and intelligent sensors in smart factory. Sensors **21**(4), 1470 (2021)
4. Larocque-Villiers, J., Dumond, P., Knox, D.: Automating predictive maintenance using state-based transfer learning and ensemble methods. In: IEEE International Symposium on Robotic and Sensors Environments (ROSE), vol. 2021, pp. 1–7 IEEE (2021)
5. Di Nuzzo, F., Brunelli, D., Polonelli, T., Benini, L.: Structural health monitoring system with narrowband iot and mems sensors. IEEE Sensors J. **21**(14), 16 371–16 380 (2021)
6. Cakir, M., Guvenc, M.A., Mistikoglu, S.: The experimental application of popular machine learning algorithms on predictive maintenance and the design of iiot based condition monitoring system. Comput. Indust. Eng. **151**, 106948 (2021)
7. Sangeetha, P., Hemamalini, S.: Dyadic wavelet transform-based acoustic signal analysis for torque prediction of a three-phase induction motor. IET Signal Process. **11**(5), 604–612 (2017)

Efficient Uploading of.Csv Datasets into a Non-Relational Database Management System

Matteo Fresta(✉), Francesco Bellotti, Alessio Capello, Marianna Cossu, Luca Lazzaroni, Alessandro De Gloria, and Riccardo Berta

Department of Electrical, Electronic and Telecommunication Engineering (DITEN), University of Genoa, Via Opera Pia 11a, 16145 Genoa, Italy
matteo.fresta@elios.unige.it, {francesco.bellotti, alessandro.degloria,riccardo.berta}@unige.it, {alessio.capello, marianna.cossu,luca.lazzaroni}@edu.unige.it

Abstract. Measurement-oriented non-relational databases often have a fixed structure schema to better manage and guarantee integrity of their data. However, this leads to a redundancy of field values into the database or does not allow storing most of the existing measurement files. We propose a solution to massively load various format.csv datasets without requiring any user modification of the original file. The core of the solution is given by a key-value pair.json file mapping the database resources to the.csv columns, and adding further context information, if not already present. The solution aims at effectiveness, efficiency and flexibility. The implemented module has been successfully tested in a couple of use cases using existing datasets.

1 Introduction

As the spreading of *Internet of things* (IoT) [1] technologies is bringing the need to deal with Big Data [2] not only in the cloud but also in the edge, therefore, it is important to develop efficient tools for end-to-end machine processing of IoT measurements already from the edge [3]. Each sensor or entity generates an impressive amount of measurements and often sends the data to cloud-based databases for future use. One problem encountered is that there are many datasets for machine learning IoT (e.g., Kaggle [4]), but poor end-to-end integration into measurement database usable for training machine learning (ML) models. In this context, we aim at investigating a solution that takes as input "any" dataset in.csv format and uploads it to a state-of-the-art schema-based, non-relational IoT database. We focus on the csv format as it is by far the most widely used across ML dataset.

2 Background

Browsing the popular dataset collection website, it appears that the vast majority of datasets used for ML (and not involving multimedia material) are contained in csv files

R. Berta and A. De Gloria (Eds.): ApplePies 2022, LNEE 1036, pp. 9–15, 2023.
https://doi.org/10.1007/978-3-031-30333-3_2

and share the same tabular format, where each row corresponds to a measurement and the columns provide the feature and label values. The main differences between one csv dataset and another can be the contents of the different fields, as they can be single or multiple elements and might contain numeric values or strings. Another aspect to take into account is when a dataset also contains descriptive values of the environment context, such as the name of the device and the location of the detection, or when it is based only on the numerical values recorded.

On the database side, the measurement-oriented DB class is emerging, that save data using a fixed non-relational schema to better describe the environment from which the measurements are generated. This mode makes it possible to better characterise the context scenarios and allows the information to be better structured for the subsequent processing by analysts and other applications.

Among such databases, we have chosen Measurify in order to test our idea of a.csv loader module. Measurify is an open source, state of the art API framework already used in a number of applications in real-world contexts [3, 5, 6], Measurify aims to represent the application context and its elements as interrelated software objects on top of which applications can be built. These objects are modeled as resources with own models and functionalities, accessible through a RESTful API interface. These resources abstract common elements in IoT environments, such as: Thing, Feature, Device, and Measurement: a *Thing* represents the subject of a Measurement; a *Feature* describes the (usually physical) quantity that is measured; each Feature consists of one or more *Items*, each one with a name and a unit; a *Device* is a tool that provides Measurements related to a Thing; finally, a *Measurement* represents the actual value of a Feature measured by a Device on a particular Thing. Each entity in the database can be tagged (i.e., specified) with one or more *Tags*.

This data storage design mode has to be addressed by any module that wants to insert measurements into it. In other words, having a fixed structure implies that all entities defined as mandatory must be respected. Particularly, Measurify organises the resources in a very generic way, adding information on the environment from which they were taken in addition to the numerical information. This complicates the insertion, but makes it possible to characterise datasets from completely different areas by adding useful information to contextualise the values.

For instance, considering the example of a sensor measuring temperature, Measurify mandates that, in addition to the simple value, the server should receive, for each measurement, such information as Device, Feature, and Thing. In a csv file, having all these elements would be redundant, and, as a matter of fact, none of the existing datasets do match this structure. So, there is the need for an adaptation, that we address in the following section.

3 Designing the Dataset Loading Module

This dataset loading module must accept a variety of datasets and must allow the mapping of the structure of each dataset to the resource structure of the database. In order to find a solution which respects the Measurify's structure, it was decided to accompany the csv file with an additional configuration file which will serve as a map for the server on how to interpret the dataset file.

The database server (i.e., Measurify) will thus receive two files, one is the original.csv dataset, the other is a.json file containing the description of how the content of the dataset should be stored in the database. This second file represents the mapping between the Measurify's resource types and the.csv columns, and also offers the possibility of adding context information not present in the original dataset. As a response to the storage request, the server returns a report indicating which rows were saved correctly and which ones ended up in an error.

The code below provides an example of a description file, in which each Measurify entity (resource type) mandatory in a Measurement has been linked to a value. The file has a key-value structure mapping each Measurify entity either to a number (representing the corresponding column number in the csv) or string value, which is assigned to all the measurements in the dataset.

```
{ "thing":1,

"feature":2,

"items":{ "feature_A":[3],

          "feature_B":[4],

          "feature_C":[3,4]

        },

"tags":[5,6],

"device": "control-unit"}
```

The example code declares that the first column of the csv contains the name of the "thing" to which the measurement is related. The type of each measurement in the dataset (i.e., its feature) is available in column 2. The example shows a significant degree of flexibility. In fact, depending on their actual feature value, measurements may be constituted of items of different dimensions. In the example, "feature_A" and "feature_B" are mono-dimensional, and their values are in columns 3 and 4, respectively, while "feature_C" is bi-dimensional, and its values are in column 3 and 4 of the.csv file.

The example also shows the possibility of creating array entities, whose values span multiple columns, such as an array of two tags (in columns 5 and 6), that are usable to specify each measurement.

Finally, a string can be inserted in place of the column number, so that this key-value pair will be set for all measurements in the csv file, which is the case of the "control-unit" device in the example. This helps especially when some mandatory entities are missing in the csv dataset.

3.1 Unit Test Dataset

For development purposes, a unit test was built, verifying the insertion of a.csv file with a single row (i.e., measurement), with the following values, in four columns:

thing-1	tag-1	device-1	10

Fig. 1. Unit test measurement

To give meaning to this line, the description file was generated:

```
{"thing":1,"tags":[2],"device":3,"feature":"feature-1",
"items":{ "feature-1":[4]}}
```

Once the mapping is computed, the server stores one measurement of type "*feature-1*", with one item having the numerical value of 10 Fig. 1.

4 Results

In this section, the case of two different datasets found online are presented, to show the functionality of the developed system, which allows seamlessly loading datasets of different structure.

4.1 Smart City

The first dataset contains air quality data for the current year from three control units located in the municipality of Bologna, Italy.

The data are made available by ARPAE Emilia Romagna [7]. An example of those measurements is provided here Fig. 2:

987792	2022-01-01T00:00:00	GIARDINI MARGH.	35	PM10
988241	2022-01-01T00:00:00	GIARDINI MARGH.	2	O3
989200	2022-01-01T00:00:00	VIA CHIARINI.	17	NO2

Fig. 2. Three lines of Bologna.csv dataset

The last column contains the pollutant type and we can consider it as the *Feature.* We can also see that the first column presents the unique code of each measurement and the last but one column contains the value measured in μg/m3. We argue that these two terms should be part of the *items* of the feature. So, each measurement will contain two numerical values, the first one is the value measured in air, the second is the identification code of the measurement. The date column will be used to fill an entity *startdate*, which records the time in which the measurement was taken, and the third column indicates the *Thing*, i.e., in this case, the location where the monitoring unit is located. A Device column is missing in this dataset, so a common name to all measurements will be assigned (e.g. "Monitor_Bo"). Thus, the.json file that will need to be attached to the.csv dataset for the Measurify server to properly store the corresponding measurements will be as follows:


```
{ "feature":5,
"items":{ "PM10":[4,1],
"O3":[4,1],
"NO2":[4,1]},
"startdate":2,
"thing":3,
"device":"Monitor_Bo"}
```


For the Thing, Startdate and Feature keys, the numbers of the corresponding column have been specified, while a string common to all measurements has been assigned as the Device. The features are specified in column 5 of the.csv file, and for each feature the.json file indicates in which column(s) the values are to be found. In this example all the item values are in the same columns (4 and 1).

4.2 Biomedical Dataset

The third dataset taken into consideration comes from the biomedical field, and takes into account the cardiac measurements of various patients, with the addition of some demographic data. This dataset was chosen because it only has "*Items*" values, and we consider it as representative of a wide set of datasets. Here we report the header and a couple of example rows Fig. 3:

age	sex	cp	trestbps	chol	fbs	restecg	thalach	exang	oldpeak	slope	ca	thal	target
63	1	3	145	233	1	0	150	0	2.3	0	0	1	1
37	1	2	130	250	0	1	187	0	3.5	0	0	2	1
41	0	1	130	204	0	0	172	0	1.4	2	0	2	1
56	1	1	120	236	0	1	178	0	0.8	2	0	2	1

Fig. 3. Header plus four lines of Heart.csv dataset

The description file adds the values of all the missing parameters:


```
{ "thing":"Patients",

"device":"Heart_monitor",

"items":{
"heartValues":[1,2,3,4,5,6,7,8,9,10,11,12,13,14]},

"feature":"Heart_disease_indicators" }
```


In this case, all the columns have been mapped as *items* values and the context information has been added as strings in order to comply with the Measurify schema.

5 Conclusions and Future Works

We have presented an end-to-end, domain-independent solution to store dataset files in.csv format inside schema-based non-relational databases. The solution relies on accompanying the file with a key-value file mapping the columns of the.csv file to the resources (i.e., the structure) of the target dataset. The solution has been designed with a view to supporting a high degree of user-configurability. Tests with two datasets, that we deem quite well representative of those typically available online for ML training and data science analysis, have been passed successfully in terms of effectiveness (all the data were successfully inserted and could be easily retrieved), efficiency (preparation of the additional mapping file was not complicate and could be done in few minutes) and flexibility (we argue that the two analyzed types of datasets cover a wide variety of datasets). On the downside, the additional information over pure.csv upload, which is due to meet the structure of the host Measurify server, introduces redundancy, thus a wider memory footprint and bandwidth.

As future work, we are interested in a more in-depth analysis on the.csv dataset files that can and that cannot be uploaded to the database, possibly finding solutions to address also more complicate cases than those considered until now. Furthermore, beside the datasets, we are also interested in extending the Measurify's support to edge-cloud ML by supporting the storing and versioning of ML models.

References

1. https://en.wikipedia.org/wiki/Internet_of_things
2. Cai, H., Xu, B., Jiang, L., Vasilakos, A.V.: IoT-based big data storage systems in cloud computing: perspectives and challenges (2017) IEEE Internet Things J. **4**(1), 75–87 (2017)
3. Berta, R., Bellotti, F., De Gloria, A., Lazzaroni, L.: Assessing versatility of a generic end-to-end platform for IoT ecosystem applications. Sensors **22**(3), 713 (2022) https://doi.org/10.3390/s22030713
4. Kaggle: Your Machine Learning and Data Science Community https://www.kaggle.com/

5. Berta, R., Kobeissi, A., Bellotti, F., De Gloria, A.: Atmosphere, an open source measurement-oriented data framework for IoT. IEEE Trans. Industr. Inf. **17**(3), 1927–1936 (2021). https://doi.org/10.1109/TII.2020.2994414
6. Bellotti, F., et al.: Managing big data for addressing research questions in a collaborative project on automated driving impact assessment. Sensors **20**, 6773 (2020)
7. Agenzia regionale per la prevenzione, l'ambiente e l'energia dell'Emilia-Romagna (ARPAE) https://opendata.comune.bologna.it/explore/dataset/centraline-qualitaaria/information/?disjunctive.agente_atm

Microcontroller Based Edge Computing for Pipe Leakage Detection

Fulvio Lo Valvo, Giacomo Baiamonte, and Giuseppe Costantino Giaconia(✉)

DEIM, Università degli Studi di Palermo, viale delle Scienze – (edificio 9), 90128 Palermo, Italy
{fulvio.lovalvo,giacomo.baiamonte02}@community.unipa.it,
costantino.giaconia@unipa.it

Abstract. In the embedded system field a correct resource management is crucial, especially in systems that use Machine Learning (ML) algorithms. The resources in that case are in terms of memory, footprint and time used to compute the tasks. The system should be able to be both accurate and compact although the precision is directly proportional to the memory used to storage data. In this paper we describe a comparison between three ML models implemented in a microcontroller, with an application scenario devoted to monitor a Water Distribution Network by using vibrations input and trying to investigate the computational complexity of each tested solution.

1 Introduction

Nowadays almost half of the drinkable water is wasted while distributing it from the source to the user. The causes of leakages are several: pipes and infrastructures aging, the sign of wear, weather etc.…

Due to the difficulties in recognizing the leak, also because the small leakages are the most common and the hardest to find, it takes a long time before the damaged pipe is detected and the problem solved. Acoustic instruments allow to find also these non-visible leaks, because the change in pipe's vibrations can be captured from an accelerometer sensor [1] and analysed by a ML hardware component.

The aim of this work is to compare the computational complexity and memory footprint on board of three ML models which are the Multilayer Perceptron (MLP), the Support Vector Machine (SVM) and the Classification Tree (CT) and evaluate the performance of the algorithms in the estimation of the leakage, when they are implemented in a constrained microcontroller-based environment.

The computation complexity is a measure of the number of compute resources, which are time and space, that the algorithm uses when running a compiled code. The time complexity measures how much time the algorithm takes to generate the output according to a predetermined input size. Space complexity, instead, measures the quantity of memory the algorithm occupies to execute and store all the data according to a fixed input size.

R. Berta and A. De Gloria (Eds.): ApplePies 2022, LNEE 1036, pp. 16–22, 2023.
https://doi.org/10.1007/978-3-031-30333-3_3

Typical pattern recognition system includes features extraction. This step is determined during the development of the ML algorithm and is important to group the information into a single vector and increase the robustness of the system.

1.1 Multilayer Perceptron

An artificial neural network (ANN) is a type of data processing which represent a model inspired by biological neuronal system [2]. Among the different types of ANN, MLP is a common choice for classification. Each element that composes the model is the neuron, which, depending on the activation threshold, sums up the inputs and generates the output. According on the depth, there will be more hidden layers between the input layer neurons and the output one.

1.2 Support Vector Machine

A SVM is a supervised ML model that uses classification algorithm for two-group of classification problems, regression and classification which is our case [3]. Compared to newer algorithms like neural networks, they have two main advantages: higher speed and better performance with a limited number of training samples; this makes the algorithm very suitable for text classification problems in case of a low number of samples.

1.3 Decision Tree

A deterministic decision tree is a rooted ordered binary tree (T). Each internal node of T is labeled with a variable x_i and each leaf is labeled with a value of 0 or 1. An input x deterministically establishes the leaf and the output. The computation will start at the root of the tree, and it will end at the final leaf. For this model the complexity is the depth of the tree, that is the number of queries made on the worst-case input. More detailed description of these ANN can be found in [4].

2 Hardware Used

Water leaks produce audible noise typically up to a few kHz. Since the noise is a vibration, it can be sensed using an accelerometer with a sampling rate higher than the senses bandwidth. The obtained signal constitutes the dataset and can then be elaborated and fed into the ML algorithm.

The dataset has been collected in a laboratory experimentation consisting of a water circuit featuring a pipe with 4 taps of different sizes and distances from the sensor acting as for simulating different leaking conditions [5]. The accelerometer used is the IIS3DWB [6] which presents an integrated digital conversion circuit capable of producing 26,6 ksps and its configurable acceleration range has been set to $\pm$ 2g. The evaluation board used is equipped with a STM32L476RG microcontroller, providing an ARM Cortex-M4 processor unit, 2 MB of Flash memory and 640 KB of RAM.

ML models have been developed and trained using sklearn [7] and keras [8] libraries in python. To train the algorithm, the dataset has been split into windows 12288 samples long (about 0,46 s) generating a training set of 3600 signal windows, evenly split between leak and no leak conditions. 3000 samples have been used during training and validation and 600 during testing.

The three selected features have been extracted automatically. All these features have been computed starting from the modulus of the acceleration and removing the gravity component. Their scattering diagram is presented in Fig. 1. As it's shown into the 3D diagram, even though the number of points is the same for the two classes, it is quite clear to distinguish between leakage and no leakage conditions.

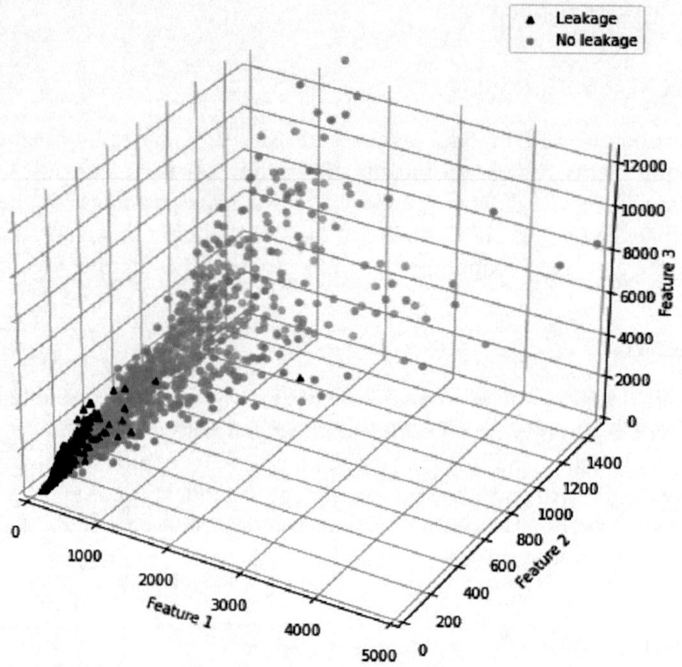

Fig. 1. Scattering diagram of the selected features extracted.

XCUBE-AI has been used as the main tool to evaluate the performance and the memory footprint of each model, directly on the microcontroller. The models from sklearn are converted to ONNX (an open format built to represent machine learning models) in order to be compatible with the STM32 utility, while the MLP network – developed using keras library - is natively compatible.

Since the purpose of this study is to highlight the differences between various ML models implemented on microcontrollers, the training of the algorithms has not been optimized for this application. Therefore, this approach aims to remove the influence of model optimization and the parameters of interest (computational complexity and hardware requirements in our hypothesis) have been evaluated in the worst case of each model type. To this purpose all models have been obtained by exploring several activation functions and neural topologies until an acceptable and comparable error was found for each algorithm.

The decision tree model has been trained without constraints in terms of possible number of branches and leaves. Since decision trees try to find rules for the whole dataset, they usually produce rules for all rare cases too, thus increasing the size of the model. For these reasons, the final trained model includes 190 leaves and 379 nodes.

Training the SVM model required 767 iterations. The points used to calculate the SVM classifier were 584 *leak* and 604 *no leak* samples.

The MLP's topology instead featured two consecutive dense layers with 512 neurons each, followed by a flatten layer. The chosen activation function for the dense layers has been a *selu* (scaled exponential linear units) whereas the last activation function was a *sigmoid*. These choices take into account the need for a small size of the MLP network and the low error during inference. During training, dropout layers have been used to reduce overfitting.

3 Results

To better show the differences in complexity between different models, they have been trained until the errors are comparable in at least one of the two classes. The characteristics of the models are presented in Table 1.

Table 1. Performance comparison among the three proposed models.

Model	Error on *leaks*	Error on *no leaks*	Number of MACCs	Ram (KB)	Memory flash (KB)
MLP	12%	8%	277515	6.09	1020
SVM	24.3%	5.3%	35640	8.13	48.16
DT	6%	8.3%	13	2.1	16.96

After the models have been trained and converted with XCUBE-AI, they have been tested on a complete record to see how they performed in a long-time scenario.

In Fig. 2 are represented the outputs of the three models. With a value of 1 the system is in a regular condition, whereas a value of 0 the system presents a leakage. The reference value contains the true classification of the recording, where a tap has been opened and then closed. The SVM model has the highest error in no leak conditions, this is reflected in the corresponding plot with the presence of many false leak classifications.

The output signal of each of the ML algorithms is usually noisy and it would be difficult to use in a real-case scenario, so it has been filtered by calculating the mean of 32 consecutive prediction values and assigning a value of 1 or of 0 if the result is higher or lower than its middle value. The resulting filtered signals are shown in Fig. 3. The filter introduces a delay of about 15 s that is relatively negligible for the use case of leakage detection, since a fast response is not required.

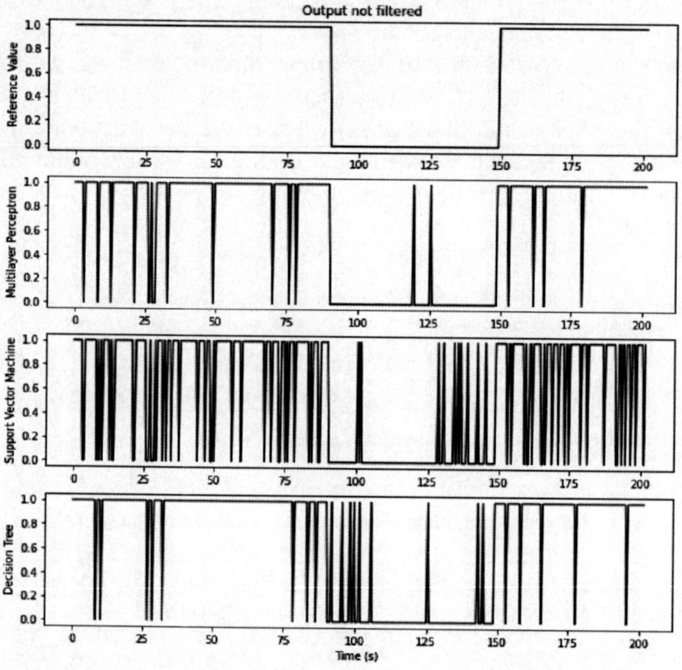

Fig. 2. Rounded inference results of the three proposed models.

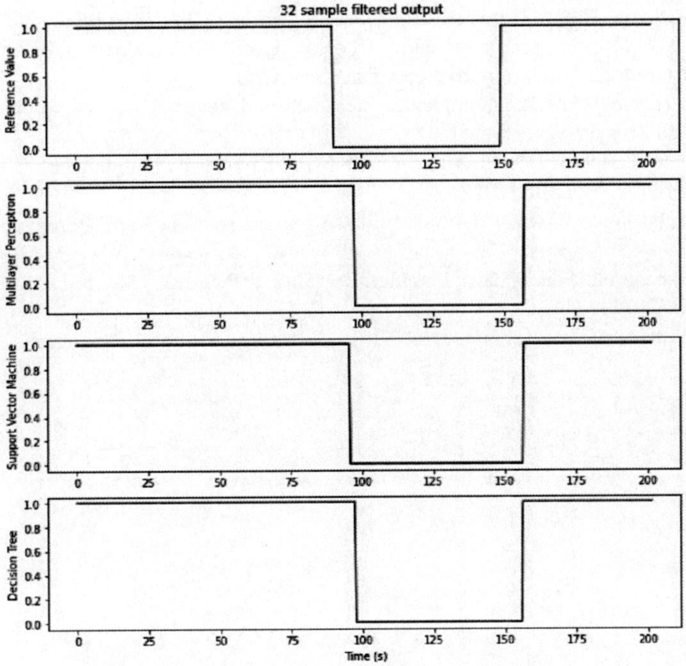

Fig. 3. Filtered inference of the three proposed models.

4 Conclusions

When trying to solve problems with a ML approach it is important to select the correct algorithm for the specific scenario. In the embedded world this choice is also influenced by the occupied memory, the features extracted and energy constraints.

For the leakage detection use case, the extracted features are very representative of the data acquired in real experiment, this allows to use low resources algorithm such as decision trees with accuracies comparable with more complex systems. The filter in the last step introduces a delay with the advantage of reducing noise on the output signal of the algorithm, increasing performances with a relatively low computation effort. Future work will include the analysis of the computational complexity of the three algorithms in terms of execution time and CPU cycles, as well as exploring new possible ML models and extracted features in order to improve the overall performances while lowering hardware resources request.

Acknowledgments. This work has been partially financed by the Project "TiSento" (Azione 1.1.5. - POC Sicilia 2014/2020 Asse 1 - PO FESR 2014/2020).

References

1. Mistretta, L., et al.: Embedding Monitoring Systems for Cured-In-Place Pipes. Applepies (2016)

2. Various, Artificial Neural Network. https://en.wikipedia.org/wiki/Artificial_neural_network
3. Stecanella, B.: Support Vector Machines (SVM) Algorithm Explained. https://monkeylearn.com/blog/introduction-to-support-vector-machines-svm/
4. Buhrman, H., de Wolf, R.: Complexity measures and decision tree complexity: a survey. Elsevier (2002)
5. Lo Valvo, F., Giaconia, G.C.: Algoritmi di Machine Learning implementati su Sistemi a Microcontrollore. Palermo (2022)
6. STMicroelectronics. Iis3dwb datasheet (2020) https://www.st.com/resource/en/datasheet/iis3dwb.pdf
7. Pedregosa, F., et al.: Scikit-learn: Machine Learning in Python. J. Mach. Learn. Res. **12**(85), 2825–2830 (2011)
8. Chollet, F. and a. others, Keras, GitHub (2015). https://github.com/fchollet/keras

eSysId: Embedded System Identification for Vibration Monitoring at the Extreme Edge

Federica Zonzini(✉), Matteo Zauli, and Luca De Marchi

Advanced Research Center on Electronic Systems for Information and Communication Technologies "Ercole De Castro", University of Bologna, 40136 Bologna, Italy
{federica.zonzini,matteo.zauli7,l.demarchi}@unibo.it

Abstract. Enabling extreme edge processing functionalities will lead a breakthrough in the development of the next generation of Structural Health Monitoring (SHM) systems, thanks to the adoption of sensor–near data analtycs which will make the structural inference process faster and more advantageous in terms of power consumption and data volume. In this work, we specifically endorse this paradigm in the context of vibration–based diagnostics by proposing a novel, intelligent accelerometer sensor combining, in an embedded device, advanced edge data analytics implementing System Identification algorithms, and energy–aware custom hardware supporting it. The effect of the bit–depth quantization of the collected signal on the quality of the retrieved structural parameters is assessed; moreover, a cost–benefit analysis is also encompassed, showing how the developed solution might be globally more advantageous from an energy point of view, reaching up to 10x power saving if compared with standard alternatives.

Keywords: Energy saving · Embedded System Identification · Extreme Edge Processing · Smart Accelerometer Sensor · Vibration Monitoring

1 Introduction

Structural Health Monitoring (SHM) aims at assessing the integrity condition of structures throughout their life cycle and in their normal operative conditions [1]. Conventionally, SHM is implemented with sensor networks transmitting raw data to central aggregators for information processing. Scaling up this approach for structures with very large and dense networks is one of the major technical challenges currently under investigation [1,2]. To overcome this issue, recent advancements in the field of micro–system design promoted the investigation of new solutions built on the extreme edge computing paradigm: i.e., information processing in strict proximity where it is sensed [3].

Such an edge–enabled approach presents three main benefits: i) it compresses all the meaningful structural parameters in a batch of damage-sensitive features

R. Berta and A. De Gloria (Eds.): ApplePies 2022, LNEE 1036, pp. 23–29, 2023.
https://doi.org/10.1007/978-3-031-30333-3_4

(scalars), thus offering a very powerful means for data compression and network load reduction, ii) it reduces the data–to–user transfer time, hence supporting real–time diagnostics, and iii) it ensures long–term functionalities by extending sensor battery life–cycle, which is one of the most crucial challenges in autonomous systems. In this scenario, the contribution of this work is to present the HW/SW co–design of a novel monitoring architecture featuring the three above–mentioned characteristics. In particular, the proposed solution consists in the extreme edge embodiment of System Identification (henceforth referred to as eSysId) algorithms on a purposely developed *intelligent*, wireless accelerometer sensor. The device is designed to make eSysId feasible, notwithstanding the complexity of the entailed algorithmic routines. The main reason for preferring eSysId among the list of available estimators is that it presents a twofold advantage: firstly, it offers an indirect means for network load reduction, which is a very desirable functionality in large–scale applications; secondly, it presents a very potent tool for dynamics analysis and, by extension, for vibration–based structural assessment, which is the prospective application scenario considered in this work.

The final objective is to extend the preliminary studies discussed in [5] and [6] by:

(i) considering the effect of the bit–depth quantization on the quality of the computed eSysId parameters and, in turn, of the computed structural parameters
(ii) quantitatively estimating the saving on the power budget while running eSysId on the developed prototype board. To this end, a comparison with alternative data compression strategies classically adopted in the context of vibration–based inspection is also encompassed

The content of the manuscript is organized as follows. In Sect. 2, the principles behind eSysId and the embedded programming strategies are explained, then introducing, in Sect. 3, the novel wireless acceleration board and its constitutive elements. The effect of the bit-width of the input data on the quality of the retrieved eSysId parameters is assessed in Sect. 4.1, while the power savings in comparison with alternative compression strategies are discussed in Sect. 4.2. Conclusions are drawn at the end.

2 The Software: eSysId Implementing ARMA Models

System identification refers to an ensemble of signal processing techniques applied in a broad range of application domains, including vibration–based structural characterization. In particular, its realization via output–only autoregressive models is compatible with extreme edge implementations, for two main reasons: 1) it is built on a filter representation which can be readily implemented via simple multiply–and–accumulate operations and 2) it does not require the measurement of the input exciting force, which is, instead, assumed equal to a zero–mean white noise Gaussian term $e(t) \sim \mathcal{N}(0, \sigma_e^2)$ with variance σ_e^2. SysId belongs to the class of *parametric* spectral analyzers, meaning that time series can be modeled via stationary processes whose spectral density function $S(f)$ is uniquely determined by a set of N_p *model parameters*, the latter corresponding to the filter taps used to model the system transfer function. By knowing $S(f)$, the identification of frequency–dependent structural features, also referred to as modal parameters (e.g., the natural frequencies of vibration) can be performed by looking at peak–related quantities in the power spectrum.

Autoregressive with Moving Average (ARMA) models are particularly effective in modeling the structural dynamics, proving better performances with respect to pure Autoregressive ones. Its mathematical representation at a generic instant $k \in \{0, \dots, K-1\}$ for a K–long discrete–time signal $s[k]$ sampled with periodicity T_s is given by:

$$s[k] + \sum_{i=1}^{q} \theta_i s[k - iT_s] = e[k] + \sum_{s=0}^{p} \gamma_s e[k - sT_s] \tag{1}$$

in which q and p indicate the number of parameters preserving memory of the past p input and q output instances, while θ_i and γ_s are the feedback and feed–forward taps of the corresponding filter. p and q are also known as the orders of the filter numerator and denominator polynomials, while their summation $N_p = p + q + 1$ equals the total amount of model coefficients to be determined. Therefore, the SysId problem turns into the estimation of the N_p parameters, a goal which can be efficiently fulfilled via ordinary least–squares algorithms. Finally, the system power spectrum can easily be estimated as the square power of the associated filter transfer function:

$$S(f) = \left| \frac{1 + \sum_{s=0}^{p} \gamma_s e^{-j2\pi f s T_s}}{1 - \sum_{i=0}^{q} \theta_i e^{-j2\pi f i T_s}} \right|^2 \tag{2}$$

Noteworthy, despite its simple algebraic formulation, porting the entire SysId workflow on resource–constrained processors as the ones typically equipped on low–cost and low–power computing architectures is a challenging task. To this end, an embedded system–oriented version, the eSysId procedure, has been presented by the authors in [5]: it leverages strategies from the dense linear algebra processing field to shrink both the algorithmic and memory complexity, proving to be portable on microcontrollers.

3 The Hardware: An Intelligent Accelerometer Sensor

A novel smart accelerometer node has been designed to support the eSysId processing. The device consists of four main components (see the prototype in Fig. 1):

1. the Digi XBee 3 wireless communication module based on the 802.15.4 protocol [7], ensuring two-way data transfer between the sensor node and a central network gateway. Among the main features of this component, the following can be listed: a consumption of 17 mA and 135 mA in reception and transmission mode, respectively, while only 50 µA are drained in sleep mode;

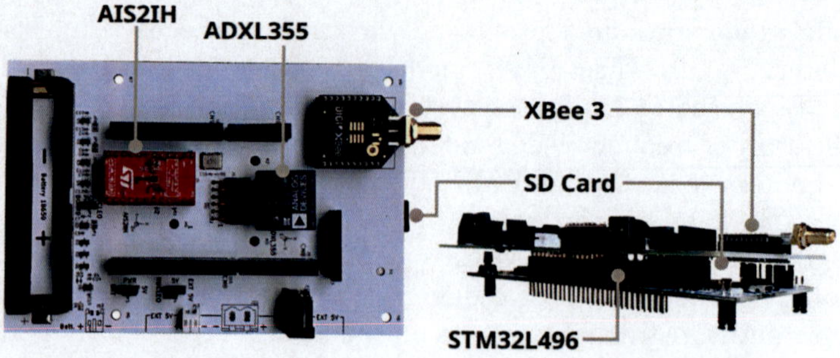

Fig. 1. Top and lateral view of the developed accelerometer sensor node.

2. a high–performance STM32l496 microcontroller unit (MCU), which integrates an ARM Cortex–M4 core operating at a frequency up to 80 MHz, with a Floating Point Unit, a Digital Signal Processing instruction set, 320 KiB of SRAM and 1 MB of Flash memory. It also provides a consistent variety of power–saving modes, which makes it appealing for ultra–low–power applications: e.g., 15 nA in shutdown mode, 115 nA in standby mode and 1.1 μA in stop mode;
3. a combination of two tri–axial MEMS accelerometers, necessary to efficiently trigger the acquisition of vibration signals based on energy thresholds while maximizing the power saving. One is the ADXL355 digital accelerometer, that is characterized by an ultra–low noise density of 22.5 $\mu g/\sqrt{Hz}$, 20–bit Analog–to–Digital Conversion (ADC) resolution and a maximum sampling rate of 4000 Hz. The second one is the AIS2IH device, featuring a sub-1 μA current consumption, 12–bit resolution in low–power mode and an output data rate from 1.6 Hz to 1600 Hz;
4. an SD card necessary to expand to storage capability.

4 Experimental Validation

4.1 Effect of Bit–Depth Quantization

Defining the format and the depth of the data resolution might have a huge impact on the required memory space necessary for temporary data storage, while also determining the dimension of the payload to be delivered over the sensor network, i.e., the lower the number of bits used to represent a single piece of information, the higher the gain in the sensor performances. On the other hand, an excessively coarse quantization might hamper the possibility to achieve an accurate structural characterization. As such, an experimental campaign has been executed in which three different types of data resolutions have been considered, namely `int8`, `int16` and `float`, whose power spectra (computed according with Eq. (2)) have been compared with the ones obtained via offline data processing performed in the MATLAB environment working with full–precision `double`.

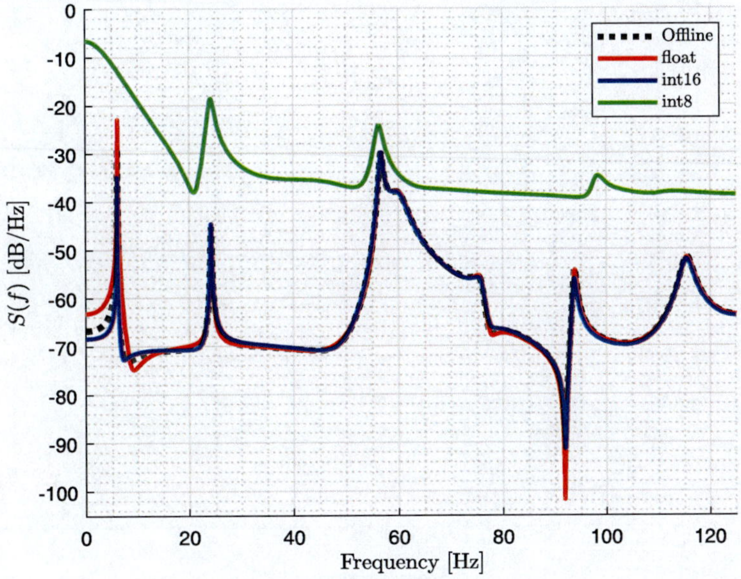

Fig. 2. Spectral estimation results for different quantization strategies.

To this end, the sensor node was installed on a laboratory steel beam under ground motion excitation to mimic real-world scenarios [6]. After sampling, data were processed directly aboard via an ARMA model requiring $N_p = 17$[1] parameters; then, these parameters were transmitted to a remote PC serving as a central gateway, where the spectral profile was reconstructed and the structural parameters were finally retrieved. Results are depicted in Fig. 2, from which it is immediate to observe that both `float` (red curve) and `int16` (green curve) quantization strategies are really effective in preserving the meaningful structural information, the latter being even more precise with respect to the target (black dotted curve) offline computation. This outcome might be due to the fact that, when reducing the bit-depth, minor details, which might be associated with high frequency noise, are implicitly filtered out by the quantization process, thus yielding to a smoother spectral profile. Conversely, despite showing a good alignment in correspondence of the second and third peak, 8 bit precision is not sufficient in pursuing the same task.

4.2 Cost–Benefit Analysis

Finally, the actual gain in running eSysId on the edge accelerometer board was numerically quantified by computing the energy expenditure assuming a duty-cycle of one hour. The latter value has been chosen since it is commonly adopted in

[1] This quantity has been estimated in a preliminary phase via the Bayesian Information Criterion, as presented in [5].

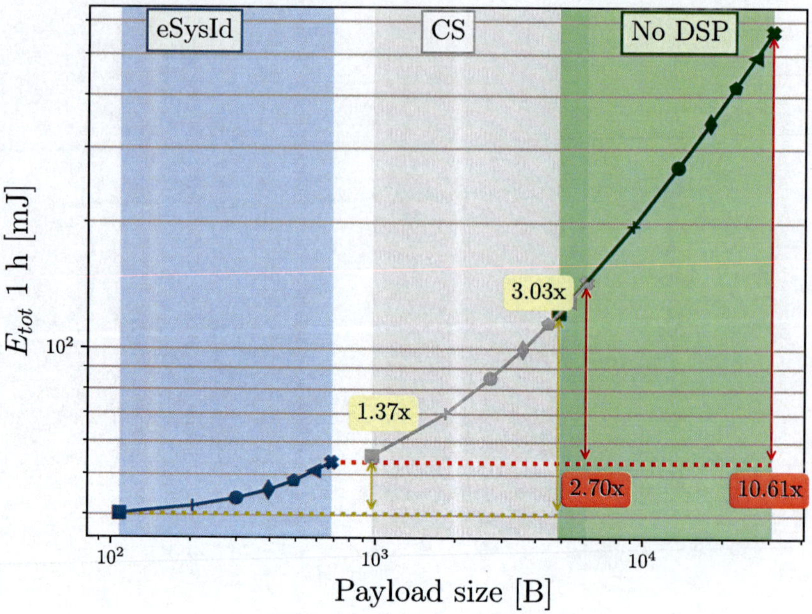

Fig. 3. Energy expenditure over one hour for different sensor node configurations: eSysId running on the MCU (blue curve), CS running aboard (grey curve) and no DSP enabled (green curve).

large–scale real–field scenario considering the long inertia of structural changes on the dynamic response of the monitored asset [5]. In order to make the analysis as much accurate as possible, the overall power consumption due to data sampling, processing and transmission are thoroughly estimated by taking into consideration the power requirements in Sect. 3 for increasing payload sizes. Beside eSysId, two additional processing frameworks are considered: compressed sensing techniques (label 'CS'), which is a well–known solution in the field [8], and the absence of any edge processing (label 'No DSP'). Concerning CS, an MCU implementation has been pursued as well assuming a static compression matrix statically loaded in the device memory at the network start up phase and consisting of random entries. The compression factor for eSysId has been fixed to 45, which corresponds to the ratio between the total number of collected samples and the number of model parameters, while a compression factor equal to 5 has been imposed for CS since higher rates are barely exceed in CS–based applications [5].

In Fig. 3, the energy curves are depicted, proving how the combination of the eSysId software and the hardware characteristics of the designed sensor node make the extreme edge implementation of eSysId energetically more efficient than CS and up to 10 times more efficient than centralized processing solutions. More specifically, even in the worst configuration, eSysId is at least 1.37x and 3.03x times more convenient with respect to its competitors, reaching a gain up to 2.70x and 10.61x for very long data payloads.

5 Conclusions

In this work, the edge implementation of system identification (eSysId) algorithms for vibration–based structural inspection is presented and validated on a novel wireless accelerometer sensor. The proposed architecture was shown to be more efficient in terms of energy consumption with respect to state–of–the–art alternatives available in the field. Furthermore, an analysis has been conducted, in which the impact of the bit–depth precision has been specifically evaluated with respect to the consistency of the spectral estimation.

References

1. Alokita, S., et al.: Recent advances and trends in structural health monitoring. Structural health monitoring of biocomposites, fibre-reinforced composites and hybrid composites, pp. 53–73 (2019)
2. Abdulkarem, M., Samsudin, K., Rokhani, F.Z., A Rasid, M.F.: Wireless sensor network for structural health monitoring: a contemporary review of technologies, challenges, and future direction. Structural Health Monit. **19**(3), 693–735 (2020)
3. Abner, M., Wong, P.K.Y., Cheng, J.C.: Battery lifespan enhancement strategies for edge computing-enabled wireless Bluetooth mesh sensor network for structural health monitoring. Autom. Constr. **140**, 10435 (2022)
4. Reynders, E.: System identification methods for (operational) modal analysis: review and comparison. Arch. Comput. Methods Eng. **19**(1), 51–124 (2012)
5. Zonzini, F., Dertimanis, V., Chatzi, E., De Marchi, L.: System identification at the extreme edge for network load reduction in vibration-based monitoring. IEEE Internet of Things J. (2022)
6. Zauli, M., et al.: A novel smart sensor node with embedded signal processing functionalities addressing vibration-based monitoring. In: European Workshop on Structural Health Monitoring, pp. 1000-1008. Springer, Cham (2023). https://doi.org/10.1007/978-3-031-07322-9_101
7. Digi: XBee 3 Zigbee 3 RF Module. https://www.digi.com/resources/library/data-sheets/ds_xbee-3-zigbee-3
8. Zonzini, F., Carbone, A., Romano, F., Zauli, M., De Marchi, L.: Machine learning meets compressed sensing in vibration-based monitoring. Sensors **22**(6), 2229 (2022)

Microcontroller Based Portable Measurement System for GaN and SiC Devices Characterization

Alberto Vella, Giuseppe Galioto, and Giuseppe Costantino Giaconia(✉)

Department of Engineering, University of Palermo, Viale delle Scienze Ed. 9, 90128 Palermo, Italy
{alberto.vella,giuseppe.galioto,costantino.giaconia}@unipa.it

Abstract. The aim of this work is to design and implement an embedded system capable to characterize some relevant figures of merit of Gallium Nitride and Silicon Carbide transistors in a wide range of frequencies. In particular, the designed system is focused on measuring the parameters involved in both the power loss phenomena and the reliability of the device during switching operations. Both the employment of a low-cost microcontroller unit and the equivalent-time sampling technique contributed to make the measurement system flexible, affordable and capable of enhanced sampling performance. As a result, different GaN and SiC devices were compared, in order to characterize the behavior of the measured quantities with respect to the switching frequency.

1 Introduction

Reducing power losses in conversion systems has always been a critical issue in most power electronics applications. nowadays, systems with low losses, and high efficiency, allow for a better power conversion with enhanced performance and lower costs. As explained in [1], main losses in power electronics systems are caused by the different phenomena related to each switching device (hemt, mosfet, bjt, igbt) but all of them can be seen as the sum of several terms:

$$P_L = P_{on} + P_s + P_{off} \tag{1}$$

where, P_L is the total power loss, P_s represents commutation losses due to switching operations, P_{on} is the conduction loss during the on-state and P_{off} is the leakage term collecting losses related to the off-state.

Since the drain current is almost zero (~μA), when a device is off, the related P_{off} term can be generally neglected, then the total power loss can be well approximated with the sum of switching and conduction losses:

$$P_L \cong P_{on} + P_s \tag{2}$$

R. Berta and A. De Gloria (Eds.): ApplePies 2022, LNEE 1036, pp. 30–38, 2023.
https://doi.org/10.1007/978-3-031-30333-3_5

Conduction losses occur during the transistor's on-state and can be expressed as:

$$P_{on} = R_{DSon} I_0^2 \frac{t_{on}}{T_S} \tag{3}$$

In which R_{DSon} is the on-resistance of the switching device, I_0 is the on-state current, t_{on} is the on-state time and T_S is the switching period. So, as it can be seen from the previous equation, conduction losses depend on the on-resistance and, consequently, they can be minimized using switching devices with low R_{DSon}, like SiC MOSFET or GaN HEMT because the theoretical limit of the on-resistance for these devices is much lower if compared with the limit of Silicon devices.

Switching power losses occur when, during the on- or off- state transition, the drain-source voltage V_d and the drain current I_0 are simultaneously different from zero.

$$P_s = \frac{1}{2} V_d I_0 f_s \left(t_{c(on)} + t_{c(off)}\right) \tag{4}$$

It is possible to notice that P_s linearly depends, as well as on the switching frequency f_s, on the switching time $t_{c(on)}$ and $t_{c(off)}$, which are obviously related to the parasitic capacitance of the switching device. For this reason, the input capacitance C_{iss} is a common parameter to be considered and its value can strongly affect power conversion performances. So, it is desirable to have a device with small input capacitance, in order to reduce the switching times and, therefore, switching power losses.

Furthermore, transistors' reliability can strongly be affected by threshold voltage instability phenomena. A negative threshold voltage shift can bring to unwanted turn-on of the device, while a positive shift can make switching times longer and affect the on-resistance, so this phenomenon can increase the device's power losses if there is an increase of R_{DSon}, this will particularly bring to larger conduction losses.

All these considerations lead to the necessity of a direct measurement of some important figures of merit of the switching devices, such as the on-resistance, the input capacitance and the threshold voltage variations. It is also important to evaluate the behavior of these quantities with respect to the switching frequency, in order to characterize the device in a wide range of frequencies.

The presented measurement system provides a low-cost portable solution compared with commercially available products, as in [2, 3], usually more than an order of magnitude more expensive. In particular, the present work employs a separated power supply, in contrast to the integrated one of the aforementioned products. Beside this, it has to be stated that the maximum frequency for the capacitance measurement reaches 10 MHz, while the present work currently reaches a maximum of 1 MHz - even if this limit may be pushed ahead in future work. Finally, thanks to the equivalent time sampling, the developed embedded platform can achieve a high equivalent sampling frequency with a simpler and cheaper design, while keeping a high grade of flexibility.

2 Embedded System Design

2.1 System Requirements

To perform all the measurements of interest, the designed system must satisfy some requirements due to the number of quantities to be measured and to the nature of the involved signals. The system, in fact, must be able to sample at the same time at least two different signals between the gate voltage, the drain current and the drain voltage. All the measurement must be performed in a wide range of switching frequency. In particular, the selected frequency range is between 10 kHz and 1 MHz, in order to investigate the behavior of high frequency devices like the Gallium Nitride HEMTs.

2.2 Designed System

In order to satisfy all the requirements, most importantly the frequency requirements, the STM32H743ZI microcontroller has been selected. This choice was due to the relatively high clock frequency (480 MHz), the presence of three independent ADCs and the presence of a high resolution timer (HRTIM), which can use the microcontroller clock as a clock source.

The diagram in Fig. 1 and the circuits' schematics in Appendix show the measurement system structure, where the drain current, drain voltage and gate voltage sensing circuit as well as the gate driver are included.

The load resistor act also as sensing element for the drain current. The R_G selection sets the gate voltage equivalent time constant. The gate voltage sensing circuit is essentially a level shifter in order to adapt the gate signal to the MCU's ADC input voltage range.

Finally, a power relay is used to decouple the on-resistance measurement from the others setup, since within the drain voltage sensing circuit, a power diode could affect the drain voltage falling time.

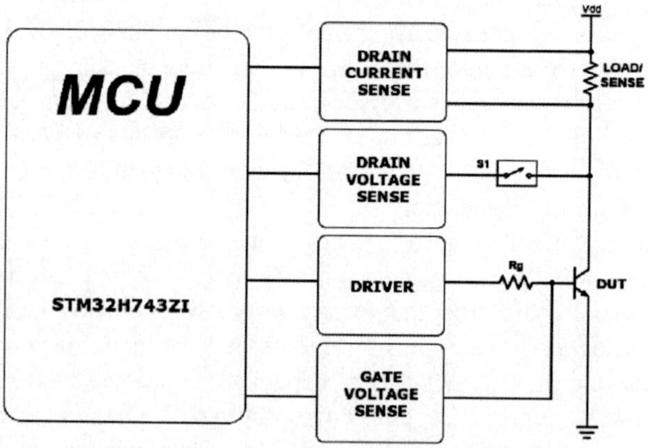

Fig. 1. Measurement system block diagram

The threshold voltage was extracted using the *constant current* method explained in [4]. So, the threshold voltage is calculated as the gate voltage corresponding to a predetermined drain current. To select the constant drain current I_C, the absolute drain current measurement uncertainty, u_{I_D}, is considered. The selected drain current value is chosen equal to 100 times u_{I_D}. So, during the OFF-ON transition the constant current threshold is I_C, while during the ON-OFF transition the current threshold is the.

ON-state drain current value, lowered by a quantity equal to I_C.

The drain-source on-resistance is measured starting from the drain current I_D and the drain voltage V_D, since R_{DSon} is equal to the ratio between these two quantities. A gate signal with 90% duty cycle is used in order to switch the device and during conduction state, an average value of V_D and I_D samples is calculated, and these averaged values are used to extract R_{DSon} measurements.

The input capacitance is instead calculated using the following equation, which relates this quantity to the rising time of the gate voltage.

$$C_{iss} = \frac{\tau}{R_g} \tag{5}$$

where τ is the time constant of the gate voltage transient and R_g is the external gate resistance.

In order to sample signals which frequency can reach 1 MHz using a microcontroller system, normal techniques would bring to use ADC with very high sampling frequency, while we wanted to keep it relatively low. In fact, the maximum sampling frequency is limited by the embedded ADC conversion time. For this reason, the *equivalent-time sampling* method has been implemented. This sampling technique can be used only if the measured signal is periodic. It consists in sampling the input signal with a sampling period equal to the signal period plus an additional time defined as $T_{ET} = T/n$, where T is the input signal period and n is the desired number of samples. So, the sampling period can be written as:

$$T_S = T + T_{ET} \tag{6}$$

Exploiting the equivalent-time sampling technique allows to sample a signal with a sampling frequency even slightly lower than the signal's frequency. In order to enhance system's sampling performances the conversion is triggered by the embedded HRTIM and the converted data is transferred to memory with the aid of the internal DMA. Moreover, the ADC conversion and DMA transfer are triggered by internal signals, achieving a reduced execution time without interrupt routines. In this way, the computational burden of the processor is lowered and the sampling period can be minimized.

3 Experimental Results

The experimental tests carried out in order to characterize GaN and SiC devices were performed on the following three products: a 900 V 15 A GaN-Cascode, a 650 V 15 A GaN E-mode and a 900 V 11.5 A SiC MOSFET.

The measurements for GaN devices extend to 1 MHz, while for SiC MOSFET the maximum achieved switching frequency is equal to 200 kHz. Moreover, the bias conditions of the tested devices are different: for GaN Cascode and E-mode the power supply is 60 V and 0.4 A, while for SiC MOSFET is 50 V and 2 A. This choice is due to the different characteristics of SiC devices, whose saturation needs a higher drain current level. Furthermore, SiC device is driven with a 0–15V gate voltage, while for GaN devices a driving voltage of ± 5V is already enough to push an pop from saturation.

In Table 1 the absolute measurement uncertainties of the considered quantities are shown. The absolute uncertainties are calculated as follows:

$$u = \sqrt{\sum_i \left(\frac{df}{dx_i}\right)^2 u_{x_i}^2} \tag{7}$$

where f is the function that defines the measurement, x_i are the measured samples, u_{x_i} is the absolute uncertainty related to each single measurement, while u is the calculated absolute uncertainty.

Since for the threshold voltage extraction the constant current method is employed, the absolute uncertainty cannot be calculated and so the relative one is presented. This was calculated as:

$$u = \sqrt{\sum_i u_{x_i}^2} \tag{8}$$

Table 1. Measurement absolute uncertainty of the devices' figures of merit

Figure of merit	Uncertainty	Type of uncertainty
Ciss	35 pF (worst case)	Absolute
RDSon	0.022 mΩ	Absolute
ΔV_{th}	0.007%	Relative

As shown in Fig. 2, the input capacitance of all the tested devices approximately doubled their low frequency value in the considered range of frequencies. The physical phenomenon behind this behavior resides in the electron trapping occurring within the switching period.

In Fig. 3 the extracted relation between R_{DSon} and the switching frequency is shown. The measured values are referred to the on-resistance at a switching frequency of 10 kHz. This choice is motivated by the difference in the R_{DSon} values between the devices under test, allowing for a better visualization of the frequency behavior. In SiC device, the variation is minimal, while GaN devices show a more consistent increase in the high-frequency region, approximately 25–30% at 1 MHz.

The last characterized figure of merit is the threshold voltage variation during a switching period, as shown in Fig. 4. The post-processed values of ΔV_{th} are calculated as difference between the positive threshold voltage variations during the on-state and the negative ones during the off-state. GaN Cascode shows an almost null variation of its threshold voltage as a function of switching frequency. GaN E-mode's threshold voltage variation increases at high switching frequencies of hundreds of mV. Finally, the highest variation occurs for the SiC device, whose threshold voltage increases over 4 V at increasing switching frequencies.

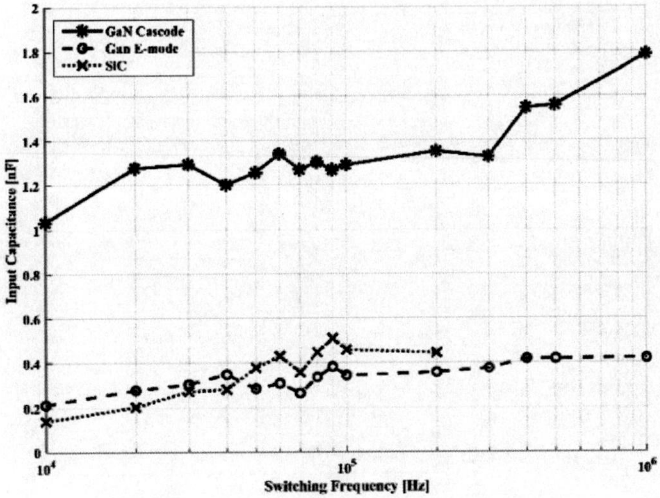

Fig. 2. Input capacitance vs switching frequency

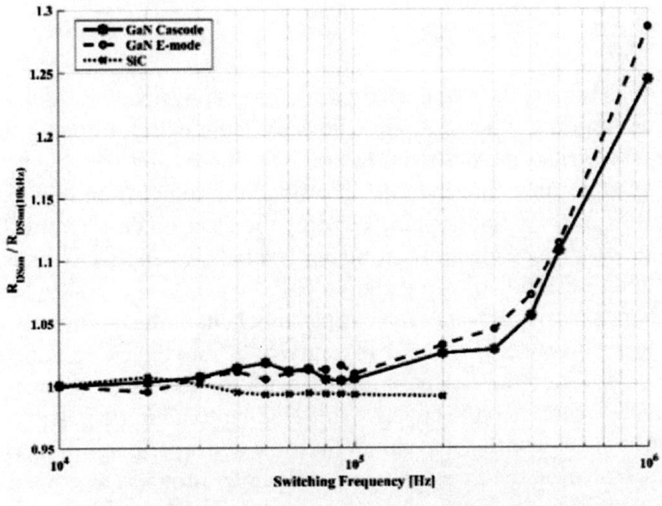

Fig. 3. $R_{DSon}/R_{DSon(10\,kHz)}$ vs switching frequency

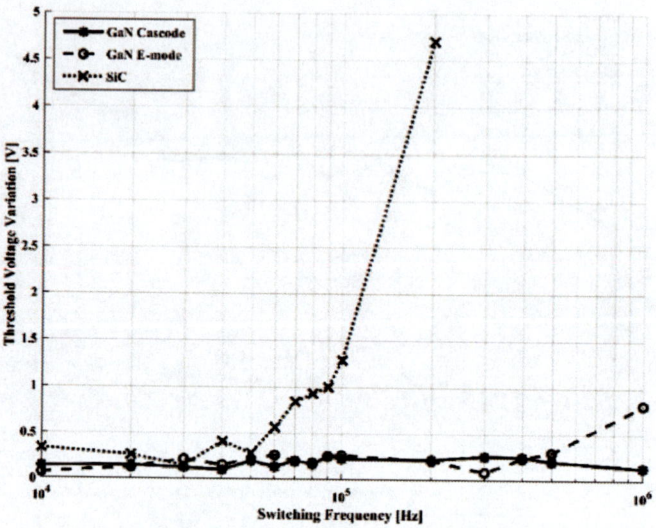

Fig. 4. ΔV_{th} vs switching frequency

4 Conclusions

The frequency behavior of the extracted figures of merit physically resides in the electron trapping and de-trapping times. For GaN devices, the electron trapping time is orders of magnitude lower than the de-trapping time [5]. Increasing the frequency, both the off-state and on-state times are reduced, but since the trapping time is much smaller, this will not affect the electron trapping but only the electron de-trapping. This leads to a bigger amount of trapped electrons at higher frequencies and so to an increase of the on-resistance and threshold voltage variation [6].

For SiC devices, during off-state, electrons tunnel out of the oxide causing a negative shift of V_{th}. During on-state, electrons can tunnel back into the oxide, causing a positive shift of V_{th} [7, 8]. Since the device is driven with 0−15V gate voltage, the positive variation is much higher than the negative one leading to a higher positive ΔV_{th}.

The present work shows the behavior of the most relevant figures of merit of GaN and SiC transistors. However, the investigation is still under progress to deepen the involved phenomena and their causes [9].

Acknowledgments. This work has been partially financed by the EC-H2020 Project "GaN4AP" (Proposal n. 101007310 - H2020-ECSEL-2020-1-IA-two-stage).

Appendix: Driving and Sensing Circuits Schematics

In this appendix paragraph, the driving circuits of the measurement setup related to Fig. 1 are briefly described. For the drain current sensing circuit, as shown in Fig. 5, in order to protect the operational amplifier (in differential configuration) from the high voltage of the DUT's power supply, two voltage dividers are applied at both terminals of the sensing resistor.

In Fig. 6 the drain voltage sensing circuit is shown. Here, the main issue when measuring the DUT's drain current is the high difference between the off-state and on-state voltage. In order to overcome this problem, as represented, the voltage across a diode, placed between the drain and ground, is measured. This solution is implemented to avoid the attenuation of the drain voltage to an acceptable value for the amplifier input, since it would make the drain current in on-state indistinguishable from the noise. Consequently, in the off-state the amplifier input voltage will be the diode threshold and it will be not considered, while in conduction state, the true drain voltage can be sampled.

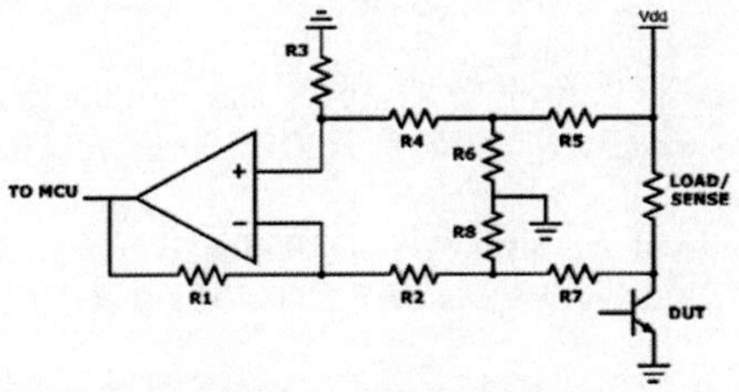

Fig. 5. Drain current sensing circuit
$R_1 = R_2 = R_3 = R_4 = 10\,k\Omega$
$R_5 = R_7 = 604\,k\Omega R_6 = R_8 = 40.2\,k\Omega$

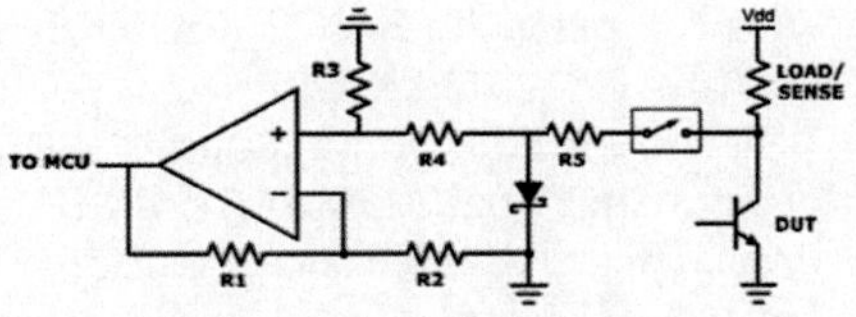

Fig. 6. Drain voltage sensing circuit
$R_1 = R_2 = R_3 = R_4 = 10\,k\Omega\ R_5 = 2.5\,k\Omega$

In Fig. 7 the gate driver is shown. It is a non-inverting summing amplifier with the aim of shifting the input level from the MCU to the driving voltage range of the DUT.

Finally, in Fig. 8, the gate voltage sensing circuitry is presented. Here, an operational amplifier is employed in non-inverting unity-gain summing configuration. The output voltage is fed to the MCU with a simple voltage divider.

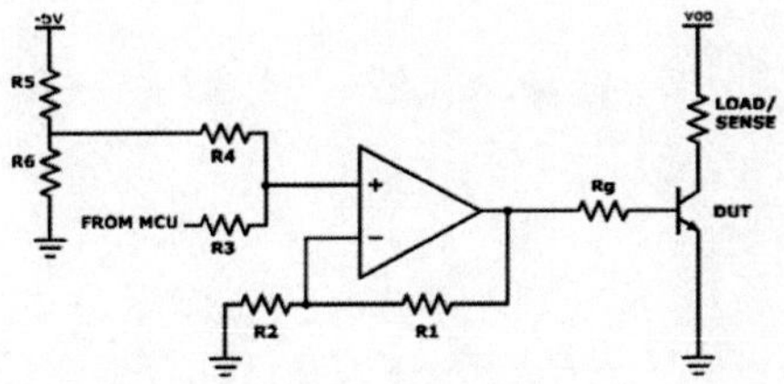

Fig. 7. Gate driving circuit (e.g. for GaN)
$R_1 = 50\,k\Omega R_2 = 10\,k\Omega\ R_3 = R_4 = 100\,k\Omega$
$R_5 = 31.6\,k\Omega\ R_6 = 22.1\,k\Omega$

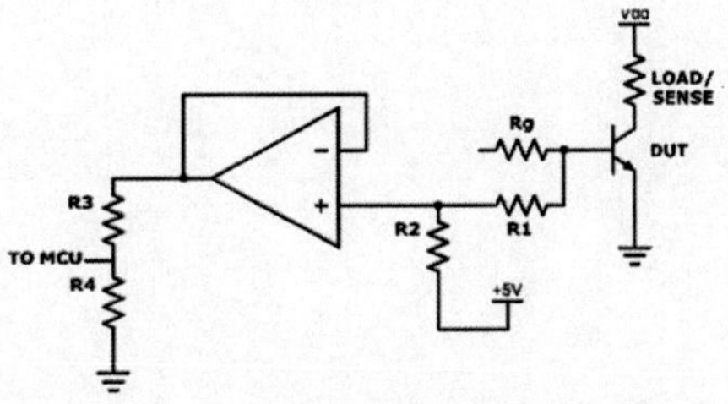

Fig. 8. Gate voltage sensing circuit (e.g. for GaN) $R_1 = 50\,k\Omega\ R_3 = 59\,k\Omega\ R_4 = 100\,k\Omega$

References

1. Mohan, N., Underland, T.M., Robbins, W.P.: Power Electronics Converter, Applications and Design, John Wiley & Sons, Inc. (2003)
2. Tektronix, 4200A-SCS Parameter Analyzer Datasheet
3. Keysight Technologies, B1505A Power Device Analyzer/Curve Tracer Datasheet
4. Ortiz-Conde, A., et al.: Revisiting MOSFET threshold voltage extraction methods, Microelectronics Reliability **53**(1), 90–104 (2013) ISSN 0026–2714, https://doi.org/10.1016/j.microrel.2012.09.015
5. Lei, J., et al.: Precise extraction of dynamic rdson under high frequency and high voltage by a double-diode-isolation method. IEEE J. Electron Devices Society **7**, 690–695 (2019). https://doi.org/10.1109/JEDS.2019.2927608
6. Meneghini, M., et al.: Charge trapping in GaN power transistors: challenges and perspectives. IEEE BiCMOS and Comp. Semicond. Integr. Circ. Technol. Symp. (BCICTS) **2021**, 1–4 (2021). https://doi.org/10.1109/BCICTS50416.2021.9682455
7. Puschkarsky, K., et al.: Threshold voltage hysteresis in SiC MOSFETs and its impact on circuit operation. IEEE Int. Integr. Reliab. Workshop (IIRW) **2017**, 1–5 (2017). https://doi.org/10.1109/IIRW.2017.8361232
8. Lelis, A.J., et al.: Basic mechanisms of threshold-voltage instability and implications for reliability testing of SiC MOSFETs. IEEE Trans. Electron Devices **62**(2), 316–323 (Feb.2015). https://doi.org/10.1109/TED.2014.2356172
9. Vella, A., Giaconia, G.C.: Embedded System for GaN Devices Characterization, Master's Degree Thesis, University of Palermo (2022)

Check for updates

Design-Time Tool for Energy Harvesting Solutions

Alessandro Bertacchini[1,2,3](✉) and Yuri Ricci[4]

[1] DISMI Dept of Sciences and Methods for Engineering, University of Modena and Reggio Emilia, 42122 Reggio Emilia, Italy
alessandro.bertacchini@unimore.it
[2] Centro Interdipartimentale InterMech MO-RE, 41125 Modena, Italy
[3] Centro Interdipartimentale En and Tech, 42122 Reggio Emilia, Italy
[4] DIEF Dept of Engineering "Enzo Ferrari", 42125 Modena, Italy
yuri.ricci@unimore.it

Abstract. A Matlab-LTSpice based tool is presented in this paper. It has been developed to help the designer to add Energy Harvesting (EH) capabilities to a generic electronic device that is being developed. The realized easy-to-use tool allows to quick evaluate several EH solutions at design time, contributing to considerably reduce both design time and costs because no hardware prototypes are needed. The tool is based on a set of pre-designed EH circuits covering all the main EH sources available in industrial applications (i.e. light, thermal gradients and vibrations) and drives the designer in the choice of the most suitable one accordingly with both the estimated environmental working conditions and the estimated power consumption profile of the specific electronic device that is being developed.

1 Introduction

In many industrial applications there is a growing interest in the use of Internet of Things (IoT) or Industrial IoT (IIoT) solutions as key element to enable functions like remote diagnostics, process monitoring, preventive maintenance, digital twins, etc.

All these solutions are based on system architectures that use sensor nodes as edge devices allowing collecting information about the status of the system under monitoring and/or the environment where they work [1]. Due to the increasing complexity of the applications and the needs for miniaturized solutions, wireless sensors are the key element because of their easy installation and integration into existing networks/infrastructures.

From an application point of view, the main limiting factor to the extensive use of wireless sensors is that they are traditionally battery powered. The adoption of Energy Harvesting (EH) solutions can help to extend the battery lifetime or even obtain batteryless autonomous devices, because the edge devices (i.e. wireless sensor node) become able to gather energy directly from the environment where they operate. Depending on the specific application it is possible to exploit two main general architectures for EH systems known as Harvest-Use and Harvest-Store-Use [1], respectively.

R. Berta and A. De Gloria (Eds.): ApplePies 2022, LNEE 1036, pp. 39–45, 2023.
https://doi.org/10.1007/978-3-031-30333-3_6

With the first approach the primary energy source of the system is a battery, and the EH system operate as secondary energy source. Its scope is to extend as much as possible the lifetime of the battery and the harvested energy is used as soon as a minimum amount of energy needed to complete a task is collected.

The main advantage of this approach is that the battery is usually non-rechargeable, allowing reducing the power management circuit complexity. The main drawback is that the battery must be replaced when it reaches the end of life, and therefore it must be properly sized at design time to guarantee an acceptable lifetime. Moreover, the seamless energy source switching (i.e. from battery to energy harvester and vice versa) must be guaranteed at operation time.

The second approach, instead, is the most common one, and it is the one targeted by the proposed tool. It allows to cover a wide range of applications, and it is suitable to handle environmental working conditions varying over time and resulting in uncontrolled or unpredictable amount of energy collected by the energy harvester. To do this a completely different energy approach is used. The energy is always provided to the load (i.e. sensor node) by a rechargeable battery and the energy gathered by the energy harvester is used to recharge the battery. In this way no load power supply source switching must be handled, but the power management circuits become more complex and expensive. Nowadays integrated battery chargers supporting multi-chemistry profiles are available on the market, allowing higher design flexibility and obtaining EH solutions suitable for more than one application.

1.1 Main Contributions of This Work

Despite of energy harvesting techniques are nowadays mature enough for several industrial applications targeting IoT or IIoT scenarios and the benefits related to introduction of autonomous devices are well known in terms of zero maintenance, reduced impact on the environment and green sustainability, industry is still introducing them slowly. The reasons are mainly related to the high development time and costs of energy harvesting systems. Indeed, to obtain reliable and effective solutions a high degree of customization is usually needed.

Moreover, to obtain effective EH solutions, it is very important to have at design time good estimations of system power consumption and amount of harvestable energy in real working conditions.

From an application point of view, this information allowing choosing optimization strategies accordingly with the amount of energy (generated or stored) available at a given time (e.g. adaptive application duty cycle, [2]), adaptive sensors sampling rate and data transmission rate [3]). Unfortunately, these kinds of data are often available only during the field tests of the prototype leading to possible redesign steps that in turn contribute to increase development time and costs.

In this context, aim of this work is provide an easy-to-use design tool helping the designer to add EH capabilities to a generic electronic device contributing to minimize design efforts. In particular, it provides an easy way to find effective and sustainable EH solutions operating in real working scenarios, by evaluating different variants and different design ideas in the very early design stages without the need for test the corresponding real prototypes.

2 The Developed Design Tool

The proposed one is a MATLAB-LTSpice based tool that can be used either to design a completely new power supply stage of a generic electronic device from scratch or to add EH capabilities to an existing battery-powered device. It is based on a set of EH solutions pre-designed to gather energy from solar irradiance, thermal gradients and mechanical vibrations under different environmental working conditions and for different ranges of load power consumption (i.e. power consumption of the device that is being developed).

The operating cycle of the tool is comprised of 4 steps as summarized in Fig. 1:

1. By exploiting an intuitive Graphical User Interface (GUI), the designer specifies: i) the estimated environmental working conditions of where the device that is being developed will operate; ii) the estimated electrical characteristics of the device.
2. During the data entry, the tool runs a decision logic algorithm and provides a sorted list of the available EH solutions supported by the tool, suggesting the EH solution that best match the combination of provided environmental and electrical estimated working conditions. The list is updated in real-time during the user's data entry and it is sorted from the most suitable EH solution to the less suitable one, depending on the overall score obtained by each supported EH solution as result of the decision logic algorithm execution.
3. A pre-designed EH circuit is associated to each EH solution shown in the list. The user selects the desired EH solution and the tool automatically runs the corresponding circuit simulation.
4. At the end of the simulation, the tool shows in the GUI qualitative and quantitative results about feasibility and estimated performances of the chosen EH solution.

Once the first iteration is completed, the designer can iterate the procedure by changing some input parameters (environmental and/or electrical) and evaluate a variant of the initial solution. For example, the designer can define the maximum application duty cycle allowing obtaining a fully autonomous device for a given power consumption and a range of real environmental working conditions.

As mentioned before, in this way, the tool allowing obtaining a quick evaluation of different solutions without the need for test real prototypes and, consequently, this leads to a significant design time and costs savings.

The simplified framework of the tool is shown in Fig. 2. The tool is comprised of a MATLAB-based[1] core application able to interact with an electrical circuit simulator. In this case the LTSpice from Analog Devices has been used. It is worth noting that it is very easy to adapt the tool to any other PSpice engine-based simulator because the main parameter passed from the MATLAB core to the electrical simulator is the EH circuit netlist and not the schematic.

The MATLAB core of the tool is divided into 4 main modules: i) an intuitive ad-hoc GUI used to provide the simulation parameters related to the estimated electrical and

[1] MATLAB has been chosen essentially due to the variety of options offered in terms of data manipulation/visualization. Other tools or programming languages (e.g. Python) able to interface with LTSpice (or any other electronic circuits simulator) can be used.

environmental working conditions of the system that is being developed; ii) an inline repository containing all the EH circuits supported by the tool; iii) an algorithm engine used to assign a weighted score to each EH circuit in the repository depending on the environmental and electrical data provided by the user; iv) a dedicated module to retrieve simulation results and perform the data processing required to provide to the user the final results through the GUI.

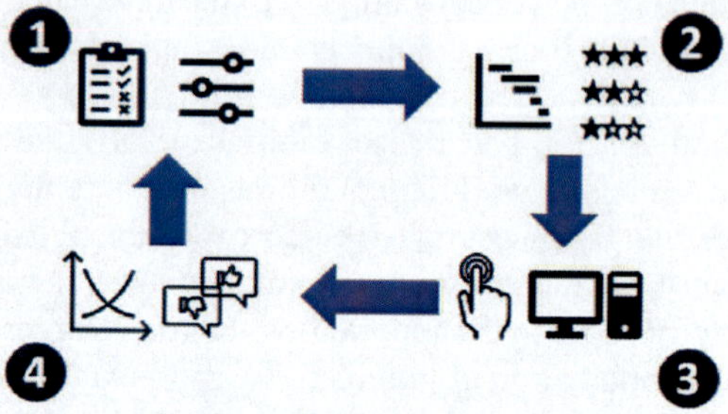

Fig. 1. Tool operating principle - Design Cycle.

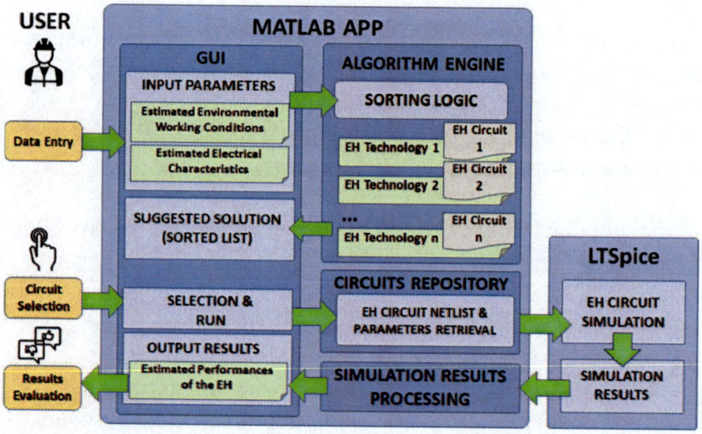

Fig. 2. Simplified Framework of the Tool

3 Decision Logic Algorithm

The choice of the most suitable EH solution for the working scenario of interest is the key element allowing saving design time and cost. Consequently, the decision logic algorithm running in background during the user's data entry plays a key role.

The operating principle is based on the assignment of a weight coefficient to each parameter provided by the user (environmental and electrical) and can be summarized by the following key points:

- Accordingly with the use case of interest, the user selects the estimated available energy sources in the corresponding GUI's tabs. Only the EH circuits related to the selected energy sources participate in the decisional process.
- The overall score of an EH solution supported by the tool is obtained by the sum of two different scores. The first one is calculated depending on the estimated average power consumption of the device that is being developed (computed from the electrical characteristics provided through the dedicated GUI's section). The second one is calculated by summing the static weight coefficients assigned to each selectable value of each selectable environmental parameter in the GUI.

Basically, the score related to the expected environmental working conditions allowing obtaining a rough estimation of the amount of harvestable energy during the operation in a real scenario, while the score related to the expected power consumption allowing obtaining a rough estimation of the output power that the energy harvester must provide to fully sustain the load. In this way a two-level sort is obtained allowing identifying the best trade-off solution among the available ones. Each EH solution supported by the tool, indeed, has been designed and optimized considering a well-defined incoming energy budget range and a well-defined load power consumption range.

4 The Tool at Work: Example of Simulations Results

The screenshot in Fig. 3 shows how the tool's GUI appears at the end of a simulation of one of the solar EH circuits supported by the tool. The upper half of the GUI is dedicated to the estimated environmental and electrical working conditions of the device that is being developing provided by the user. Noticeably the GUI accepts environmental parameters in form of both quantitative and qualitative data, with available options changing dynamically accordingly with user's selection (e.g. in case of solar EH circuit simulation with estimated irradiance conditions unknown a priori, by selecting the "unknown" option, the energy source profile used in the simulation resembles a typical irradiance profile over the day, as shown in Fig. 3.

The lower half of the GUI is reserved to the simulation output results. While typical electrical output data are independent from the simulated EH circuit, results show in the "Harvester" tab vary accordingly with the simulated EH circuit. They basically provide information about the capability of the selected EH circuit to sustain or not the device under the provided working conditions, the estimated battery lifetime, and the rate of variation of the battery state of charge, as shown in Fig. 3. In this case, under the provided environmental and electrical information the simulated EH circuit is able to fully sustain the load with a minimum irradiance of about 110 W/m^2.

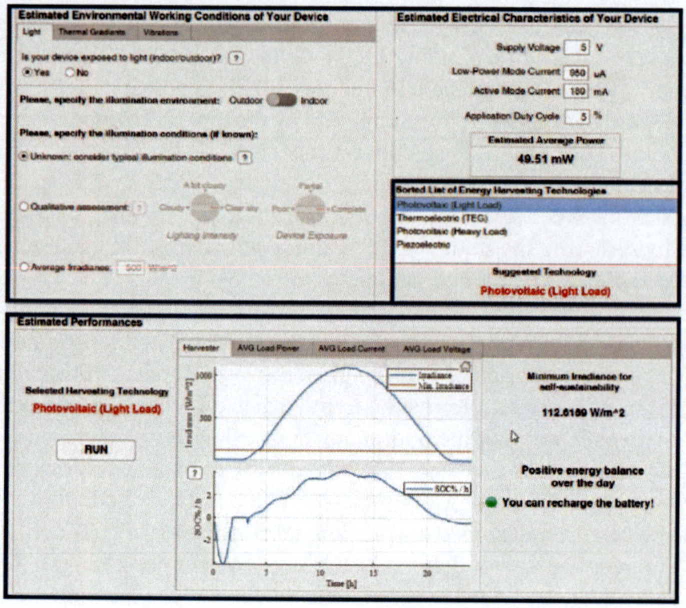

Fig. 3. Example of Simulation Results.

5 Conclusions

A simulation software tool developed to help the designer in adding Energy Harvesting (EH) capabilities to a generic electronic device under development has been presented. The realized easy-to-use tool allows to quickly evaluate several EH solutions at design time reducing both development time and costs because it is based on a set of pre-designed EH circuits covering all the main combinations of EH sources (i.e. light, thermal gradients and vibrations) and power consumption ranges (from μW to W) of industrial applications targeting IoT and IIoT scenarios. The tool drives the designer in the choice of the most suitable one accordingly with both the estimated environmental working conditions and the estimated power consumption profile of the specific electronic device under development.

Acknowledgments. This research is funded by ECSEL, the Electronic Components and Systems for European Leadership Joint Undertaking under grant agreement No 826452 (Arrowhead Tools), supported by the European Union Horizon 2020 research and innovation programme and by the member states. The authors thank Proff. P. Pavan and L. Selmi for the discussions and their contributions to the Arrowhead Tools project.

References

1. Sanislav, T., Mois, G.D., Zeadally, S., Folea, S.C.: Energy Harvesting Techniques for Internet of Things (IoT). IEEE Access **9**, 39530–39549 (2021). https://doi.org/10.1109/ACCESS.2021.3064066

2. Zhang, Z., Shu, L., Zhu, C., Mukherjee, M.: A Short Review on Sleep Scheduling Mechanism in Wireless Sensor Networks. In: Wang, L., Qiu, T., Zhao, W. (eds.) QShine 2017. LNICSSITE, vol. 234, pp. 66–70. Springer, Cham (2018). https://doi.org/10.1007/978-3-319-78078-8_7
3. Abella, C.S., et al.: 'Autonomous energy-efficient wireless sensor network platform for home/office automation.' IEEE Sensors J. **19**(9), 3501–3512 (2019). https://doi.org/10.1109/JSEN.2019.2892604

Design and Validation of an Electronic Unit for Monitoring Water Distribution in Plastic Pipes

Christian Riboldi, Daniele M. Crafa, and Marco Carminati(✉)

Dipartimento di Elettronica, Informazione e Bioingegneria, Politecnico di Milano, 20133 Milano, Italy
marco1.carminati@polimi.it

Abstract. A modular IoT node for monitoring water distribution with non-invasive sensors (ultrasonic flow rate and leakage) applied to plastic pipes is presented. It combines miniaturization and low-power design (380 μWh for a 1-h transmission period) with the potential to connect multiple remote units to the central one with powering and serial communication through the same 4 copper strip electrodes used for impedimetric leakage detection.

1 Introduction

A more efficient distribution of water, in both urban (drinking) and rural (irrigation) applications, is of key importance for the increased sustainability at which society is aiming. The ongoing consolidation of electronic IoT technologies, combining sensors miniaturization, low-power devices, radio networks and Cloud-based automatic data processing provides excellent tools to address the challenges of distributed real-time monitoring of large infrastructures, in particular of pipelines [1]. So far, the majority of IoT units applied to water distribution have been focused on the measurement of flow rate (for smart metering). Here we present a compact and low-power electronic unit combining traditional sensors, both in contact with the liquid (temperature, pressure) and contactless (ultrasonic flow rate) with a novel impedance sensor for the detection of water leakage outside of the pipe [2]. In order to minimize data transmission, embedded processing allows smart monitoring and rejection of false alarms. The main novelty of this solution is the connection of the main unit, endowed with a battery, solar panel and radio, with ancillary nodes placed along the plastic pipeline measuring local water leakage by means of the same longitudinal band electrodes used for sensing. This sensor node focuses on accurately measuring water "quantity", targeting the primary goal of reducing water loss in distribution network like similar efforts [3]. It can be however combined with other sensors for measuring water "quality" (such as pH, conductivity and fouling of the internal pipe surface [4]).

2 Electronics Design

The system architecture is shown in Fig. 1: the central unit interfaces with thermal, pressure (up to 6.6 bar) and ultrasonic transducers (resonating at ~1 MHz) for flow rate

R. Berta and A. De Gloria (Eds.): ApplePies 2022, LNEE 1036, pp. 46–53, 2023.
https://doi.org/10.1007/978-3-031-30333-3_7

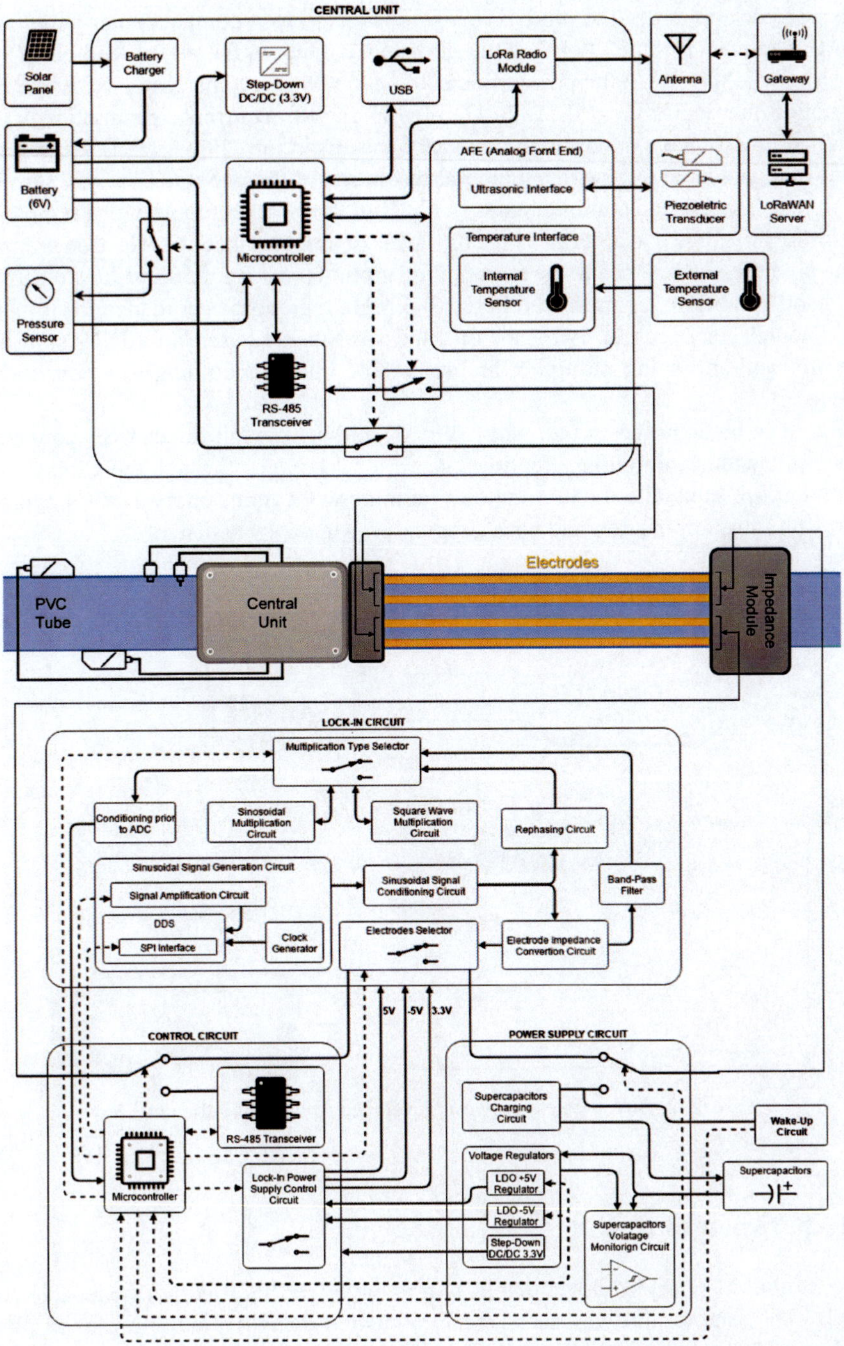

Fig. 1. Architecture of the central and auxiliary units connected by 4 longitudinal sensing electrodes switched between supply, serial communication and multiplexed impedance sensing.

measurement applied to the pipe. It is also connected to a compact solar panel and a 432 MHz antenna for LoRaWAN communication. It contains the sealed-lead-acid (SLA) battery (6 V, 1300 mAh) the power management and battery charging circuit, the main low-power microcontroller (SMT32L4 family with sub-μA current consumption in sleep mode) and the sensors conditioning and acquisition circuits. Flow rate is measured by means of a dedicated chip with a temporal resolution of the internal TDC of 22 ps. Furthermore, it powers and communicates (with RS-458 serial protocol) with a remote unit measuring impedance to detect leakage by means of the same 4 copper electrodes applied on the pipe. The impedance of the surrounding ground (~50 Ω for 50-cm-long electrodes in dry soil conditions) is measured in the 2–4 MHz region, in accordance to a previous experimental investigation by means of a lock-in scheme based on a DDS sinusoidal generator and an analog multiplier and selectable filters, according to a consolidated scheme [5].

Data can be acquired in two ways (Fig. 2), either through direct USB connection to the unit (with a minimum sampling interval of 70 ms) or through the Cloud server, where data are transmitted with the LoRa radio network (min. Interval of 4 s, typically set to 1 h in this application and with a typical ~1 km gateway distance).

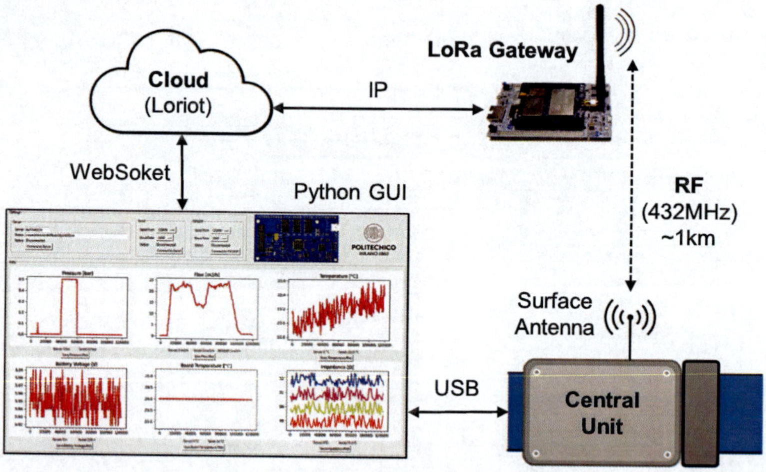

Fig. 2. IoT wireless network developed to validated the unit.

3 Experimental Results

The central unit is very compact thanks to a miniaturized PCB with a swappable radio module. The remote unit was packaged in a custom-designed and 3D-printed plastic annular collar to be clamped directly on the pipe (Fig. 3). The electrical contact with the copper electrodes is achieved by means of spring-loaded pins (pogo pins), while the vertical connection between the PCBs in the two jaws is granted by several robust banana plugs.

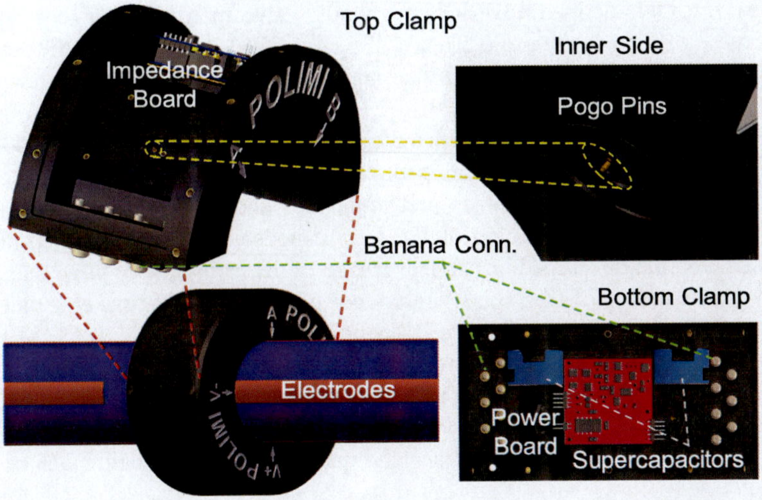

Fig. 3. Annular plastic case (180 mm in diameter and 90 mm in width) hosting the remote unit directly clamped on the pipe and connected to the strip electrodes by means of pogo pins.

The prototype was tested in laboratory by means of a 7 m-long loop realized for this purpose a with 4mm-thick rigid PVC pipes of 90 mm diameter equipped with all the sensors (see Fig. 4, where the battery and LoRaWAN antenna are also visible in the same standard waterproof box)). The water flow is driven by an immersion pump able to produce a water speed above 1 m/s (and a nominal prevalence of 12 m for a ~1 kW power consumption).

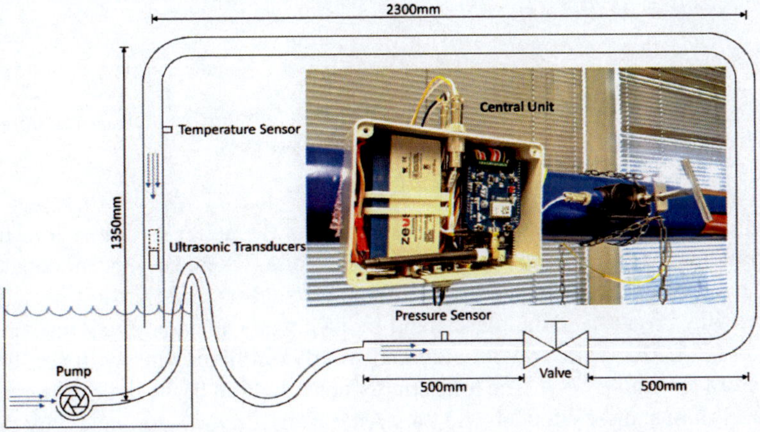

Fig. 4. Laboratory validation of the units with a closed water loop in which water (~30 L in total) circulates at 1.1 m/s (22 m^3/h) speed and with a Reynolds number > 10^5.

An example of complete time tracking of the acquired sensors (water pressure, flow rate, temperature, soil impedance, board temperature and battery voltage) of a typical test cycle (OFF, ON full open valve at 22 m^3/h, 50% valve open at 15 m^3/h corresponding to an increase of the pressure to 0.5 bar, full again, OFF) is reported in the screenshot of Fig. 3 collecting data from both USB serial link and via a LoRaWAN gateway to the Cloud (Loriot IoT server).

The flow sensor was calibrated with a commercial instrument (Riels RIF600P), connected to the same type of ultrasonic transducers. Different digital filters were compared to reject fluctuations due to bubbles and impurities and preserve a response time of ~1 s. As shown in Fig. 5, zooming on the step response of the sensor when the pump is switched ON, the developed instrument is free of the error at zero flow. Samples are acquired at a frequency of 14.2 Hz. A digital low-pass filter with a time constant of 2.2 s is applied to the samples, achieving a RMS noise of 0.64 m^3/h (i.e. 0.6% of the FSR of 5 m/s). In the remote unit, the optimal sensing frequency to detect leak is identified at 2 MHz. The impedance measuring unit was tested separately achieving a RMS noise of 0.04% for a 50 Ω test resistor measured from 0.1 to 4 MHz. All the sensors offer a resolution (combination of noise sources and quantization noise) below 1% of the full scale range.

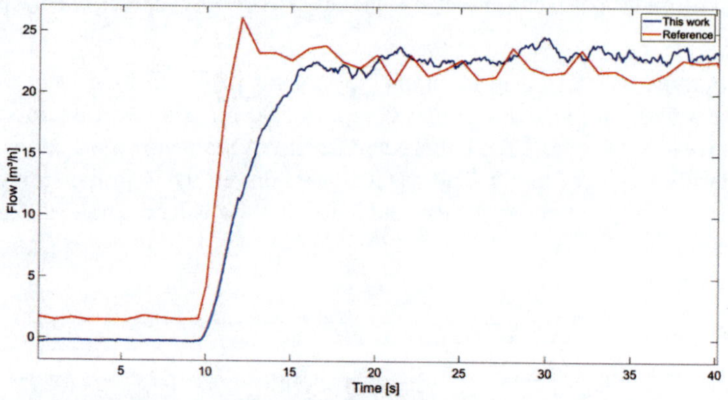

Fig. 5. Validation of the flow rate sensor with a commercial reference during a step in the flow rate from 0 to the maximum (22 m^3/h) achievable in the test loop.

As for all IoT devices, the major figure of merit of the unit is its power consumption. The central unit is optimized to sleep in stand-by mode (where the current consumption is 30 μA) and wake-up every hour, acquire all the sensors and transmit data. The run cycle takes 40 s (Fig. 4) and includes: stand-by (A), wake up and radio join (B), sensors acquisition (C, D, E) radio transmission and serial communication with auxiliary unit (F) and return to stand-by (G). The total energy consumption of the main unit during this cycle (32 μAh in stand-by for 1 h and 24 μAh during run time dominated by the radio transmission) is 56 μAh, matching the battery self-discharge (~3%/month). It would give an autonomy (without the solar panel whose min. Area should be ~2 cm^2 considering the average efficiency of amorphous silicon and the average insolation in Italy) of 488 days. This corresponds to a power consumption of 380 μWh that is 100 times lower than what achieved by a previous system powered with an invasive small turbine [4].

The current consumption of the central unit is comparable with that of a single Application Specific Integrated Circuit (ASIC) for the readout of ultrasonic flow sensors (13.5 μA at 3 V with a sampling period of 30 ms [6]). When looking at similar IoT nodes measuring only flow rate and communicating via LoRaWAN, the average current consumption is more than twice higher (132 μA at 3 V with the same transmission period of 1 h [7]) than our system measuring additional parameters (Fig. 6).

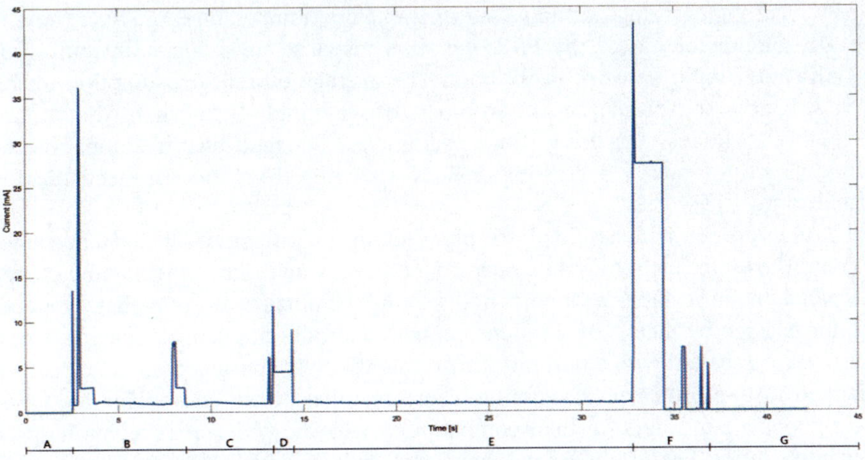

Fig. 6. Measured instantaneous current consumption of the central unit during the ON cycle.

Finally, we can compare the performance of the flow rate measuring circuits with state-of-the-art commercial instruments based on clamp-on sensors. As reported in Table 1, the range of measurable flow velocities and accuracy are similar, while the

Table 1. Performance comparison of the realized flow meter with commercial instruments compatible with clamp-on ultrasonic transducers. The advantage in terms of power consumption is evident.

Model	This Work	UD2100	U1000 MKII-FM	PORTA FLOW D550	UF25B
Max. Speed	>5 m/s	12 m/s	10 m/s	12 m/s	5.32 m/s
Min. Speed	0.1 m/s	0.1 m/s	0.1 m/s	0.03 m/s	0.05 m/s
Pipe Diameter	90–200 mm	16–4500 mm	22–180 mm	12.5–4500 mm	9.5 mm
Accuracy [% of FSR]	±1%	±2% (>0.3 m/s)	±1% to ±3% (>0.3 m/s)	2%	3% to 5%
Power Consumption	340.8 μW	10 W (max)	7 W (max)	N. A	<90 mW
Producer		GF Piping Sys	Micronics		Sensata

sensor here proposed offers an improvement of several orders of magnitude in terms of power dissipation.

4 Conclusions

We have presented the design, realization and laboratory validation of a solar-powered, credit-card-sized IoT electronic system to pervasively monitor water distribution in real-time by locally measuring multiple water parameters (temperature, pressure, flow rate with a resolution of 1% of the FSR and ~1 s response time) and transmitting data via LoRaWAN radio network every hour. The average current consumption matches the SLA battery self-discharge, is two orders of magnitude better than what archived by other WSN nodes measuring similar parameters [4] and, thus, is compatible with a broad variety of energy harvesting devices, spanning from solar to mechanical and thermal sources [8].

The units are coupled to rigid plastic pipes relying on a modular scheme that connects the central module (including the underground battery and radio antenna and compact solar panel on the surface) with other nodes along the pipeline for impedance detection of water leakage by means of the same copper electrodes applied for sensing. Copper electrodes, passivated with a thin (50 μm) Kapton® layer and glued on the pipes, were employed in this prototypical activity. These materials were chosen for the ease in shaping copper bands and for their compatibility with the widespread technology used for manufacturing flexible PCBs, i.e. copper on polyimide (Kapton®). Flexible PCBs are commonly limited to a maximum length of ~1 m. However, the choice of metal is not critical and, given the increasing cost of copper, other cost-effective metals could be adopted. Furthermore, conductive polymers could be used too, to be directly embedded in the pipe and, thus, potentially simplifying the manufacturing of the sensorized pipe. Simulations have shown that the impedance sensor is able to reliably detect small leaks (with a volume of a few liters) with an electrodes length up to ~10 m. Longer electrodes produce a smaller relative variation ($\Delta R/R$), causing a compromise between sensitivity and density of nodes. Also considering financial constraints, we expect the installation of a central unit (covering in total ~20 m of pipe, 10 on each side) in most critical sections of the network, with an initial density in the order of ~1 unit/km.

Acknowledgments. The early contribution to this development by Ilenia D'Adda, Danilo A. Carnevale Castillo and Rebeka Lukacevic is acknowledged, as well as initial support from FITT S.p.A.

References

1. Moubayed, A., et al.: Water leak detection survey: challenges & research opportunities using data fudion & federated learning. IEEE Access **9**, 40595–40611 (2021)
2. D'Adda, I., Battaglin, G., Carminati, M.: A low-cost flexible pipe sheath for multi-parameter monitoring of water distribution. In: IEEE ISCAS Proceedings, pp. 1–5 (2021)
3. Fabbiano, L., Vacca, G., Dinardo, G.: Smart water grid: a smart methodology to detect leaks in water distribution networks. Measurement 151, 107260/1–7 (2020)

4. Carminati, M., et al.: A self-powered wireless water quality sensing network enabling smart monitoring of biological and chemical stability in supply systems. MDPI Sensors **20**, 1125 (2020)
5. Carminati, M., et al.: A smart sensing node for pervasive water quality monitoring with anti-fouling self-diagnostics. In: IEEE ISCAS Proceedings, pp. 1–5 (2018)
6. Kaga, Y., Johguchi, K.: A 180-nm CMOS smart ultrasonic water flow meter circuit for IoT smart society. Jpn. J. Appl. Phys., SBBL05 (2021)
7. Slany, V., et al.: An integrated IoT architecture for smart metering using next generation sensors for water management based on LoRaWAN technology: a pilot study. MDPI Sensors 20, 4712 (2020)
8. Virk, M.A., et al.: Review of energy harvesting techniques in wireless sensor-based pipeline monitoring networks, 112046 (2022)

Hardware Acceleration

Design and Implementation on FPGA of a HW Accelerator for Post-Quantum RLWE Polynomial Operations

Stefano Di Matteo[1](✉), Sergio Saponara[1], and Riccardo Locatelli[2]

[1] Department of Information Engineering, University of Pisa, Via G. Caruso, 16, Pisa, Italy
stefano.dimatteo@phd.unipi.it

[2] Intel Corporation, Pisa, Italy

Abstract. Ring learning with errors (RLWE) is largely adopted in Post-Quantum cryptography and Homomorphic encryption schemes. RLWE cryptosystems are defined over polynomial quotient rings, where polynomial additions/subtractions and multiplication are required. In this paper we propose the implementation of a hardware accelerator for polynomial operations required in RLWE cryptosystems. Such hardware accelerator includes a Arithmetic Logic Unit (ALU) able to perform polynomial additions/subtractions and multiplications. The latter adopts Negative Wrapped Convolution (NWC) and Number Theoretic Transform (NTT) techniques. The hardware accelerator has been integrated in a hardware system for accelerating the NewHope-1024 algorithm, and a demo on a Intel DE4 FPGA board has been developed to test the functionality of the system. Synthesis results on different FPGA technologies are proposed too, showing competitive performances respect to the State of Art.

Keywords: Ring Learning With Errors (RLWE) · Hardware Accelerator · Post-Quantum Cryptography · Security

1 Introduction

Currently most of the communications protected by asymmetric cryptography use RSA [1], Diffie-Hellman [2] or ECC [3] to guarantee security services as confidentiality and authentication. In the year 1994, Peter Williston Shor [4] announced the discovery of a quantum algorithm that could break most of the currently used asymmetric cryptographic schemes, including Diffie-Hellman and those based on RSA and ECC. Even though large-scale quantum computers are still not available right now, in the future they would seriously compromise the security of digital communications on Internet and elsewhere. Lattice-based cryptography (LBC) is a good candidate to replace the classical public-key cryptography because no known quantum algorithms exist for solving the lattice problem and at the same time lattice problem remains hard for classical computers as

R. Berta and A. De Gloria (Eds.): ApplePies 2022, LNEE 1036, pp. 57–64, 2023.
https://doi.org/10.1007/978-3-031-30333-3_8

well. Moreover, some LBC schemes can be used to implement advanced security services such as Identity-Based Encryption, Fully Homomorphic Encryption (FHE) [5] and "Somewhat" Homomorphic Encryption (SHE) [6]. Learning with Errors (LWE) [7] problem is probably the most known lattice-based scheme, which relies his security on the impossibility of extracting the vector s from the equation $b = A \cdot s + e$ where A is a matrix and b, s and e are vectors. However, LWE requires the large matrix A to be stored and expensive matrix-vector multiplications, resulting in high memory resources consumption and expensive computation cost. To overcome these drawbacks, Ring-Learning With Errors (RLWE) was proposed by Lyubashevsky et al. [8] in 2010. In RLWE the matrix A becomes a vector a and the arithmetic operations are defined over a polynomial ring. Therefore, the computational and storage costs of RLWE cryptosystems are severely reduced respect to LWE ones. The arithmetic operations included in RLWE cryptosystems are additions, subtractions and multiplications over polynomial rings, and polynomial multiplication is considered the most complex and time-consuming among them. This paper will focus on the design of a RLWE HW accelerator for polynomial operations. The proposed design is able to perform polynomial addition and subtraction, and polynomial multiplications adopting the NTT technique. The main contributions of this paper include the architectural design of a RLWE cryptoprocessor for polynomial operations, synthesis on Xilinx FPGAs and Intel Stratix IV FPGA, and the implementation of a demo on DE4 board equipped with Stratix IV FPGA of a hardware system for NewHope-1024 algorithm. The rest of the paper is organized as follows: Sect. 2 gives an introduction on RLWE, Sect. 3 presents our hardware architecture and the demo on FPGA, Sect. 4 shows the results and a comparison with previous works and Sect. 5 draws the conclusion of this work.

2 Preliminary

2.1 Ring Learning with Errors

Cryptography schemes based on LBC exploit the hardness of computational problems on lattices. The LWE problem serves as the foundation for different modern lattice-based cryptography schemes. It states that given the equation $b = A \cdot s + e$ where A is a matrix and b, s and e are vectors, it is difficult to find the vector s if A and e are sampled from appropriate distributions. An efficient derivation of the LWE scheme is the RLWE, which is based on ideal lattices and defined over quotient ring $R_q = \mathrm{Z}_q[x]/(x^n + 1)$. All the polynomials contained in the quotient ring have degree at most $n - 1$ and the polynomial coefficients are defined in Z_q. In RLWE n is selected a power of 2 and q is a prime number such that $q = 1 mod 2n$. Degrees of the polynomials in RLWE usually vary between 256 and 1024 [9]. As a use case, we considered the Key Encapsulation Mechanism (KEM) in NewHope-1024 [10], and we fixed $n = 1024$ and $q = 12289$.

2.2 Multiplication over Polynomial Rings and Number Theoretic Transform

Polynomial multiplication in RLWE cryptosystems represents one of the most onerous part of the computation. Number Theoretic Transform (NTT) is a specialized version of the Discrete Fourier Transform (DFT), which allows to reduce the complexity of polynomial multiplication from $\mathcal{O}(n^2)$ of the schoolbook algorithm to $\mathcal{O}(n \log n)$. NTT transformation can be denoted as:

$$NTT(a) = \sum_{i=0}^{n-1} \breve{a}_i x^i,\ where\ \breve{a}_i = \sum_{j=0}^{n-1} a_j \omega^{ij} \tag{1}$$

where a is a polynomial and ω is the n-th primitive root of unity in Z_q. Computing the NTT for a polynomial multiplication $c = a \cdot b\ mod(x^n + 1)$ requires the zero padding of each input in order to obtain the length of polynomials a and b equal to $2n$ and a reduction modulo $(x^n + 1)$. To avoid this overhead the Negative Wrapped Convolution (NWC) [11] technique can be used, which involves a pre-scale and post-scale phase on the polynomials. Given the $2n$-th primitive root of unity ψ such that $\psi^{2n} \equiv 1 mod q$, given the vectors $\Psi = (1, \psi, ..., \psi^{n-1})$, $\Psi^{-1} = (1, \psi^{-1}, ..., \psi^{-(n-1)})$ and given the polynomials a and b, the polynomial multiplication in the quotient ring becomes $c = \Psi^{-1} \circ INTT(NTT(a \circ \Psi) \circ NTT(b \circ \Psi))$, where $\circ$ is the coefficient-wise multiplication between vectors. NTT computation can be executed through butterfly operations in Cooley-Tukey (CT) or Gentleman-Sande (GS) configurations [12].

3 Hardware Implementation

3.1 Post-Quantum Arithmetic Logic Unit (ALU) and Overall Hardware Architecture

As reported in the previous chapters, RLWE-based cryptosystems are built using polynomial operations on quotient rings, which are polynomial additions, subtractions and multiplications. Polynomial multiplications have been implemented using the technique reported in 2.2, which decomposes the computation in butterfly operations and coefficient-wise multiplications. Figure 1 reports the CT (i.e. on the left side) and the GS (i.e. on the center) configurations for butterfly. The right side of Fig. 1 shows the schematic that allows to unify the capabilities of both CT and GS configurations. The hardware architecture of the proposed ALU for polynomial operations is reported in Fig. 2. Modular additions and subtraction are implemented using the standard approach [13], modular multiplications adopt the multiply-then-reduce approach. The Barret algorithm has been employed for modular reduction after multiplication, and the hardware implementation follows the one reported in [14]. The ALU has six stages of pipeline, and four of them are dedicated to the modular multiplier. The proposed configuration allows to execute the following operations: CT and GS butterfly computation, multiplication between polynomial coefficients and vectors Ψ and Ψ^{-1}, coefficient-wise multiplication, addition and subtraction between two different polynomials.

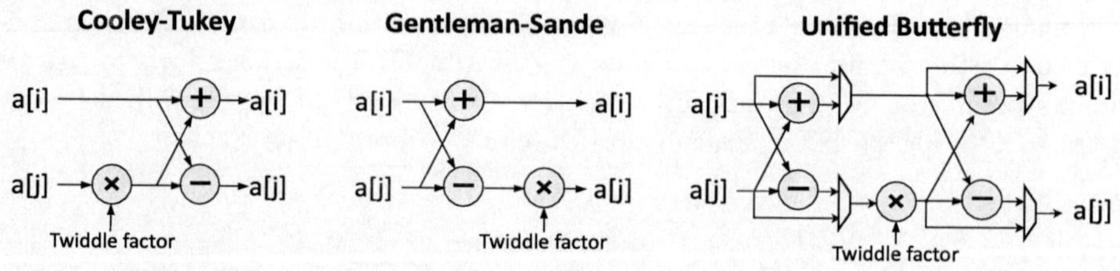

Fig. 1. Butterfly configurations. CT on the left side, GS on the center and a unified CT-GS configuration on the righ-side. Twiddle factor refers to the n-th primitive root of unity ω and $2n$-th primitive root of unity ψ.

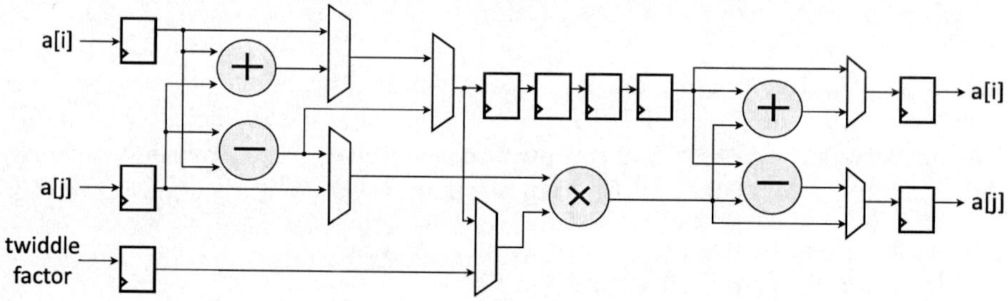

Fig. 2. Hardware architecture of the six stages ALU. Four stages of pipeline are dedicated to the modular multiplier.

Figure 3 shows the overall architecture of the presented HW accelerator for polynomial operations. Two different memory banks have been used to store the intermediate results and polynomials, the ROM bank instead stores the precomputed values (i.e. n-th primitive root of unity and $2n$-th primitive root of unity) for NWC and NTT operations. The F.S.M manages the data flow and controls the ALU and memories. As reported in Sect. 2.1, as use case we are considering the RLWE protocol NewHope-1024; then, n is set to 1024 and q to 12289. The parameter q affects both the modular operations inside the ALU (including the size of data) and the precomputed values in the ROM memory. The parameter n affects the size of the memories and the precomputed values in ROM. Each RAM bank is composed of four single-port RAM inspired by the work reported in [15]. For the selected parameters (i.e. n and q), the proposed architecture consumes 7KB for RAM memories and 7KB for the ROM memory.

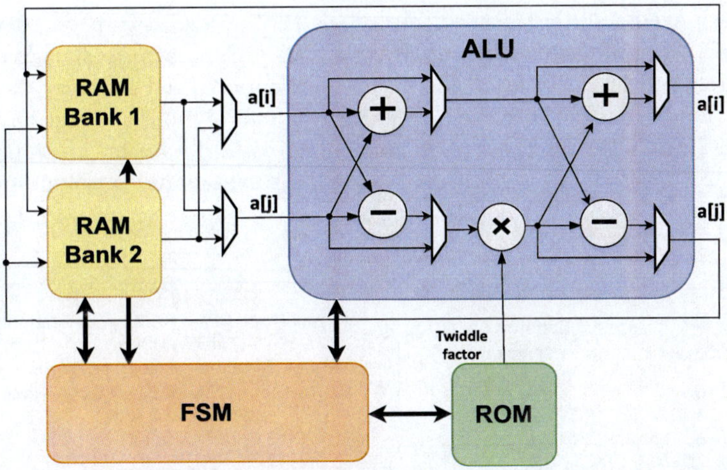

Fig. 3. Overall architecture of the HW accelerator for polynomial operations. The detailed architecture of the ALU block is represented in Fig. 2.

3.2 Demoboard on Intel DE4: NewHope-1024 Use Case

As use case for the proposed hardware accelerator we selected the NewHope-1024 algorithm, which is a RLWE KEM. Further details and specifications can be found in the official documentation [10]. The main functions of NewHope algorithm are: key generation, encryption and decryption. These functions require in addition to the RLWE polynomial operations also HASH functions (i.e. actually eXtendable Output Function based on SHAKE128/256), sampling of polynomial coefficients over different distributions (i.e. Gaussian and Uniform distributions) and other encoding/decoding functions. We implemented in hardware a dedicated module for SHAKE128/256 based on the architecture reported in this work [16], and a hardware module for both Gaussian and Uniform sampling of polynomial coefficients based on respectively binomial and rejection sampling as reported in the work [15]. In order to test the functionality of the proposed hardware accelerator we implemented the "NewHope-CPA-PKE Key Generation" function reported in the official supported documentation of NewHope [10]. The steps from line 3 to line 11 of the algorithm have been executed in hardware. In particular, steps 8 and 10 (NTT) and step 11 (polynomial multiplication and addition) are executed by the proposed hardware accelerator for polynomial operations. The steps 3 is computed by the dedicated module for SHAKE128/256 algorithm and steps 6,7 and 9 are executed by the hardware distribution sampler. Figure 4 reports the architecture of the system and the achieved results. The system is implemented in a Intel Stratix IV board, the CPU is a soft-core NIOS II in the fast-version, the interface is a standard Avalon BUS interface. The peripherals are a on-chip memory (192 KB), interval timer and UART for communication with the host PC, and the hardware accelerator (i.e. in the green box of Fig. 4). The clock frequency is 120 MHz. Figure 4 reports also the achieved results: the speed-up of the HW accelerator respect to the SW (from line 3 to 11 of the NewHope-CPA-PKE Key Generation) is around a x534, which is reduced to a x17.5 for the entire NewHope-CPA-PKE Key Generation function. This because the EncodePK and EncodePolynomial functions (line 12 of the NewHope-CPA-PKE Key Generation algorithm) have not been performed in hardware, resulting in a high overhead due to the b and s polynomial reads from the accelerator RAM banks.

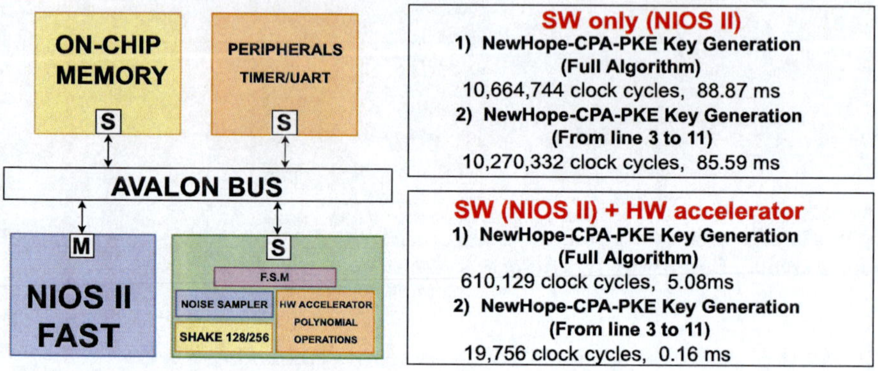

Fig. 4. On the left side the NIOS-II system implemented on a Intel Stratix IV FPGA Board. On the right side the achieved results of the NewHope-CPA-PKE Key Generation algorithm with and without the HW acceleration.

4 Results and Comparison with State of Art

Table 1 reports the synthesis results for the proposed hardware accelerator and a comparison with the State of the Art. In order to make a fair comparison we selected works in literature that adopt the same parameter set (i.e. $n = 1024$ and $q = 12289$), and we evaluated the performance respect to the NTT. Our results in Table 1 refers only to the hardware accelerator for polynomial operations (i.e. Figure 3) and do not include the resources consumption due to the logic described in Sect. 3.2. The work in [17] is a hardware accelerator for NewHope algorithm that employs two different butterfly units for polynomial operations. It takes in fact less cycles for NTT at the price of more logic resources. The authors adopt the SAMS2 [18] technique instead of the Barret for modular reduction, which does not require additional multiplications, but can be used only for special form moduli. In our work we preferred the more generic Barret reduction algorithm that can be easily customized for different moduli. The authors in [19] implemented a high performance hardware accelerator for NewHope where the NTT module is composed of four butterfly units. The NTT unit is faster than ours but it consumes around x10 of registers and x5 of LUTs respect to our work.

The work in [20] is a high-speed NTT-based polynomial multiplier with a 2×2 butterfly core. Also in this case our architecture requires fewer logical resources at the price of higher execution time.

Table 1. Synthesis results on FPGA technologies and comparison with the State of the Art. The reported results of Our work refer to the hardware accelerator for polynomial operations only (the hardware reported in Fig, 3).

Work	FPGA	NTTc.c.	Frequency [MHz]	Time[μs]	CLB LUTs	**CLB** Regs	DSPs	BRAMs
Our	Xilinx xczu7ev	5240	310	16.9	543	175	3	6
Our	Intel Stratix IV	5240	120	43.6	554	257	8	15
Our	Xilinx Artix 7	5240	100	52.4	567	173	6	4
[17]	Xilinx Artix 7	2569	244	10.5	847	375	2	6
[19]	Xilinx Artix 7	2616	150	17.4	2832	1381	8	10
[20]	Xilinx Artix 7	1591	234	6.8	798	715	4	10

5 Conclusion

This paper presented the the implementation of a hardware accelerator for RLWE polynomial operations. Negative Wrapped Convolution (NWC) and Number Theoretic Transform (NTT) techniques have been used for polynomial multiplications. The paper showed the hardware architecture of the Arithmetic Logic Unit and the memory organization. The proposed accelerator has been implemented in a Stratix IV FPGA and tested with the NewHope-1024 Key Generation algorithm. Synthesis results on different FPGA technologies showed competitive performances respect to the State of Art.

Acknowledgments. This work has been supported by an Intel Research Grant and by MIUR Dipartimento di Eccellenza program.

References

1. Rivest, R.L., Shamir, A., Adleman, L.: A method for obtaining digital signatures and public-key cryptosystems. Commun. ACM **21**(2), 120–126 (1978)
2. Diffie, W., Hellman, M.: New directions in cryptography. IEEE Trans. Inf. Theory **22**(6), 644–654 (1976)
3. Koblitz, N.: Mathematics of computation. Elliptic Curve Cryptosys. 48n **1**(77) (1987)
4. Shor, P.W.: Algorithms for quantum computation: discrete logarithms and factoring. In: Proceedings 35th Annual Symposium on Foundations of Computer Science, pp. 124–134, IEEE (1994)
5. Gentry, C.: Fully homomorphic encryption using ideal lattices. In: Proceedings of the Forty-first Annual Acm Symposium on Theory of Computing, pp. 169–178 (2009)

6. Naehrig, M., Lauter, K., Vaikuntanathan, V.: Can homomorphic encryption be practical?. In: Proceedings of the 3rd ACM Workshop on Cloud Computing Security Workshop, pp. 113–124 (2011)
7. Regev, O.: New lattice-based cryptographic constructions. J. ACM (JACM) **51**(6), 899–942 (2004)
8. Lyubashevsky, V., Peikert, C., Regev, O.: On ideal lattices and learning with errors over rings. In: Gilbert, H. (ed.) EUROCRYPT 2010. LNCS, vol. 6110, pp. 1–23. Springer, Heidelberg (2010). https://doi.org/10.1007/978-3-642-13190-5_1
9. Nejatollahi, H., Shahhosseini, S., Cammarota, R., Dutt, N.: Exploring energy efficient quantum-resistant signal processing using array processors. In: ICASSP 2020-2020 IEEE International Conference on Acoustics, Speech and Signal Processing (ICASSP), pp. 1539–1543. IEEE (2020)
10. de la Piedra, A.: Erdem alkim, roberto avanzi, joppe bos, léo ducas, antonio de la piedra, thomas pöppelmann, peter schwabe, douglas stebila version 1.0-(updated june 14, 2018) (2018)
11. Pöppelmann, T., Güneysu, T.: Towards efficient arithmetic for lattice-based cryptography on reconfigurable hardware. In: Hevia, A., Neven, G. (eds.) LATINCRYPT 2012. LNCS, vol. 7533, pp. 139–158. Springer, Heidelberg (2012). https://doi.org/10.1007/978-3-642-33481-8_8
12. Longa, P., Naehrig, M.: Speeding up the number theoretic transform for faster ideal lattice-based cryptography. In: Foresti, S., Persiano, G. (eds.) CANS 2016. LNCS, vol. 10052, pp. 124–139. Springer, Cham (2016). https://doi.org/10.1007/978-3-319-48965-0_8
13. Di Matteo, S., Baldanzi, L., Crocetti, L., Nannipieri, P., Fanucci, L., Saponara, S.: Secure elliptic curve crypto-processor for real-time iot applications. Energies **14**(15), 4676 (2021)
14. Nannipieri, P., Di Matteo, S., Zulberti, L., Albicocchi, F., Saponara, S., Fanucci, L.: A risc-v post quantum cryptography instruction set extension for number theoretic transform to speed-up crystals algorithms. IEEE Access **9**, 150798–150808 (2021)
15. Banerjee, U., Ukyab, T.S., Chandrakasan, A.P.: Sapphire: A configurable crypto-processor for post-quantum lattice-based protocols, arXiv:1910.07557 (2019)
16. Nannipieri, P., et al.: Sha2 and sha-3 accelerator design in a 7 nm technology within the european processor initiative. Microprocess. Microsyst. **87**, 103444 (2021)
17. Zhang, N., Yang, B., Chen, C., Yin, S., Wei, S., Liu, L.: Highly efficient architecture of newhope-nist on fpga using low-complexity ntt/intt. IACR Trans. Cryptograph. Hardware Embedded Syst., 49–72 (2020)
18. Rentería-Mejía, C.P., Velasco-Medina, J.: High-throughput ring-lwe cryptoprocessors. IEEE Trans. Very Large Scale Integrat. (VLSI) Syst. **25**(8), 2332–2345 (2017)
19. Kuo, P.-C.: High performance post-quantum key exchange on fpgas. Cryptology ePrint Archive (2017)
20. Bisheh-Niasar, M., Azarderakhsh, R., Mozaffari-Kermani, M.: High-speed ntt-based polynomial multiplication accelerator for post-quantum cryptography. In: 2021 IEEE 28th Symposium on Computer Arithmetic (ARITH), pp. 94–101 (2021)

A Novel Front-End Circuit for the Digital Conversion of QCM-D Responses for FPGA-Based Frequency Measurements

Tommaso Addabbo, Ada Fort, Elia Landi, Riccardo Moretti(✉), Marco Mugnaini, and Valerio Vignoli

Department of Information Engineering and Mathematics, University of Siena, Via Roma 56, 53100 Siena, Italy

{addabbo,ada,landi,moretti,mugnaini,vignoli}@diism.unisi.it

Abstract. This work presents a front-end circuit aimed at enhancing the performance of counter-based digital frequency meters for Quartz Crystal Microbalances with Dissipation Monitoring (QCM-D) applications. The proposed electronics converts the QCM-D vanishing sinusoidal transient response into a square wave signal with voltage levels compliant with standard digital I/O ports. The front-end electronics is designed to preserve the QCM-D response oscillation frequency during the analog-to-digital conversion by means of an Automatic Gain Control (AGC) amplifying stage.

1 Introduction

A Quartz Crystal Microbalance (QCM) is a piezoelectric electromechanical resonator, whose resonance frequency varies according to the formation of adlayers on its surface, or more in general due to interactions with the environment. This characteristic allows to perform different kind of in-gas and in-liquid measurements, such as gas sensing, humidity sensing, biosensing and so on [1, 2].

To evaluate a quartz resonance frequency, the state of the art proposes several approaches, such as oscillation monitoring or the QCM impedance evaluation [3, 4]. Among the first approaches, one of the most established is the Quartz Crystal Microbalance with Dissipation Monitoring (QCM-D) technique [5, 6], which allows the simultaneous measurement of the QCM resonance frequency and of the dissipation factor during an exponentially decaying sinusoidal transient response induced by an applied excitation burst. With respect to other QCM measurement techniques, QCM-D provides a good trade-off between accuracy and cost, also providing the possibility to perform contactless measurements, coupling the front-end electronics and the quartz by means of two coils. However, its main limitation is that it typically requires laboratory instrumentation, analog-to-digital converters, and advanced processing systems to excite the QCM and to extract its response signal features.

To reduce the complexity of architectures employed to measure the QCM resonance frequency, over the years numerous compact electronic interfacing systems have been proposed in literature, e.g., relying on the design of oscillators [7], zero-phase lock-in

R. Berta and A. De Gloria (Eds.): ApplePies 2022, LNEE 1036, pp. 65–72, 2023.
https://doi.org/10.1007/978-3-031-30333-3_9

circuits [8], or compact impedance meters based on mixed-signal architectures, realized through Programmable Logic Devices (PLDs) and Direct Digital Synthesis (DDS) generators [9]. Anyhow, no relevant solution has yet emerged regarding the QCM-D technique; for this reason, the development of low complexity systems for frequency measurement with the QCM-D technique currently represents an open research field, for which a reference state of the art has not yet been defined.

In this paper, we address the problem of reducing the QCM-D measurement systems complexity, proposing an optimized front-end circuit for the 1-bit analog-to-digital conversion of the QCM-D response signal, such to allow its acquisition by a digital frequency measurement architecture implemented in FPGA.

2 Background and Motivation

The measurement of a signal frequency in digital hardware is typically performed using frequency counters. These well-known devices are characterized by a very simple structure, as they are basically composed only by two counters. The input signal sets the measurement time T_x, defined as an integer multiple of its period; calling f_x the input frequency and n_g the number of counter periods, the measurement time is given by $T_x = n_g/f_x$. The device output is the number of counted edges n_p of a reference clock with frequency f_{clk} during T_x. Accordingly, the frequency measure $\widehat{f}_x$ is nominally given by:

$$\widehat{f}_x = \frac{n_g}{n_p} f_{clk} \leq f_x \tag{1}$$

As indicated in (1), neglecting the stochastic contribution of electronic noise, jitter, clock instability and the phase displacement between the input signal and the reference clock, the frequency counter introduces a systematic underestimation of the input signal frequency if the measurement time T_x is not a multiple integer of the clock period $1/f_{clk}$; when this condition is verified, a residual time going from the last observed clock edge to the end of T_x is not measured. The maximum underestimation value is therefore given by the time corresponding to one clock count, or, in terms of frequency:

$$e_{max}\left(\widehat{f}_x\right) = \frac{n_g}{n_p} f_{clk} - \frac{n_g}{n_p + 1} f_{clk} \approx \frac{n_g}{n_p^2} f_{clk} \approx \frac{1}{n_g} \frac{f_x^2}{f_{clk}} \tag{2}$$

It can be observed from (2) that the systematic error is directly proportional to the input frequency f_x and inversely proportional to the clock frequency f_{clk} and to the measurement time n_g/f_x. As also shown in Fig. 1, this means that, to achieve high frequency resolution, the frequency counter must be designed using a reference clock with much higher frequency than the target signals, especially when the measurement time is limited, such as in the QCM-D application case, in which the target signal is, in best conditions, few milliseconds long. Moreover, a QCM resonance frequency can be in the order of tens of MHz, thus requiring reference clock frequency values which are unfeasible for the used hardware to get an acceptable measurement resolution.

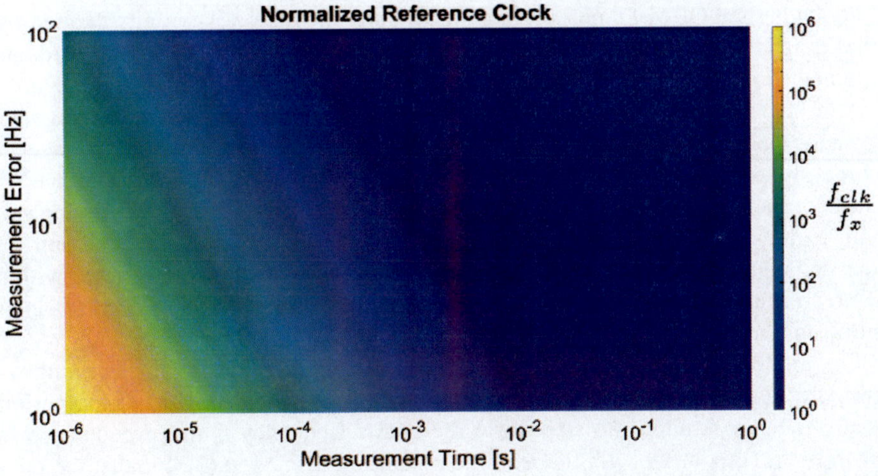

Fig. 1. Graphical representation of the minimum ratio between a digital frequency meter reference clock frequency f_{clk} and the input signal frequency f_x required to expect a certain minimum measurement error in a fixed measurement time.

Furthermore, the 1-bit analog-to-digital conversion of the QCM-D signal should preserve its oscillation frequency, avoiding the introduction of spurious transitions (e.g., due to noise), considering the vanishing dynamical behavior of the resonating sensor.

To approach these aspects, apart from maximizing the frequency counter performance [10], a possible solution is to design a proper analog signal conditioning circuit, which converts the QCM-D transient response into a square wave processable by the digital hardware, maximizes the square wave duration, and steps down the QCM-D frequency, allowing to recover the original frequency at the end of the measurement.

3 Proposed Front-End Electronics

The proposed front-end electronics is composed by different blocks, as shown in Fig. 2.

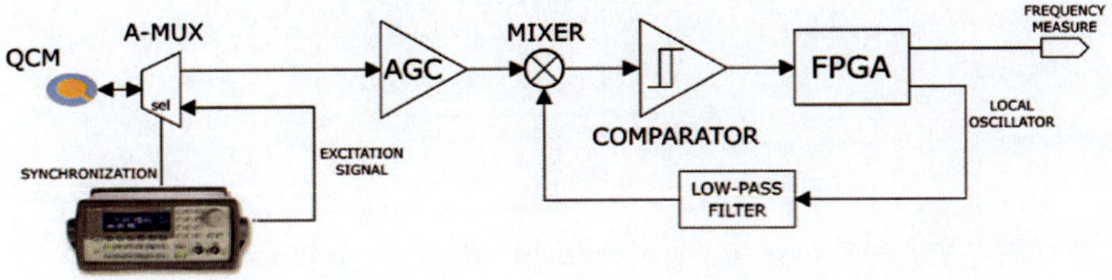

Fig. 2. Functional representation of the proposed front-end electronics.

The quartz resonator is connected to a high-speed analog multiplexer, which alternatively interfaces the QCM either with an excitation source or with the proposed read-out circuit.

Following the QCM-D measurement procedure, the quartz is initially excited by a periodic signal; the excitation frequency is approximately equal to the quartz resonance frequency, to maximize the transient response amplitude and duration.

After the excitation, the resonance sensor is connected to an amplifying stage with Automatic Gain Control (AGC). The AGC automatically tunes the amplifier voltage gain accordingly to the input signal envelope, providing as output a sinusoidal voltage with almost constant amplitude, until the vanishing input signal fades. The amplifying stage circuit includes a Variable Gain Amplifier (VGA), whose gain is feedback-controlled with a demodulating logarithmic amplifier, whose input and output are coupled with the VGA by means of filtering stages, aiming at reducing noise interferences and enhancing control stability.

The amplified signal is then provided as input to a low-distortion mixer, along with a periodic signal generated by a local oscillator implemented in FPGA; the periodic signal frequency f_0 allows to step-down the quartz frequency f_q to a frequency value $f_x = |f_q - f_0|$ compatible with the required resolution defined in (2). Finally, the mixed signal is passed through a high-speed hysteresis comparator to generate a square wave processable in digital hardware. The comparator is interfaced to the digital hardware via a high-speed optocoupler coupling the analog section with the FPGA I/O port.

Finally, the FPGA performs the digital frequency measurement and generates the oscillating signal provided as input to the mixer. As shown in Fig. 3, the oscillating signal is generated by a Mixed-Mode Clock Manager (MMCM) module taking as input the system clock of the FPGA. The same module is also used to generate the clock signal employed as reference clock with frequency f_{clk} employed by the frequency meter. The frequency meter is designed as an equal precision reciprocal frequency counter, i.e., a modified version of the classical frequency counter architecture, targeting, through the pre-set flip-flop, the synchronization of the measurement start command with the input signal [11].

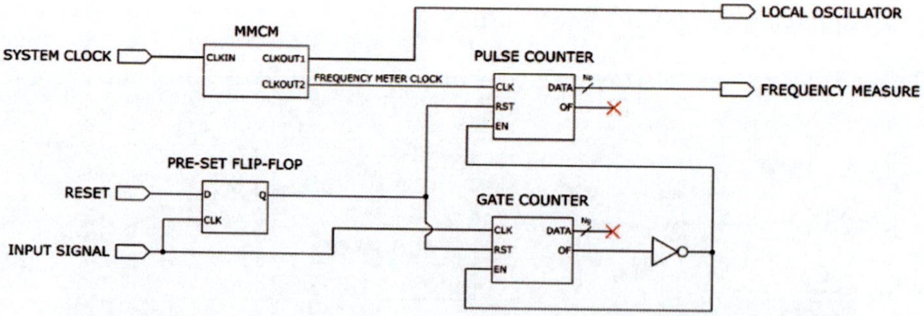

Fig. 3. Functional representation of the digital architecture implemented in FPGA.

The use of the automatically tuned VGA allows to preserve the frequency of the resonance response during the analog-to-digital conversion, eliminating the phase shift error introduced by the combined effect of the comparator thresholds and the decaying behavior of the sensor signal.

4 Measurement and Results

The proposed system was tested on an AT-cut quartz with nominal resonance frequency of 10 MHz, providing the digitally converted signal as input to a digital frequency meter implemented on a Xilinx Artix 7 xc7a45 FPGA, configuring the MMCM to generate a reference clock of 462.5 MHz starting from the system clock set at 100 MHz. The FPGA was exploited also as local oscillator, generating a 9.968 MHz signal which was fed to the mixer, with expected output frequencies ranging from 30 kHz to 40 kHz. The excitation signal was a 0–3.3 V, 10 MHz square wave generated by a Rigol DG4162 waveform generator. The AGC high-pass and low-pass filters cut-off frequencies were set to 1 MHz and 500 kHz, respectively. Both in-air and in-water tests were performed, with measurement times set at $T_{g,air} = 15/f_x$ and $T_{g,water} = 4/f_x$, respectively, according to the transient response duration in the two considered setups.

Figure 4 shows the experimental results obtained from the in-air tests, comparing the performance of the proposed solution enabling and disabling the AGC on the amplifier.

Focusing on the frequency measurement (Fig. 4a), it can be noticed that, disabling the AGC, the phase noise due to the hysteresis comparator threshold introduces a bias in the measured frequency, as the time duration of the intervals between subsequent threshold crossings increases at each period (Fig. 4b). Conversely, the square wave generated enabling the ACG is more stable in frequency, as the phase noise contribution

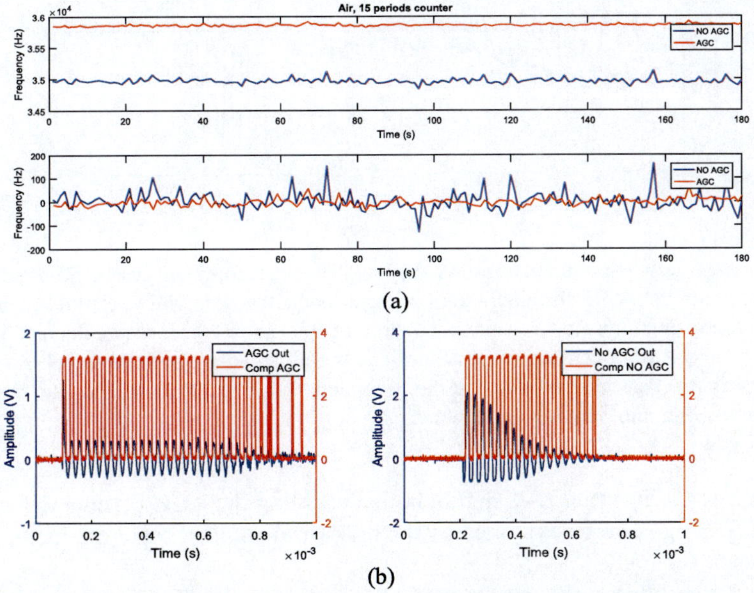

Fig. 4. In-air tests experimental results. (a) Frequency measurement results. The red signal is the measure with the AGC disabled, the blue signal is the measure with the AGC enabled. The upper plot shows the frequency measurement, the lower plot shows the measure fluctuation around its mean value (measurement error). (b) Front-end output signals. The blue signal is the output of the amplifying stage, the red signal is the generated square wave. In the left plot the ACG is enabled, in the right plot the AGC is disabled.

becomes evident only on the last periods, for which the AGC is not able to provide a sufficient gain and must be therefore discarded. Additionally, it can also be observed a larger measurement error, being the measurement standard deviation in steady-state measurement conditions equal to 15.4 Hz when the AGC is enabled and to 40.7 Hz for the fixed gain measurement.

The phase noise sensitivity of the two tested electronics becomes more evident considering the in-water measurements, for which the QCM-D response time duration becomes shorter with respect to the in-air measurements, as shown in Fig. 5.

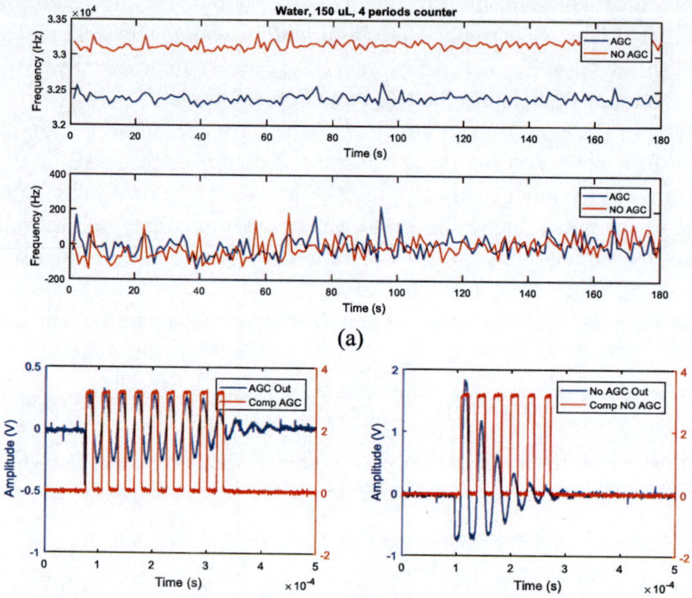

(a)

(b)

Fig. 5. In-water tests experimental results. (a) Frequency measurement results. The red signal is the measure with the AGC disabled, the blue signal is the measure with the AGC enabled. The upper plot shows the frequency measurement, the lower plot shows the measure fluctuation around its mean value (measurement error). (b) Front-end output signals. The blue signal is the output of the amplifying stage, the red signal is the generated square wave. In the left plot the ACG is enabled, in the right plot the AGC is disabled.

Moreover, the in-water tests highlight that enabling the AGC permits to have more periods of the signal to be exploited by the reciprocal counter, therefore increasing the measurement accuracy.

Table 1 summarizes the results presented in Fig. 4 and Fig. 5, evaluating for the measurement system both the standard deviation and the maximum deviation from the measurement mean value for the in-air and in-liquid scenarios, comparing the performance when the AGC is enabled and disabled. As confirmed by the reported data, the use of the AGC allows, starting from the same operating conditions, to increment the measurement performance, as it increases the measurement precision.

Table 1. Measurement setups comparison evaluated for the whole measurement system, composed by the QCM, the front-end electronics and the FPGA architecture.

Measurement Setup	Standard Deviation	Maximum Deviation
In-air AGC OFF	40.7 Hz	162.7 Hz
In-air AGC ON	15.4 Hz	62.9 Hz
In-liquid AGC OFF	54.6 Hz	224.6 Hz
In-liquid AGC ON	53.1 Hz	185.4 Hz

This result shows how, in the design of digital architectures for frequency measurement of QCM-D natural responses, it is necessary to focus not only on the optimization of the digital part but also on the front-end analog electronics. In fact, as shown in this work, proper design of analog pre-processing stages aimed to mitigate noise effects can improve the overall measurement performance.

5 Conclusions

In this work an optimized front-end circuitry for the conversion of QCM-D response signals into digital signals for frequency counter-based measurements has been presented. The proposed circuit converts the QCM-D transient response into a square wave processable by the digital hardware, maximizing its duration, and minimizing the output frequency instability due to noise, undesired spurious transitions, and phase noise. Both in-air and in-water experimental tests have been performed, comparing the performance of the system with a classical sine-to-square converter architecture. The obtained results showed that the presented circuit allows to reach lower frequency measurement error with respect to the classical one, confirming the validity of the proposal.

References

1. Fauzi, F., Rianjanu, A., Santoso, I., Triyana, K.: Gas and humidity sensing with quartz crystal microbalance (QCM) coated with graphene-based materials – a mini review. Sens. Actu. A **330**, 112837 (2021)
2. Noi, K., Iwata, A., Kato, F., Ogi, H.: Ultrahigh-frequency, wireless MEMS QCM biosensor for direct, label-free detection of biomarkers in a large amount of contaminants. Anal. Chem. **91**(15), 9398–9402 (2019)
3. Fort, A., et al.: An adaptive measurement system for the simultaneous evaluation of frequency shift and series resistance of QCM in liquid. Sensors **21**(3), 678 (2021)
4. Fort, A., et al.: QCM measurements in newtonian liquids: problems and performance analysis. IEEE Trans. Instrum. Meas. **71**, 1–13 (2022)

5. Su, X., Chen, D., Li, N., Stevenson, A.C., Li, G., Hu, R.: A wireless electrode-free QCM-D in a multi-resonance mode for volatile organic compounds discrimination. Sens. Actu. A **305**, 111938 (2020)
6. Shoaib, S., Tabrizian, M.: A QCM-D sensing strategy for investigating the real-time effects of oxidative stress on the viscoelastic properties of pre-osteoblast cells. Sens. Actu. B Chem. **293**, 235–246 (2019)
7. Rodriguez-Pardo, L., Cao-Paz, A.M., Fariña, J.: Design and characterization of an active bridge oscillator as a QCM sensor for the measurement of liquid properties and mass films in damping media. Sens. Actu. A **276**, 144–154 (2018)
8. Park, J.Y., Pérez, R.L., Ayala, C.E., Vaughan, S.R., Warner, I.M., Choi, J.W.: A miniaturized quartz crystal microbalance (QCM) measurement instrument based on a phase-locked loop circuit. Electronics **11**(3), 358 (2022)
9. Wudy, F., Multerer, M., Stock, C., Schmeer, G., Gores, H.J.: Rapid impedance scanning QCM for electrochemical applications based on miniaturized hardware and high-performance curve fitting. Electrochim. Acta **53**(22), 6568–6574 (2008)
10. Addabbo, T., Fort, A., Landi, E., Moretti, R., Mugnaini, M., Vignoli, V.: Strategies for the accurate measurement of the resonance frequency in QCM-D systems via low-cost digital techniques. Sensors **22**(15), 5728 (2022)
11. Fang, Y.Y., Chen, X.J.: Design of equal precision frequency meter based on FPGA. Sci. Res. Publ. **4**(10), 696–700 (2012)

A FPGA HardWare Architecture for AZSPWM Based on a Taylor Series Decomposition

Andrea Donisi(✉), Luigi Di Benedetto, Rosalba Liguori, Gian Domenico Licciardo, and Alfredo Rubino

Department of Industrial Engineering, University of Salerno Fisciano (SA), Salerno, Italy
{adonisi,ldibenedetto,rliguori,gdlicciardo,arubino}@unisa.it

Abstract. The parer illustrates a new efficient FPGA hardware architecture for the Active Zero State Pulse Width Modulation, which exploits the Taylor Series to decompose the dwell-times expressions. This decomposition avoids further resources, like external reference signals or Digital Signal Processor, as well as specific architectures, like CORDIC core or Look Up Table-based approaches, which are all solutions provided by the state of the art. All the calculations are done by a fixed-point Arithmetic Logic Unit, which can perform a real time variation of the output amplitude.

The architecture has been implemented on a Xilinx Artix VII FPGA XC7A35T1CPG236C requiring 1973 LUTs and 501 Flip Flops, respectively, the 9.49% and 1.20% of the overall resources, and a dynamic power consumption of 1 mW.

1 Introduction

One of the targets to reduce the air pollution emission can be a significative electrification of transportation vehicles [1] and it is supported by new semiconductor technologies, like silicon carbide [2,3] and gallium nitride [4,5]. However, wide-band gap power devices introduce new constrains for their driving due to the high switching frequency [6], which forces to use controller with high computation speed. In this scenario, the digital controllers based on Programmable Logic Devices, PLD, like micro controller [7], Digital Signal Processor, DSP, [8] or Field Programmable Gate Array, FPGA, [9] become pervasive [10], but new hardware architecture are designed in order to optimize the resources and the performances of the processor.

In this context, one of the applications is the Active Zero State Pulse Width Modulation, AZSPWM, control used in the three-phase inverters. Due to the high computational cost involved in the calculation of the dwell-times [11], it is exclusively managed through digital controllers. The calculation of trigonometric functions is a demanding task and to deal with this problem different approaches are found in the literature, such as resorting to CORDIC algorithm, [12,13], EPROM reference table [14–16] needing an external reference signal, or with

R. Berta and A. De Gloria (Eds.): ApplePies 2022, LNEE 1036, pp. 73–81, 2023.
https://doi.org/10.1007/978-3-031-30333-3_10

a Look-Up Table, LUT, approach [17–20]. The main differences between the state-of-art and the proposed work are as follows:

- the LUT approach requires a pre-calculation stage which is performed by an external numeric computing environment, like Matlab;
- the EPROM reference table technique needs a huge amount of resources, which can require external hardware, like memory, and an external signal reference from a personal computer or Digital Signal Processor;
- the CORDIC algorithm is comparable with our proposal, but it strongly depends on how it is designed and implemented on the target devices.

In this paper we propose a hardware architecture of an AZSPWM controller fully described in VHDL and implemented on a FPGA. It generates the gate drivers control signals of a three-phase inverter with a fixed carrier and switching frequencies and a variable amplitude of the output voltages. The novelty of the work is based on the approach used to calculate the dwell-times: indeed, instead of using the technique previously mentioned, the main equations of the dwell-times are decomposed through the Taylor Series and are carried out in a fixed-point Arithmetic Logic Unit, ALU. The paper is organized in the following sections: Sect. 2 shows the theory of the proposed AZSPWM control technique; in Sect. 3 the description of our HW architecture is reported; the experimental results of its implementation is in Sect. 4; finally, the conclusions are in Sect. 5.

2 Theory of the Proposed AZSPWM

Active Zero State Pulse Width Modulation is a three-phase inverter control technique. When a three-phase inverter generates the sinusoidal output voltages, a common mode voltage (CMV) is induced due to the discrete output voltage levels. It has been demonstrated that the CMV is $\pm V_{DC}/6$ for active voltage vectors (i.e. V_1 to V_6 from Fig. 1) and $\pm V_{DC}/2$ for zero voltage vectors (i.e. V_0 and V_7 from Fig. 1) [21]. It should be noted that the numbers in brackets under the vectors in Fig. 1 refer to the state of the top transistors, for example (110) means that the first two transistors are ON and the third one is OFF. The zero voltage vectors keep all three top transistors ON or OFF, meaning that there is no contribution to the value of θ but they contribute only to the value of V_{REF}. In the traditional Space Vector Pulse Width Modulation (SVPWM) one zero vector is at least fed to the inverter, instead in the AZSPWM two active states, which are opposite to each other (for example V_1 and V_4 for the second sector), are operated with equal time period in order to give no contribute over the time integral of the tensions, reproducing the effect of the zero voltage vector. Hence, the AZSPWM technique produces an output with a reduced CMV compared to the SVPWM [22,23]. An example of the control signals generated through the AZSPWM is shown in Fig. 2, where for a single switching period there are the gate signal sequences for a vector positioned in the second sector.

By using the Taylor Series [24] we linearized the dwell-time equations, avoiding all the problems related to the performing of trigonometrical functions in

FPGA. Moreover, by using the properties of the dwell-times highlighted in [19] we obtained the following equations:

$$T_{\mathrm{A}} = \frac{2V_{\mathrm{M}}\, f_{\mathrm{CLK}}}{V_{\mathrm{DC}} f_{\mathrm{SW}}^2}\left[0.75 f_{\mathrm{SW}} - 2.72 f_{\mathrm{C}} i - \frac{14.8 f_{\mathrm{C}}^2}{f_{\mathrm{SW}}} i^2 + \frac{17.9 f_{\mathrm{C}}^3 i^3}{f_{\mathrm{SW}}^2}\right] \tag{1a}$$

$$T_{\mathrm{B}} = \frac{2V_{\mathrm{M}}\, f_{\mathrm{CLK}}}{V_{\mathrm{DC}} f_{\mathrm{SW}}^2}\left[5.44 f_{\mathrm{C}} i - \frac{35.8 f_{\mathrm{C}}^3 i^3}{f_{\mathrm{SW}}^2}\right] \tag{1b}$$

$$T_0 = \frac{f_{\mathrm{CLK}}}{f_{\mathrm{SW}}} - T_{\mathrm{A}} - T_{\mathrm{B}} \tag{1c}$$

where i is from 0 to $(\frac{f_{\mathrm{SW}}}{6 f_{\mathrm{C}}} - 1) = i_{\mathrm{MAX}} - 1$, and f_{C}, f_{SW}, f_{CLK} are the carrier, the switching and the clock frequency, respectively, V_{M} the output amplitude voltage and V_{DC} the DC-link voltage. The modulation index is defined as $m_{\mathrm{a}} = 2V_{\mathrm{M}}/V_{\mathrm{DC}} \leq 2/\sqrt{3}$.

3 HW Architecture

Figure 3 shows the proposed architecture for an open loop control of the inverter. As long as an additional block is added to control the variation of the modulation index m, this architecture can be used for a closed loop control.

At every switching period the architecture is triggered to calculate the new dwell-times. Its operation can be outlined as follows:

1. The COUNTER block communicates with the $i\,CALCULATOR$ to update the value of i and ACTUAL SECTOR;

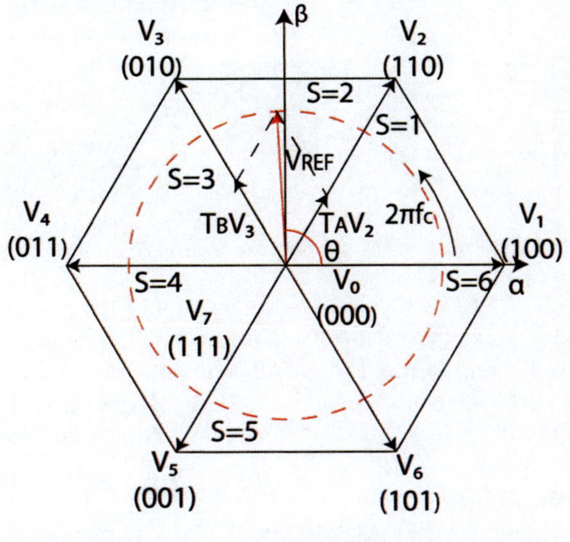

Fig. 1. Polar diagram of the SVPWM representation.

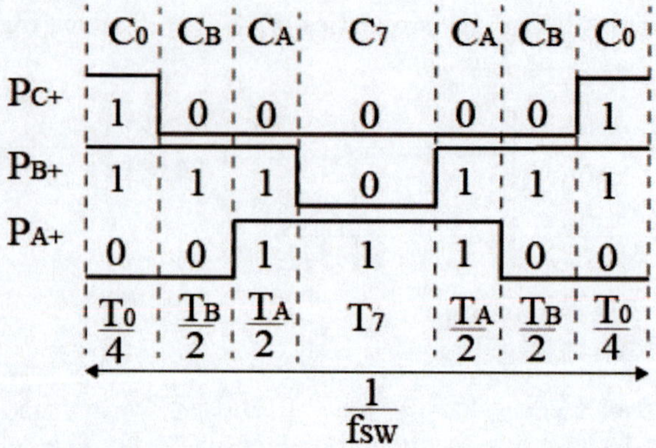

Fig. 2. Example of top transistor gate control signals for one switching period when the V_{REF} is in the 2nd Sector.

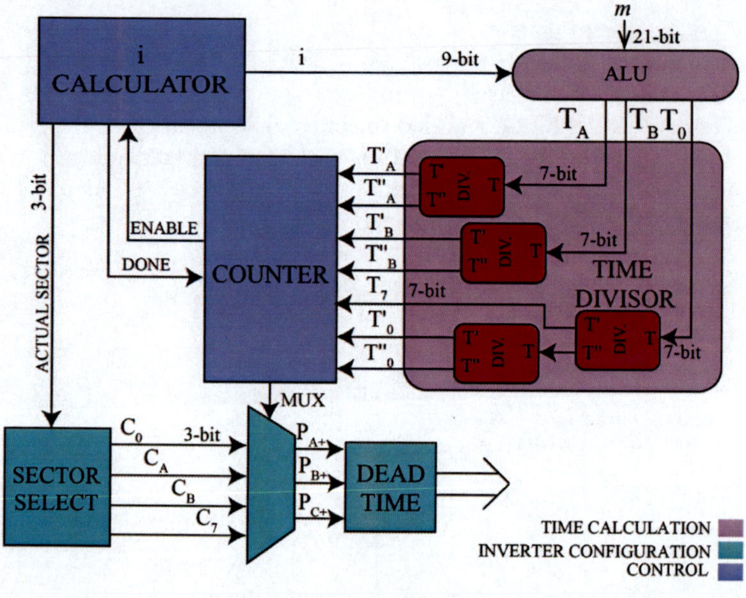

Fig. 3. Block diagram of the proposed HW architecture.

2. The ALU block, which is a 32-bit fixed-point Arithmetic logic unit, reads the values of m and i and starts the calculation. To optimize the architecture, no division is performed in the ALU as all the divisions and multiplications between fixed quantity (e.g. $\frac{2V_M f_{CLK}}{V_{DC} f_{SW}^2}$) are done offline and saved as constants;
3. The TIME DIVISOR divides the times to obtain a sequence of dwell-times as the one outlined in Fig. 2;
4. Based on the value of ACTUAL SECTOR, C_A, C_B, C_7 and C_0 are chosen as shown in Table 1.

Table 1. Top Transistor Sate for each Sector

	Sector					
	1	2	3	4	5	6
C_0	010	011	001	101	100	110
C_A	100	110	010	011	001	101
C_B	110	010	011	001	101	100
C_7	101	100	110	010	011	001

5. The COUNTER block controls the MUX, deciding for how long each control sequence has to be fed to the DEAD TIME block.
6. The DEAD TIME block generates the six gate signals and, also, implements a delay between the falling edge and the rising edge of the gate signals for the transistors belonging to the same branch.

To perform all the calculations needed to compute the new dwell-times, ten clock periods are required. As it can be seen in Fig. 2 every time a change into the configuration of the inverter occurs, only one branch changes its state, reducing the switching loss.

4 Experimental Results

The proposed architecture has been fully described in VHDL and, then, implemented on a Xilinx Artix VII FPGA XC7A35T1CPG236C [25]. In the implementation phase the carrier and the switching frequencies are equal to $f_C = 50$ Hz and $f_{SW} = 100$ kHz, meanwhile the FPGA clock frequency to $f_{\mathrm{CLK}} = 12$ MHz. The minimum angle variation is $\Delta\theta = 2\pi f_C/f_{SW} = 10^{-3}\pi$.

To be implemented the architecture needs 1973 LUTs and 501 FFs, which are, respectively, the 9.49% and 1.20% of the total resources available on the device, and has a dynamic power consumption of just 1 mW. The power has been calculated through a SAIF estimation in Vivado. For the sake of comparison, two CORDIC and LUT based architectures, which are described in [20] and [19], have been implemented on the same FPGA. The results of such comparisons are shown in Table 2, where the power estimations for the different architectures are made on the same time span. The comparison with CORDIC Cascade architecture reveals that our proposal needs less resources and dynamic power consumption; instead, although the CORDIC Iterative architecture needs less resources, which are almost comparable with ours, one more clock period is required making it disadvantageous for high switching frequency applications. Instead, from the comparison with the LUT architecture, ours has a less number of resources and dynamic power, but without the possibility to change the carrier and the switching frequencies of the sinusoidal output.

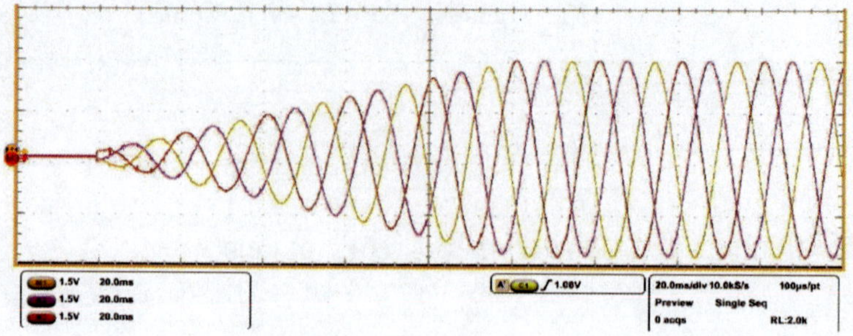

Fig. 4. Concatenate voltage with m_a variation from 0 to 1. Vertical axis: 1.5 V/div, time 20 ms/div sampling rate 10 kS/s

Table 2. Comparison of Resources Utilization between CORDIC Algorithm, LUT architecture and Proposed Architecture

	CORDIC		LUT	Proposed HW
	Iterative	Cascade		
LUT	1769(−10.3%)	2584(+31%)	9108(+362%)	1973
FF	432(−13.8%)	382(−23.8%)	1275(+155%)	501
Dyn.Pow [mW]	1	2.1(+110%)	7(+600%)	1
Clock Iteration	11	1	1	10
Energy per OP [nJ]	0.913	0.175	0.58	0.83

In Fig. 4, 5 and 6, the experimental results are shown and have been acquired by monitoring the FPGA output pins. By using a sampling frequency lower than the switching frequency of the output signals, the waveform are filtered and only the fundamental harmonics are shown. The oscilloscope configuration is written in the image captions. The modulation index can vary in real time, as shown in Fig. 4, where it changes in an interval of 100 ms from 0 to 1. This can be used to perform a soft start at the start-up of an induction motor to reduce the inrush current of an asynchronous motor. In Fig. 5 the top transistor control signals from the FPGA are shown during multiple switching periods for the 2nd sector, as in Fig. 2. The concatenate and phase voltages are reported for a modulation index $m = 1$ in Fig. 6, where the typical third harmonic injection of the control technique is evident in the phase voltage.

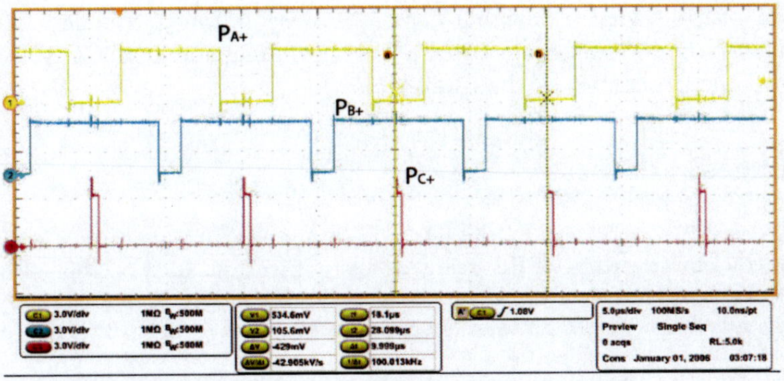

Fig. 5. Top gate control signal. Vertical axis: 2 V/div, time 5 μs/div.

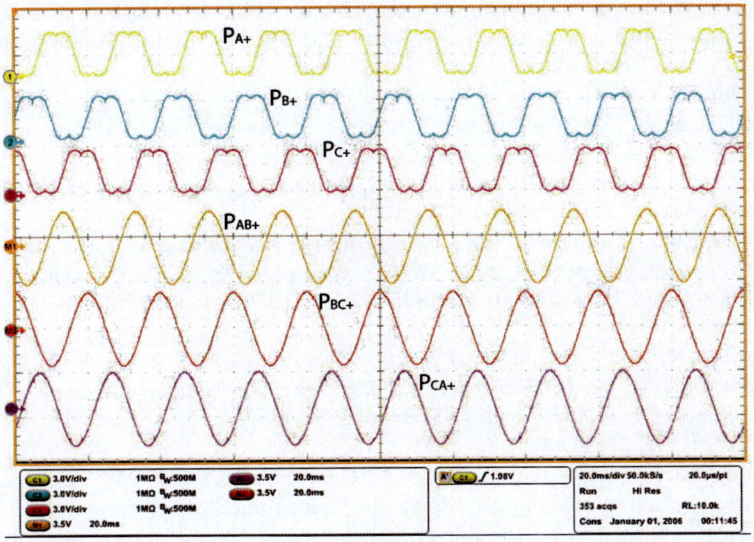

Fig. 6. Concatenate voltage and Phase voltage from the FPGA pin with $m_a = 1$. $P_{AB+} = P_{A+} - P_{B+}$, $P_{BC+} = P_{B+} - P_{C+}$, $P_{CA+} = P_{C+} - P_{A+}$ Vertical axis: 3 V/div phase voltage,: 3.5 V/div concatenate voltage, time 20 ms/div sampling rate 50 kS/s

5 Conclusions

In this paper a promising new hardware architecture for digital controllers able to implement AZSPWM technique for three-phase power converters is presented. From the comparison with other state-of-the-art architectures we can see that our proposal is well matched with CORDIC iterative and can be a good alternative to it. It works without the use of DSP, Cordic core, LUT-based architectures and does not need an external signal reference. The output amplitude can be adjusted in real time, but future works will be focused on the possibility of obtain an output with a variable carrier and switching frequency. Moreover, other improvements will come from a better management of CONTROL sub blocks and the properties of the dwell-times.

References

1. Li, X., et al.: High-voltage hybrid igbt power modules for miniaturization of rolling stock traction inverters. IEEE Trans. Industr. Electron. **69**(2), 1266–1275 (2022)
2. Di Benedetto, L., Licciardo, G., Erlbacher, T., Bauer, A., Rubino, A.: Optimized design for 4h-sic power dmosfet. IEEE Electron Device Lett. **37**(11), 1454–1457 (2016)
3. Di Benedetto, L., Licciardo, G.D., Erlbacher, T., Bauer, A.J., Liguori, R., Rubino, A.: A model of electric field distribution in gate oxide and jfet-region of 4h-sic dmosfets. IEEE Trans. Electron Devices **63**(9), 3795–3799 (2016)
4. Barth, C.B., et al.: Design, operation, and loss characterization of a 1-kw gan-based three-level converter at cryogenic temperatures. IEEE Trans. Power Electron. **35**(11), 12040–12052 (2020)
5. Naradhipa, A.M., Kim, S., Yang, D., Choi, S., Yeo, I., Lee, Y.: Power density optimization of 700 khz gan-based auxiliary power module for electric vehicles. IEEE Trans. Power Electron. **36**(5), 5610–5621 (2020)
6. Henn, J., et al.: Intelligent gate drivers for future power converters. IEEE Trans. Power Electron. **37**(3), 3484–3503 (2021)
7. Ruiz, G.E.M., Muñoz, N., Cano, J.B.: Design methodologies and programmable devices used in power electronic converters-a survey. In: IEEE Workshop on Power Electronics and Power Quality Applications (PEPQA), vol. 2015, pp. 1–6. IEEE (2015)
8. Pop, O., Chindris, G., Dulf, A.: Using dsp technology for true sine pwm generators for power inverters. In: 27th International Spring Seminar on Electronics Technology: Meeting the Challenges of Electronics Technology Progress, 2004, vol. 1, pp. 141–146. IEEE (2004)
9. Monmasson, E., Idkhajine, L., Bahri, I., Charaabi, L., et al.: Design methodology and fpga-based controllers for power electronics and drive applications. In: 2010 5th IEEE Conference on Industrial Electronics and Applications, pp. 2328–2338. IEEE (2010)
10. Rashid, M.H.: Power electronics handbook: devices, circuits and applications. Elsevier (2010)
11. m-Mode SVPWM Technique for Power Converters. CPES. Springer, Singapore (2019). https://doi.org/10.1007/978-981-13-1382-0_9
12. Wiśniewski, R., Bazydło, G., Szcześniak, P.: Svm algorithm oriented for implementation in a low-cost xilinx fpga. Integration **64**, 163–172 (2019)
13. Shreyas, S.G., Vachhani, L.: Area efficient reconfigurable architecture for current control loop of a servo controller. In: 2012 IEEE 7th International Conference on Industrial and Information Systems (ICIIS), pp. 1–6 (2012)
14. Tzou, Y.Y., Hsu, H.J.: Fpga realization of space-vector pwm control ic for three-phase pwm inverters. IEEE Trans. Power Electron. **12**(6), 953–963 (1997). https://doi.org/10.1109/63.641493
15. Neacu, D.O., Kim, J.C., Lehman, B.: A three-phase multioptimal pwm implemented on 2-gbit flash memory integrated circuits. IEEE Trans. Power Electron. **32**(7), 5813–5826 (2017)

16. Neacsu, D.O.: Novel microcontrollers with direct access to flash memory benefit implementation of multi-optimal space vector modulation. IEEE Trans. Industr. Inf. **8**(3), 528–535 (2012)
17. Donisi, A., Di Benedetto, L., Licciardo, G.D., Rubino, A., Piccirilli, E., Lanzotti, E.: A fully fpga implementation of svpwm for three-phase inverters without external reference signals. In: 2020 IEEE International Conference on Environment and Electrical Engineering and 2020 IEEE Industrial and Commercial Power Systems Europe (EEEIC / I CPS Europe), pp. 1–5 (2020)
18. Donisi, A., Di Benedetto, L., Licciardo, G.D., Rubino, A., Piccirilli, E., Lanzotti, E.: Design of digital controller for svpwm algorithm with real-time control of the output amplitude and switching frequency. In: 2021 IEEE International Symposium on Circuits and Systems (ISCAS), pp. 1–5 (2021)
19. Di Benedetto, L., Donisi, A., Licciardo, G.D., Rubino, A.: A hardware architecture for svpwm digital control with variable carrier frequency and amplitude. IEEE Trans. Industrial Informat. (2021)
20. Di Benedetto, L., et al.: Implementation of hardware architecture for svpwm with arbitrary parameters. IEEE Access **10**, 32381–32393 (2022)
21. Ali, S.M., Reddy, V.V., Kalavathi, M.S.: Simplified active zero state pwm algorithms for vector controlled induction motor drives for reduced common mode voltage. In: International Conference on Recent Advances and Innovations in Engineering (ICRAIE-2014), pp, 1–7 (2014)
22. Hou, C.C., Shih, C.C., Cheng, P.T., Hava, A.M.: Common-mode voltage reduction pulsewidth modulation techniques for three-phase grid-connected converters. IEEE Trans. Power Electron. **28**(4), 1971–1979 (2012)
23. Lai, Y.S., Chen, P.S., Lee, H.K., Chou, J.: Optimal common-mode voltage reduction pwm technique for inverter control with consideration of the dead-time effects-part ii: applications to im drives with diode front end. IEEE Trans. Ind. Appl. **40**(6), 1613–1620 (2004)
24. In: Taylor expansions and applications, pp. 223–255. Springer Milan, Milano (2008)
25. Xilinx inc.: Artix 7. (2018). https://www.xilinx.com/products/silicon-devices/fpga/artix-7.html (Accessed 27 Feb 2018)

Analysis, Design and Synthesis of an Execution Tracing Unit (ETU) Based on AUTOSAR Run-Time Interface (ARTI)

Francesco Cosimi[1,2](✉), Fabrizio Tronci[1,2](✉), Sergio Saponara[1,2], and Paolo Gai[1,2](✉)

[1] Huawei Research Center, Via Pietro Nenni 24, 56124 Pisa, Italy
{francesco.cosimi,fabrizio.tronci,paolo.gai}@huawei.com
[2] Department of Information Engineering, University of Pisa, Via Girolamo Caruso, 16, 56122 Pisa, Italy
sergio.saponara@unipi.it

Abstract. Complex vehicle applications and the need of reducing the number of processing units are giving audience to virtualization on multi-core devices. Such solution offers flexibility, but it is challenging in terms of timing predictability and resources partitioning. This explains why timing analysis is a crucial topic for Functional Safety. A knowledge of tasks and their behavior, applied at debug or runtime, allows balancing resources, refining scheduling and understanding the issues of a safety critical system. In the paper, we present the Execution Tracing Unit (ETU), a hardware support to AUTOSAR Run-Time Interface (ARTI), whose scope is to define an interface between tracing and analysis tools. The hardware element introduces a limited overhead if synthetized for an xczu9eg FPGA device. Meanwhile, simulations demonstrated that the peripheral is able to manage occurring software items with a very fast activation rate (an average of 4 clock cycles between two subsequent activations).

Keywords: execution tracing · timing analysis · autosar · functional safety

1 Introduction

Functional Safety (FuSa) is defined as "the absence of unreasonable risk due to hazards caused by malfunctioning behavior of electrical or electronic systems". To analyze the potential risk of malfunctioning on Electronic Controller Units (ECUs), electronic sensors and communication structures for automotive systems, from December 2018, a new version of ISO 26262 regulates safety implementation of electrical/electronic devices [1]. It classifies risk through Automotive Safety Integrity Level (ASIL). There are four ASILs: A, B, C, and D.

R. Berta and A. De Gloria (Eds.): ApplePies 2022, LNEE 1036, pp. 82–89, 2023.
https://doi.org/10.1007/978-3-031-30333-3_11

ASIL-A represents the lowest degree of automotive hazard, while ASIL-D represents the highest one [2]. As specified in [3], automotive systems are becoming more and more complex, and applications with diverse ASIL have to coexist in the same hardware in a Mixed Criticality System (MCS), sharing the same hardware resources, and running real-time tasks with certain time and safety constraints [4].

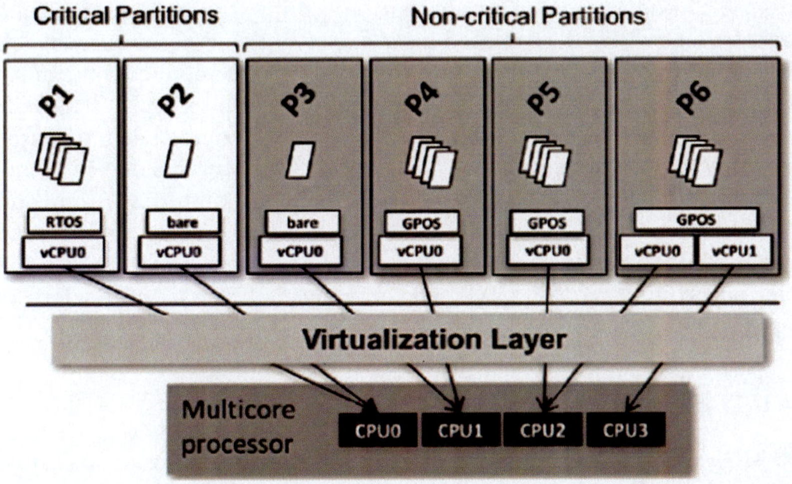

Fig. 1. Example of virtualization for a multicore ECU

Figure 1 shows how an hypervisor on a multicore device allows the co-execution of more operating systems (e.g. Real Time OS, General Purpose OS, with different ASIL) in a safe and isolated manner. However, timing Validation & Verification (V&V) becomes harder to get only through the analysis of end-to-end execution times. The interference generated by cores affects the performance and influences critical tasks' execution, with the impossibility to estimate the Worst-Case Execution Time (WCET) [5]. For this reason, several automotive systems are designed to tolerate the WCET, but to guarantee the safe functionality within the Fault Tolerant Time Interval (FTTI). To reach an adequate level of predictability the developer turns to Timing Analysis tools, which collects performance data (from a Performance Monitoring Unit), software-tracking results and log of error recovery actions. Such instrument highlights timing and scheduling issues, their causes, and defines the way to mitigate interference. Nowadays, the market offers many solutions in terms of performance monitoring and tracking of execution. Software Tools (i.e. Eclipse TraceCompass [6], SEGGER SystemView [7], Rapita RapiTask [8], Gliwa T1 [9], Lauterbach TRACE32 [10], etc.) can collaborate with the automotive application to get results. In this portfolio of multiple solutions, even AUTOSAR specified its proposal aimed to extract run-time information from executed code: AUTOSAR Run-Time Interface (ARTI) [11].

This publication proposes a prototype of an Execution Tracing Unit (ETU) hardware element designed for ARTI, which can also collaborate with a PMU to provide an efficient instrument to analyze the behavior of a safety critical application.

1.1 State of Art for Execution Tracing

Software tools cited above perform execution tracing, but at the same time they cause interference with safety related tasks executing. Furthermore, some of them need additional hardware support, and they offer only limited compatibility with ISAs, Operating Systems or standard communication interfaces. SEGGER SystemView and Eclipse tool TraceCompass are the only ones that have a relevant position regarding performance and compatibility. However, none of them offers the possibility of using ARTI as tracing and debugging interface. Implementing a core independent on-chip tracing unit frees the ECU from wasting execution time to perform tracing features, it allows to standardize the interface with the controller according to AUTOSAR, and moreover, it is part of a safety related hardware library.

1.2 AUTOSAR Run-Time Interface (ARTI)

AUTOSAR (AUTomotiveOpen System ARchitecture) is an open and standardized automotive software architecture, developed by automobile manufacturers, suppliers and tool developers. AUTOSAR Run-Time Interface (ARTI) standardizes a way to link build tools to debugging tools, and to exchange information with timing analysis tools. The tools need to have knowledge about tasks, interrupts and runnables, in other words to have the so-called "OS awareness". An ARTI solution should support distributed systems and multi-core, connect with other AUTOSAR modules, integrate instrumentation-based tracing and measurement solutions, interface with TIMEX (TIMing EXtensions to describe and specify timing information), and debugging and tracing tools should be able to read the AUTOSAR XML, and link it with the traces acquired from the ARTI backend on the target. A hardware unit optimized for ARTI helps providing an overview over running applications, for an analysis of software's behavior [12]. It accelerates the annotations sampling, minimizing the time overhead introduced by the use of traditional software solutions. Debug data are useful to estimate WCET, to optimize the scheduling of running applications and to enhance the performance of a complex system. Moreover, debugging and tracing tools should be able to read the ARTI files and be "AUTOSAR aware", giving additional debugging and tracing features to the developer. These tools should be able to read AUTOSAR XML, and link it with the traces acquired from the ARTI backend on the target.

1.3 Innovation Idea of This Work

Our research, currently targeting RISC-V technologies, has the objective of guaranteeing the proper integrity of the hardware platform and the corresponding environment where safety related software (up to ASIL D) is executed.

To achieve our goal, we might introduce some core independent peripherals, a Performance Monitoring Unit, an Error Management Unit and an Execution Tracing Unit aimed to respectively trace the performances, errors and execution information, to analyze them and then adapt the software behavior according to the defined functional and safety requirements. In fact, the trace of hardware events and software items (running virtualizations, task activations, task ends, interrupts, etc.) is essential for resources budgeting and performance analysis.

This paper focuses on a subset of the work currently carried out, and in particular we will describe the Execution Tracing Unit (ETU), a hardware support to AUTOSAR Run-Time Interface (ARTI). The paper contains five sections. Section 1 gives an introduction to the topic, a description of ARTI and of the research idea. Section 2 describes the design of a preliminary architecture of an ARTI compliant ETU. Section 3 reports an analysis of resources used for the synthesis (Vivado 2020.2 for Xilinx Zynq ZCU102 Ultrascale+ FPGA). Section 4 is about ETU behavioral simulation. Section 5 concludes the paper.

2 Design of ETU

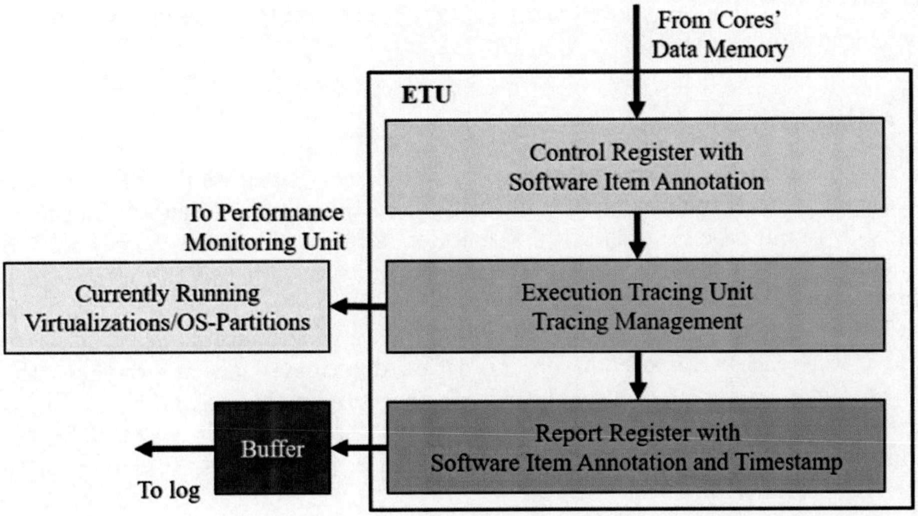

Fig. 2. ETU Block Scheme with main in/out signals

The ETU's block diagram is showed in Fig. 2. The device has to be integrated in a multicore ECU, together with other safety peripherals. The architecture has been sized for an 8-core device and integrated to the RISC-V OpenPulp platform [13], which allows to have up to 8 cores in the same cluster. The ETU collects information from a memory mapped register (ETU Control Register with Software Item Annotation). Then the ETU Tracing Management block transfers data to the ETU Report Register and associates a timestamp (to correctly reconstruct the trace). Moreover, this part of the device notifies to the PMU what are the Virtualization/OS-Partitions that are currently running on each core. This feature is necessary to implement the hardware events filtering mechanism described in [3] for the PMU.

2.1 ETU Registers

The Control Register is a 3×32 bits memory-mapped register. The first and the second row are dedicated to peripheral enable and reset controls. The third row contains the ARTI compliant annotation. This last row is divided in 4 sections. Starting from the most significant bit we have: 3 bits dedicated to core ID; 2 bits for Virtualization/OS-Partition ID; then 11 bits are reserved to the Event ID (i.e. starting point, ending point, interrupt, exception etc.). Finally, last 16 bits contains the identifier for the software item which has to be logged.

The Report register is internal and its size is 3×32 bits. First row is the same of last one of Control Register. The other two rows together contains the 64 bit timestamp, necessary to correctly reorder the flux of events.

The Register for Running Virtualizations/OS-Partitions associates to each core a 4 bit sector. Each bit corresponds to a specific Virtualization/OS-Partition on the core; if the flag is high the partition is active, otherwise it is waiting to be executed or is unused.

2.2 Buffer

This block is organized in 5 sections, divided in more subsections (20). The size of each section is the size of the subsection multiplied for the size of the report register. In our case the buffer is 5×20 x 3×32 bit, which means nearly 1.2 KB of buffer memory. Every time a section is completely filled, it is sent to a log device through a 32 bit serial bus.

While the number of sections is useful to have a wider margin in case of delays and interferences, the size of the subsection determines how many times the transmission is enabled. Larger the size of subsections is, longer and less frequent are the transmission intervals are; instead, lower the size of subsections is, briefer and more frequent transmission intervals are. The solution 5×20 seemed to be the right balance point for our system.

3 Hardware Synthesis of ETU

We synthetized the ETU on a Xilinx Zynq ZCU102 Ultrascale+ FPGA device (xczu9eg-ffvb1156-2-e), and is performed using Xilinx Vivado 2020.2 tool. Table 1 shows the amount of CLB Registers (Configurable Logic Block Registers), CLB LUTs (Look Up Tables), CARRY8s (LUTs organized as a column with an 8-bit carry chain per CLB), F7 MUXes (combine the output of two 6-input LUTs to create a 7 input function) and F8 MUXes (to combine the output of two F7 MUXes to create an 8 input function) we need to implement the octa-core ETU and the buffer. The synthesis does not involve memory blocks or Digital Signal Processing (DSP) units.

Table 1. Logic Elements of Xilinx Ultrascale+ FPGA for ETU and Buffer synthesis

Part	CLB Reg	CLB LUT	Carry8	F7Mux	F8Mux
ETU	273	298	2	0	0
Buffer	9 856	12 902	46	480	224

4 Behavioural Simulation of ETU and Buffer

To validate our design and to evaluate the performance and to predict the behavior of RTL devices description in Vivado, we realized a Python behavioral simulation of logging mechanism. While the ETU hardware introduces a low latency in detecting and managing new software items, on the other hand, the buffer constitutes the bottleneck of the architecture. With a 5 sections and 20 subsections implementation we expect to have a good margin (more than 60%) to manage unexpected overproduction of data, in a condition where the span between two consecutive activations is 4 clock cycles.

Simulations show results for a 100 MHz clock, which is a reasonable reference frequency for a RISC-V core, implemented in a Xilinx FPGA device. Figure 3's left graph presents the behavior of average and maximum percentages reached by the buffer, increasing the number of sections from 2 to 10. As we can see, more sections we have and more the buffer is relaxed in terms of occupation. These simulations help in buffer sizing, in order to customize buffer's safety margin, to ensure the correct behavior of the architecture. The number of buffer's subsections, instead, does not affect buffer's occupation. However, this influences the output rate towards log system. In fact, larger is the number of subsections and more data transmission will last.

Hardware behavioral simulations resulted to be in line with Python ones, with only a few (nearly 3%) difference.

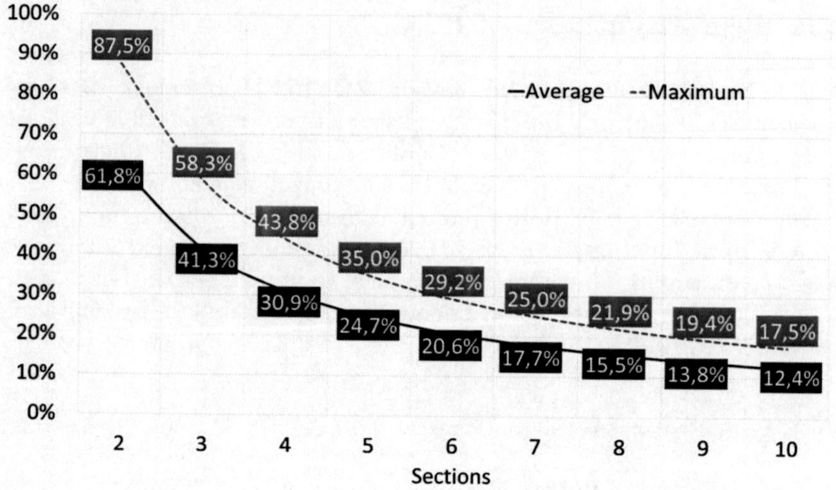

Fig. 3. The graph shows the average and maximum ETU Buffer % with a new Software Item annotations every 40 ns, for simulations lasting up to 50 ms. Here we swapped the number of buffer's sections (2–10) and fixed the number of subsections (20).

5 Conclusions

This paper defines hardware specs and details of an ARTI compliant ETU. An analysis of the architecture is given and simulation of its behavior are presented. The ETU produces data to be sent to a log unit and offers the possibility of interfacing with other peripherals, or a Timing Analysis tool. The device has a reduced hardware overhead in terms of FPGA's programmable logic. As regards performance, features, monitoring capabilities, flexibility and portability the ETU seems to be comparable with software state of art. Next step will be debugging, optimization of the hardware, and integration in a more complex system. This will include the RISC-V ECU implemented in the FPGA, as proof of concept, and other functional safety peripherals. Moreover, we will focus on the deployment of a RISC-V software library to simplify the inclusion of tracing features in automotive applications.

References

1. Ismail, A., Jung, W.: Research trends in automotive functional safety. In: 2013 International Conference on Quality, Reliability, Risk, Maintenance, and Safety Engineering (QR2MSE), Chengdu, China (2013)
2. Synopsis ASIL. www.synopsys.com/automotive/what-is-asil.html
3. Cosimi, F., Tronci, F., Saponara, S., Gai, P.: Analysis, "Hardware Specification and Design of a Programmable Performance Monitoring Unit (PPMU) for RISC-V ECUs". In: SMARTCOMP2022. Aalto University, Espoo, Finland (2022)

4. Caforio, F., Iannicelli, P., Paolino, M., Raho, D.: VOSySmonitoRV: a mixed-criticality solution on Linux-capable RISC-V platforms. In: 10th Mediterranean Conference on Embedded Computing (MECO), Budva, Montenegro (2021)
5. Crespo, A., Balbastre, P., Simó, J., Coronel, J., Gracia Pérez, D., Bonnot, P.: Hypervisor-Based Multicore Feedback Control of Mixed-Criticality Systems. IEEE Access **6** (2018)
6. Lousa Marques, A.: Monitoring Architecture for Real Time Systems. University of Beira Interior, Covilhã (2016)
7. Segger SystemView. www.segger.com/products/development-tools/systemview
8. Rapitask. www.rapitasystems.com/products/rapitask
9. Gliwa T1. www.gliwa.com
10. Lauterbach. www.lauterbach.com
11. Bo, H., Hui, D., Dafang, W., Guifan, Z.: Basic Concepts on AUTOSAR Development. In: 2010 International Conference on Intelligent Computation Technology and Automation, pp. 871-873 (2010). https://doi.org/10.1109/ICICTA.2010.571
12. ARTI. www.autosar.org/fileadmin/user_upload/standards/classic/19-11/AUTOSAR_SWS_ClassicPlatformARTI.pdf
13. Pullini, A., Rossi, D., Loi, I., Tagliavini, G., Benini, L.: Mr.Wolf: an energy-precision scalable parallel ultra low power SoC for IoT edge processing. IEEE J. Solid-State Circ. **54**(7), 1970–1981 (2019). https://doi.org/10.1109/JSSC.2019.2912307

A Side Channel Attack Methodology Applied to Code-Based Post Quantum Cryptography

Kristjane Koleci(✉), Lorenzo Cecchetti, Guido Masera, Maurizio Martina, and Massimo Ruo Roch

DET, Politecnico di Torino, Corso Duca Degli Abruzzi 24, Torino, Italy
{kristjane.koleci,guido.masera,
maurizio.martina,massimo.ruoroch}@polito.it,
s276870@studenti.polito.it

Abstract. The present work proposes a Side Channel Attack that targets the multiplier of a code-based Post Quantum Cryptography primitive. The Secret Key has been recovered with the Correlation Power Analysis obtained with the use of a methodology that simulates the power consumption profile of a design and then validates the method with the real device. The methodology is proposed as a useful tool to study weaknesses of designs during their design phase.

Keywords: Post-Quantum Cryptography · Side Channel Attack · VLSI · FPGA

1 Introduction

The security of Public Key Cryptography (PKC) is based on the hardness of recovering the Secret Key (SK) from the Public Key (PK), this is considered infeasible for classical computers, but quantum computers can bridge the gap and put the security of current cryptographic systems at risk. Post Quantum Cryptography (PQC) algorithms are one solution. The primitives have been proposed to a competition launched by the National Institute of Standards and Technology (NIST) in 2016 [1].

The adoption of NP-hard problems is not enough to guarantee the security of the whole system. Indeed, it is important to analyze possible information leakages of either the device where the algorithm is running or the hardware implementation of the primitives. Despite a Side Channel Attack has been successfully applied to a PQC primitive running on a device [2], the case of hardware implementation of such primitives has not been studied yet.

The present work analyzes the power consumption profile of a multiplier employed in PQC primitives. In Sect. 2, LEDAcrypt/BIKE [3,4] code-based cryptosystems and the multiplier are described together with the model of the attack. In Sect. 3, the attack and experimental methodology are described. Finally, in Sect. 4, the results of the validation are reported and the conclusions are drawn.

R. Berta and A. De Gloria (Eds.): ApplePies 2022, LNEE 1036, pp. 90–96, 2023.
https://doi.org/10.1007/978-3-031-30333-3_12

2 Architecture for Code Based Cryptosystem

Code-based cryptosystems are based on the idea originally proposed by McEliece [5]: the encryption adds redundancy to the message to hide ($mathbfm$), with Public Key (PK) $\mathbf{G}$, and a error pattern $\mathbf{e}$ to obtain the ciphertext ($\mathbf{x}$), while the decryption removes the errors with the Secret Key (SK) $\mathbf{H}$. In the present work, the focus is on LEDAcrypt/BIKE cryptosystems, where codes are Quasi-Cyclic Low/Moderate Density Parity Check (QC-LDPC/MDPC) codes, as a consequence, $\mathbf{H}$, the SK, is a Quasi Cyclic matrix.

A QC binary matrix is a block matrix with n_0 cyclic blocks and size p ranging from 10^4 to 10^5. In cyclic matrices only the first row (or column) is required to completely describe the matrix, as each row is obtained by cyclic-shifting the first one. The SK is: $\mathbf{H} = [\mathbf{H_0},\ \mathbf{H_1},\ \dots,\ \mathbf{H_{n_0-1}}]$. Moreover, the blocks are sparse and the number of 1 s is referred to as d_v, which value is ≈ 100; these 1s are the target of the attack. The encoding and decoding algorithms require to calculate the product between a vector and a cyclic matrix LEDAcrypt [3] [4]. As a consequence, any multiplier at the decoder side, being in charge of computing the product between a vector and the $\mathbf{H}$ matrix (the SK), is a critical block. Indeed, fundamental information about the positions of the ones can be derived by monitoring the power trace of the multiplier.

The multiplier proposed in [6] is referred to as `Vector By Circulant` (VbC). It takes as inputs a vector ($\mathbf{v}$) and a set of positions ($\mathbf{P}_v$) (the asserted 1s in the first row of the matrix) to compute the multiplication as a sum of cyclic shifts of the input vector, v_s (partial products). The shift amount is read from $\mathbf{P}_v$ as shown in Algorithm 1. This is provided as `Position` in the multiplier unit of Fig. 1. According to Algorithm 1, the module generates one partial product v_s per iteration. The single v_s is generated n_b bits per clock cycles with the `Rotate unit`. The new v_s is summed to the previous result with the `Element wise sum`.

Algorithm 1. `VbC`

Input: $\mathbf{v}(1,n)$,$\mathbf{P}_v(1,d_v)$;
Output: $\mathbf{r}(1,n)$;
$\mathbf{r} \Leftarrow \mathbf{0}$;
for $i_P = 0$ **to** d_v **do**
  $k = \mathbf{P}_v(i_P)$;
  $\mathbf{v}_s = [\mathbf{v}(k:n), \mathbf{v}(1:k-1)]$
  ;
  $\mathbf{r} \Leftarrow \mathbf{r} + \mathbf{v}_s$
end for

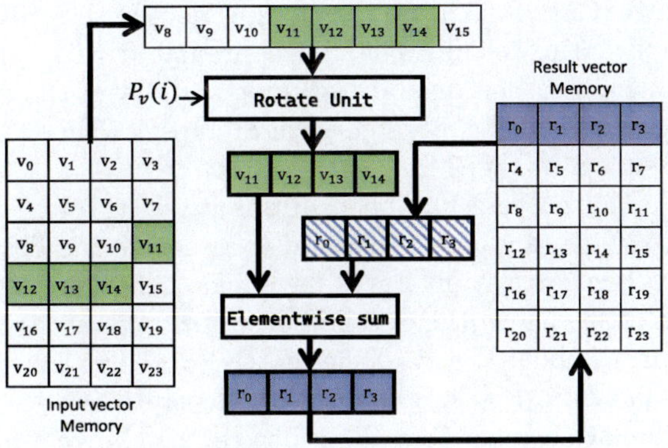

Fig. 1. VbC with $n_b = 4$.

The described multiplier is then analyzed in order to find its weaknesses. In general, the target of Side-Channel Attacks are the steps of the algorithms that employs the Secret Key (SK) to evaluate an intermediate variable, this is the case of the Syndrome (**s**) which is the result in Algorithm 1, **H** is the set of position, **x** is the input vector.

3 Attack Methodology

The present work analyzes the dynamic power consumption to predict possible information leakage of the architecture. The study presents an initial analysis of the multiplier by relying on simulations, with an approach similar to the one adopted in [7]. Then, results are validated on an FPGA device.

The Correlation Power Analysis (CPA) attack described in [8] is applied to the dynamic power consumption (P_{dyn}) of the multiplier in the code-based decoder. The idea is to retrieve the Secret Key (SK), i.e. the decoding matrix **H**, by exploiting the leakages in the computation of the Syndrome, **s**.

The attack iteratively guesses the asserted positions in the first row of **H**. The number of elements to search depends on the type of code that is employed in the algorithm, and for BIKE [4]/LEDAcrypt [3] is in the order of 10^2.

The model used to predict P_{dyn} is the Hamming Distance (HD) [8], then the *intermediate values* of the method are selected, two consecutive sequences of $\mathbf{r}_i$ after the `Elementwise sum` unit, and targeted by the attack.

3.1 Power Traces Derivation

The Power Traces, required by the CPA attack, have been obtained via post-synthesis simulation of the multiplier netlist, during the computation of the quasi-cyclic multiplication to evaluate the Syndrome **s**.

The netlist has been obtained with Synopsys Design Compiler using the UMC CMOS 65 nm technology. Then, with a SK and a random message as the inputs, the decoding has been simulated with Questa Sim by Mentor, to derive the information on the switching activity of the design. Finally, the single power trace has been generated with Synopsys PrimeTime. An example of power trace is shown in Fig. 2.

The process has been automatized through a Python script to launch the tool with random inputs and to generate the desired number of traces.

The correlation traces have been computed from the power traces and power prediction with the detailed description presented in [8] for all the asserted positions in the first row of the matrix (d_v). The example of a single correlation trace is shown in Fig. 3. The analysis clearly shows the presence of a peak: this is the correlation peak that identifies the clock cycle in which the transition between the two intermediate values occurs.

In detail, the *intermediate values* for the HD power model are computed, for this purpose two matrices are required, IV_1 and IV_2, which contains all the possible consecutive intermediate values we can have, the number of rows of the

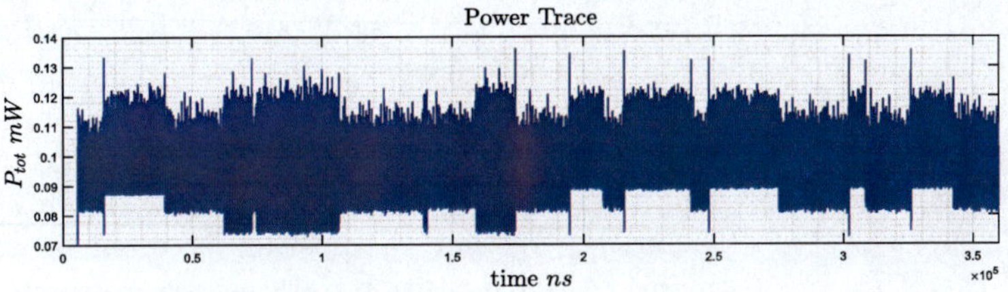

Fig. 2. Example of a Power Trace derived with Synopsys Prime time during the execution of a complete product.

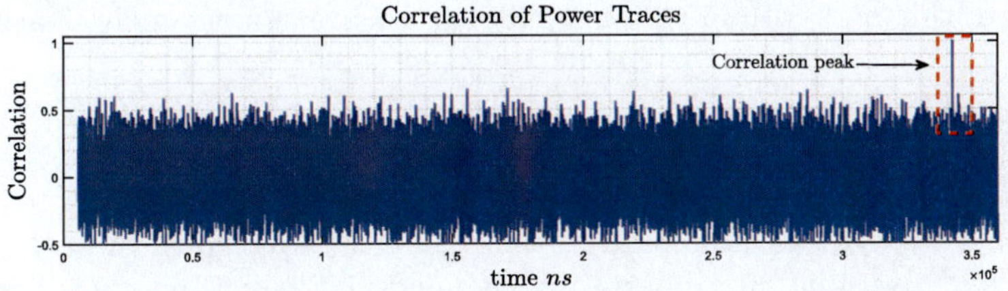

Fig. 3. Correct Correlation trace of the position 11 in d_v in SK. The correlation peak is highlighted in the dashed box.

matrices corresponds to all possible input of the `Rotate Unit` and the number of columns to the possible $\mathbf{P_v}$ selection of the unit. The hamming distance of IV_1 and IV_2 is evaluated (HD model) to derive the modeled power matrix **PM**. The measured power traces (for a fixed value of the $\mathbf{P}_v$) are collected to derive matrix **PT**, with the number of rows that corresponds to the number of traces and the columns that corresponds to the number samples. The information in **PM** and **PT** are combined with the Pearson Correlation Coefficient to derive the Correlation Matrix **CP**, in our case the formula becomes:

$$cp_{i,j} = \frac{\sum_{n=1}^{N}(pm_{n,i} - \bar{pm}_i) \cdot (pt_{n,j} - \bar{pt}_j)}{\sqrt{\sum_{n=1}^{N}(pm_{n,i} - \bar{pm}_i)^2 \cdot \sum_{n=1}^{N}(pt_{n,j} - \bar{pt}_j)^2}} \tag{1}$$

The i, j lower case variables represent the i, j value in the corresponding upper case matrix.

3.2 Secret Key Recovery

The CPA attack is structured as an algorithm that iteratively recovers one position in $\mathbf{P}_v$. In general, the candidate values for the position coming out from the CPA algorithm are associated to correlation traces that show a peak very close to 1. However, multiple candidate positions for each iteration may be found, due to the rotations in `VbC`, depending on the position we are trying to guess. In particular, this is the case of the first position in $\mathbf{P}_v$. In the first position the correct value has to be computed from the first peak that appears in the considered range of time. The guess on the second position is simpler, because only one correlation trace presents a peak, thus there is only one possible candidate. On the contrary, the third position has the same behaviour as the first one. In this case, we can have two possible scenarios: three candidates or two candidates. In the second case, the candidates are $\mathbf{P}_v(2)$ and $\mathbf{P}_v(3)$. The candidates are three if $\mathbf{P}_v(2) - \mathbf{P}_v(1)$ is a multiple of n_b, with $\mathbf{P}_v(1) < \mathbf{P}_v(2)$. In this case the candidates are $\mathbf{P}_v(1)$, $\mathbf{P}_v(2)$ and $\mathbf{P}_v(3)$, then the position is found by exclusion.

In all the other cases, the $i - th$ (with $4 \leq i \leq d_v$) position approximated by the CPA attack corresponds to a choice between $\mathbf{P}_v(i)$ and $\mathbf{P}_v(i-1)$, and again the correct position if found by exclusion.

In the end, the attack is successful and the asserted positions in SK are discovered; this makes the `VbC` multiplier not secure against such attacks.

3.3 Threshold Frequency Limit Detection

A study on the operating frequency has been carried out in order to find the limit of the CPA attack presented in the previous paragraphs. In particular, we want to find the minimum operating frequency for which no correlation peak can be detected.

The study on this limit is motivated by the fact that the P_{dyn} of the correlation traces strongly depends on the operating frequency. If the dynamic power has a contribution too negligible to the static power, it is not possible to compute a reliable correlation value. The results shown in Fig. 4 have been obtained guessing the second value of the key and represent the behaviour of the correlation peak of the correct key as a function of the operating frequency.

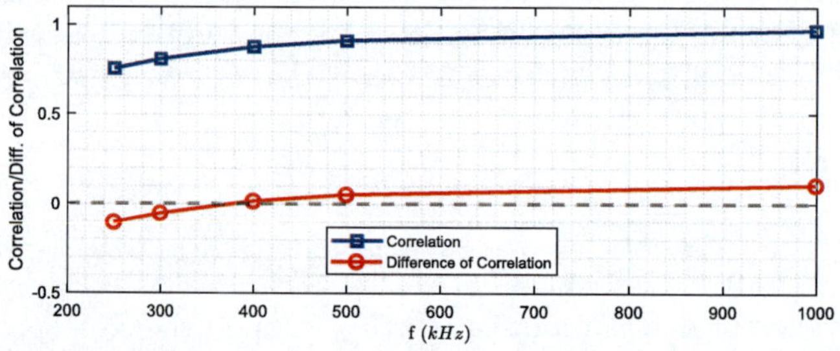

Fig. 4. Graph representing the behaviour of the correlation peak of the correct key value and an indication on the accuracy of the attack depending on the working frequency of the multiplier.

The blue curve represents the highest correlation value found in the correct key and it is directly proportional to the frequency, as expected. The orange curve shows the difference between the correlation peak in the correct key and the highest correlation value found from all the wrong key values. If the difference is positive, the attack was successful. On the contrary, if the difference is negative it means that the highest correlation value belongs to one of the incorrect keys.

As it can be observed, the threshold is around 400 kHz but since the data were obtained through simulations, it is expected that in a real case this value would be higher.

4 Results Validation and Conclusions

The same CPA attack, in order validate the results obtained via simulations for the ASIC implementation, has been conducted on a real implementation of the multiplier. Despite the resulting value of the power is different, the results of the power model remains unchanged and the difference is in the value of the collected power traces.

To this purpose, the VirtLab board [9,10], equipped with a Cyclone 10 LP 10CL025YE144C8G FPGA and an STM32L496VET6 microcontroller, was used. The multiplier architecture has been implemented on the FPGA and the microcontroller is used to stimulate the circuit and collect the measures, working as a Digital Storage Oscilloscope (DSO). The maximum DSO sampling frequency is 500 kHz, thus limiting the working frequency of the multiplier to 250 kHz in order to record two samples per clock cycle. The attack applied un such a lower frequency tourned out to be unsuccessful since SK has not been guessed. One proof of it could be the maximum working frequency analysis performed in 3.3, the value of 250 kHz is lower than the threshold we found. Moreover, the Static Power has been examined from the report of the tool employed to program the FPGA and it showed that the P_{static} was comparable to the measured P_{tot}, then the result of the correlation showed no peak since the comparison was among to uncorrelated quantities.

The proposed method of analysis is useful to prevent the security issues of an architecture during the design phase by identifying the effects of a CPA attack on a PQC multipler. The method can be applied to the whole decoder and for different types of Side-channel attacks. The future work will be dedicated to make the architecture immune to this type of attacks.

References

1. https://csrc.nist.gov/projects/post-quantum-cryptography/post-quantumcryptography-standardization
2. Rossi, M., et al.: A side-channel assisted cryptanalytic attack against QcBits. Cryptology ePrint Archive, Paper 2017/596 (2017). https://eprint.iacr.org/2017/596
3. Baldi, M., et al.: LEDAcrypt Home. https://www.ledacrypt.org/
4. https://bikesuite.org/
5. McEliece, R.J.: A public-key cryptosystem based on algebraic coding theory. DSN Prog. Rep. **44**, 114–116 (1978)
6. Koleci, K., et al.: Efficient hardware implementation of the LEDAcrypt decoder. IEEE Access **9**, 66223–66240 (2021). https://doi.org/10.1109/ACCESS.2021.3076245
7. Barenghi, A., et al.: On the efficiency of design time evaluation of the resistance to power attacks. In: 2011 14th Euromicro Conference on Digital System Design, pp. 777–785 (2011). https://doi.org/10.1109/DSD.2011.103

8. Brier, E., Clavier, C., Olivier, F.: Correlation power analysis with a leakage model. In: Joye, M., Quisquater, J.-J. (eds.) CHES 2004. LNCS, vol. 3156, pp. 16–29. Springer, Heidelberg (2004). https://doi.org/10.1007/978-3-540-28632-5_2
9. Ruo Roch, M., Martina, M.: vrLab: a virtual and remote low cost electronics lab platform. In: Saponara, S., De Gloria, A. (eds.) ApplePies 2020. LNEE, vol. 738, pp. 213–220. Springer, Cham (2021). https://doi.org/10.1007/978-3-030-66729-0_24 ISBN: 978-3-030-66729-0
10. Roch, M.R., Martina, M.: VirtLAB: a low-cost platform for electronics lab experiments. Sensors **22**(13), 4840 (2022). ISSN: 1424-8220. https://doi.org/10.3390/s22134840

A 2 GHz Wide Tuning Range LC-Tank Digitally Controlled Oscillator in 28 nm CMOS Technology

G. Ciarpi(✉), G. Puccioni, M. Mestice, D. Monda, D. Rossi, and S. Saponara

Department of Information Engineering, University of Pisa, Pisa, Italy
gabriele.ciarpi@unipi.it

Abstract. This work presents the design, in 28 nm CMOS technology, of an integrated LC-tank Digitally Controlled Oscillator (DCO) operating around 2 GHz with a wide tuning range between 1.24 GHz and 2.51 GHz. The DCO, implemented with N- and P-cross-coupled MOSFETs, uses two capacitor banks, one fine and the other coarse, to cover all the frequency range with a minimum resolution of about 10 MHz. A digital gain control solution was also implemented to guarantee the startup and sustainment of the oscillation even in the worst conditions. The reported post-layout simulation results confirm compliance with the DCO constraints.

1 Introduction

Many of today's processors implement Dynamic Voltage and Frequency Scaling (DVFS) algorithms for the reduction of power consumption and/or the effort required for its cooling [1–3]. One of the elements that enable the implementation of DVFS solutions is the oscillator integrated in Phase-Locked Loops (PLL), which is used to change the processor clock frequency depending on its workload [4, 5]. In addition, the same capability to shift the clock frequency could be used to control the switching frequency of integrated DC/DC converters, enabling the dynamic control of the processor supply voltage [6, 7]. For this purpose, different kinds of oscillators with different performances can be employed, such as ring or LC-tank oscillators [8, 9]. For its better noise performance, an LC-tank oscillator was implemented in this work. This kind of oscillator can be further classified into voltage-controlled oscillator (VCO) and digital-controlled oscillator (DCO). The former generates an oscillation with a frequency depending on the value of the tank capacitor, controlled by an analog voltage applied to varactor devices. Instead, the oscillation frequency of the latter is controlled by capacitor banks that are connected or disconnected according to a digital control word [10], following the relationship given in Eq. 1, where L and C are the inductor and the whole capacitance of the LC-tank, respectively.

$$f_{oscillator} = \frac{1}{2\pi\sqrt{LC}} \tag{1}$$

In this work, we present an LC-tank oscillator designed to be digitally controlled to operate in a wide frequency range between 1.25 GHz and 2.5 GHz with a frequency

R. Berta and A. De Gloria (Eds.): ApplePies 2022, LNEE 1036, pp. 97–105, 2023.
https://doi.org/10.1007/978-3-031-30333-3_13

resolution of 10 MHz. In addition, the DCO is designed to work properly in Process-Voltage-Temperature (PVT) corners consisting of slow, typical, and fast technology corners, a temperature range between 0 °C and 125 °C, and a supply voltage variation of ±10%. Section 2 presents the architecture of the proposed DCO with the design of the main blocks. Sections 3 and 4 show the layout of the proposed DCO and post-layout simulation results, respectively. Conclusions are drawn in Sect. 5.

2 DCO Architecture

The proposed DCO architecture is based on a CMOS LC-tank structure with a digitally controlled capacitors bank. Compared to the use of LC-tank structures with only N- or P-type cross-coupled devices, the use of both types of devices results in a higher loop-gain and, thanks to its symmetry, also lower phase noise in the low-frequency range [11]. In terms of area, different architectures can be exploited, but the most area-hungry devices are the passive ones, inductors and capacitors. They are chosen to obtain the required oscillation frequency and tuning range independently from the selected architecture. Figure 1 shows the whole DCO architecture.

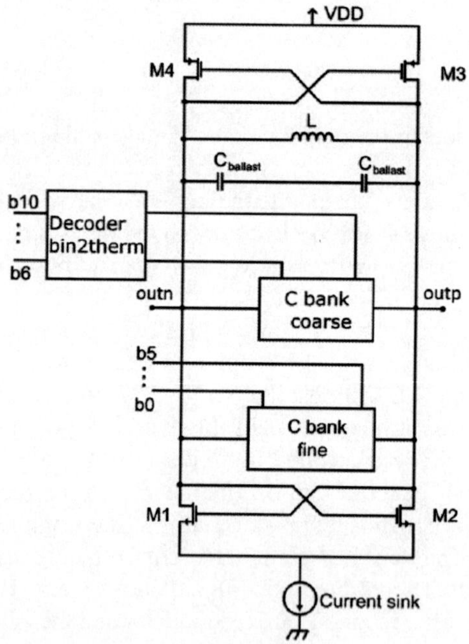

Fig. 1. Schematic of the proposed DCO architecture. It is mainly composed of two cross-coupled MOSFET pairs, one with n-MOSFETs and another with p-MOSFETs, an integrated inductor, two capacitance banks to control the operative frequency, and a current sink to bias the system.

Considering the wide tuning range required, from 1.25 GHz to 2.5 GHz and the 10 MHz frequency resolution, a coarse and a fine capacitor banks are implemented in the DCO. The idea is to connect one or more couples of capacitors to the nodes of the oscillator to increase the total capacitance of the circuit and reduce the operative frequency, and vice versa. In particular, an 11-bit control word is chosen as a good trade-off between circuit complexity and frequency control; 5 bits and 6 bits are assigned to the coarse and fine banks, respectively. In addition, considering the use of a limited number of bits, the frequency range, and the frequency resolution constraints, a pair of ballast capacitors is added to decrease the nominal central frequency.

2.1 Coarse and Fine Banks Design

Considering the use of the proposed DCO in a Frequency Locked Loop device, in order to increase the efficiency of the control algorithm, an increasing or decreasing DCO characteristic should be achieved. For this aim, the coarse bank is designed to be controlled with a thermometric encoding to ensure the required characteristic even in matching error cases [12]. Therefore, equal-value pair capacitors are connected to the DCO using a couple of N-MOSFET devices, enabled by the i-th bit of the coarse control word, as depicted in Fig. 2a. Compared with the use of a single N-MOSFET device connected between the capacitor pair, this solution reduces the equivalent switch resistance since the transistors work with a higher overdrive voltage. On the line to improve the overdrive voltage, the switches are realized using ultra-low-voltage-threshold MOSFETs, sized as a tradeoff between on-resistance and area (width and length equal to 64 μm and 0.24 μm, respectively).

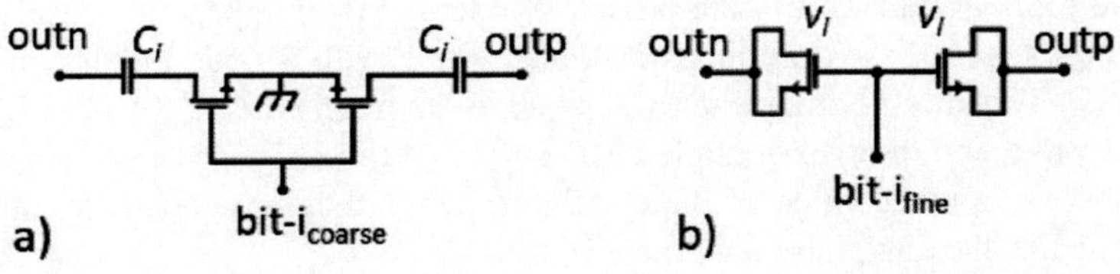

Fig. 2. Schematic connections of the capacitors of the coarse bank (a) and of the varactors of the fine bank (b).

Binary encoding is instead implemented for the fine bank. In this case, varactors with a power of two increasing capacitance are controlled by the fine digital word. This means that each fine bit is associated with a capacitance that is twice the value of the previous one. The fine bank, implemented to improve the DCO frequency resolution, consists of couples of MOSFETs connected as varactors with their gate connected to the i-th fine bit, see Fig. 2b. From simulations, a minimum capacitance value of 15.5 fF is chosen to achieve the required frequency resolution.

2.2 Inductor Sizing

Considering a trade-off between the inductor area, its quality factor, and technology rules, a symmetrical central tapered inductor with a value of about 2.5 nH is chosen for the DCO. It consists of three turns with a width of 10.7 μm, spaced of 3 μm, and with an inner radius of 80 μm. In the operating range, its inductance value variation is below 1%, whereas its quality factor variation reaches 58% yet remaining above 10 at 1.25 GHz, as shown in Fig. 3.

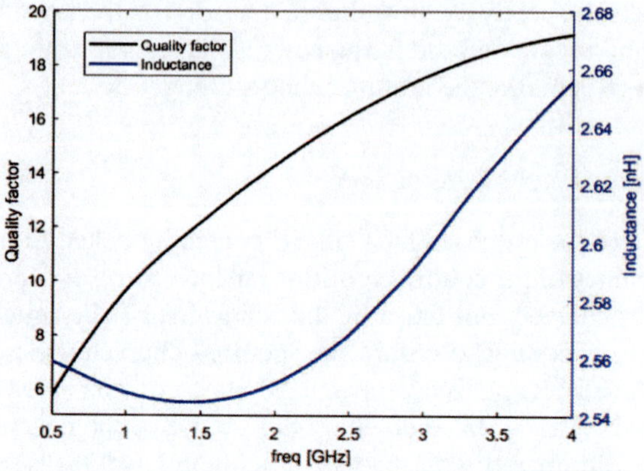

Fig. 3. Inductance and quality factor values of the used inductor as a function of frequency.

2.3 Current Sink Design

A simple current sink is used to bias the DCO core with a constant current flowing through the four MOSFETs and the passive elements that compose it. Its schematic is shown in Fig. 4. It is based on a master-slave couple with a scale factor equal to two and a MOSFET in diode configuration to provide the main current (M0 in Fig. 4). The tail current value and the cross-coupled MOSFETs of the DCO are mainly responsible for the transconductance value of the oscillator, hence they are chosen to amplify the oscillation in the starting phase and to sustain it.

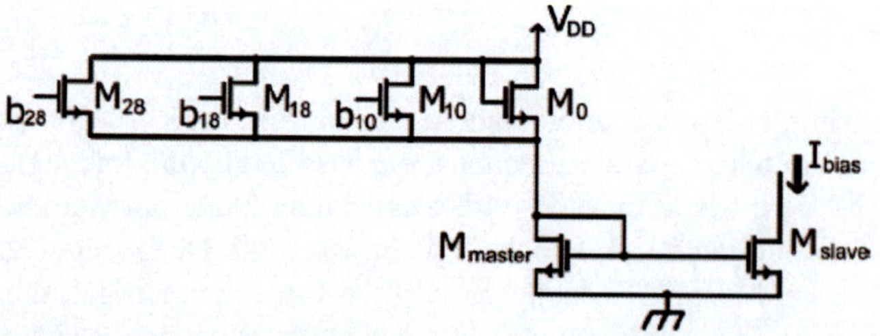

Fig. 4. Schematic of the current sink based on a simple current mirror. The MOSFETs M_{10}, M_{18}, and M_{28} are added to improve the DCO transconductance for low oscillation frequencies. The MOSFETs sizes (width – length) are: M_{master} (2.8–0.24 μm), M_{slave} (5.6–0.24 μm), M_0 (1.5–0.15 μm), M_{10} (2–0.15 μm), M_{18} (2.5–0.15 μm), M_{28} (3–0.15 μm).

The required DCO transconductance changes with the oscillation frequency. Indeed, for lower frequency, a higher transconductance is required due to two main aspects:

firstly, as shown in the previous section, the inductor Q-factor reduction at low-frequency values requires a higher transconductance to compensate for the increase in the equivalent parasitic resistance of the inductor; secondly, the greater number of capacitances connected to lower the working frequency requires to turn on switches whose on-resistance is different from zero, decreasing the capacitor Q-factor and consequently requiring a higher transconductance.

The solution proposed to compensate for these effects is the increment of the bias current for low operating frequency using additional MOSFETs driven by bits 10, 18, and 28 of the coarse thermometric encoding. This solution ensures the low-frequency oscillation start-up and sustainment conditions for all PVT corners. In addition, since the output voltage swing is related to the equivalent conductance of the oscillator and to the bias current, this solution also enables compensation for the voltage swing reduction. The MOSFETs used for this stage are the ultra-low-voltage-threshold devices sized with the values reported in the caption of Fig. 4, according to post-layout simulations and PVT corner simulations.

3 Layout

The main approach followed for the DCO layout design involved the realization of symmetrical structures to minimize mismatch effects and the use of low resistance high-layer metals to reduce parasitic elements. All metal connections are sized considering the electromigration phenomena boosted at high temperatures. The final layout area of the whole DCO is $300 \times 650\ \mu m^2$. A reduction in area could be achieved with optimal block placement, however, the main space occupation is due to the passive elements, inductors and capacitors, whose area is related to the oscillation frequency and tuning range [13, 14] (Fig. 5).

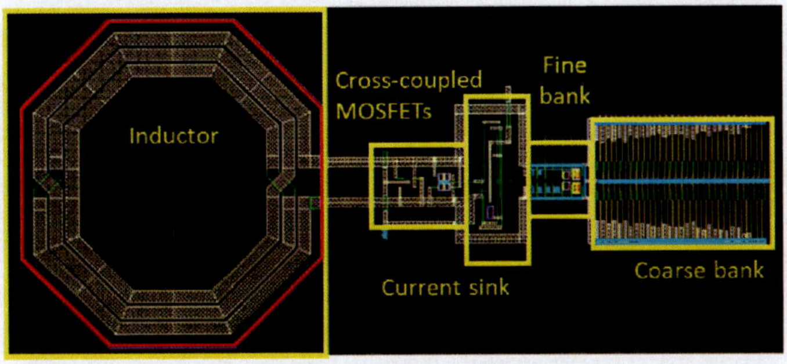

Fig. 5. DCO Layout. From left to right: the integrated symmetrical inductor, the P- and N- cross-coupled MOSFETs, the current sink, the fine bank, and the coarse bank.

4 Post-layout Simulations Results

Post-layout simulations show the ability of the designed DCO to cover a frequency range from 1.24 GHz to 2.51 GHz with a minimum frequency resolution of 9.7 MHz in the first coarse band (low-frequency) and a maximum frequency resolution of about 1.4 MHz in the last coarse band (high-frequency), as shown in Fig. 6.

Table 1 summarizes the post-layout simulation results for some frequency settings identified with the coarse control word. In particular, it is shown the master current of the tail mirror, which is chosen to be about 1 mA and increased for low-frequency values by bits 10, 18, and 28, as described in Sect. 2.3. As shown in the fourth column, this current increase limits the reduction of the oscillation amplitude at low frequencies, but, on the other hand, it increases the DCO power consumption reported in the sixth column. Concerning phase noise, as expected, the phase noise simulated at 1 MHz offset from the main frequency shows a worsening at high frequency, but it remains below −100 dBc/Hz.

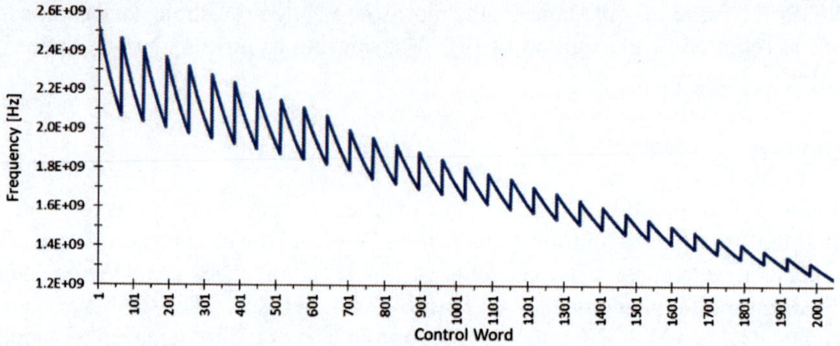

Fig. 6. Frequency behavior of the proposed DCO as a function of the digital control word.

Table 1. Summary of the main post-layout results for some frequencies.

Coarse Frequency	Frequency [GHz]	Master Current [mA]	Voltage Swing [mV]	Phase Noise @ 1 MHz [dBc/Hz]	Total Power [mW]
f0	2.514	1.098	818.8	−106.9	2.77
f4	2.322	1.097	765.4	−108.3	2.82
f8	2.150	1.097	715.1	−109.2	2.85
f12	1.958	1.632	802.0	−111.4	4.05
f16	1.810	1.631	759.0	−112.1	4.13
f20	1.671	1.999	780.9	−114.0	4.93
f24	1.538	1.998	729.6	−115.2	5.04
f28	1.415	1.998	670.1	−116.3	5.11
f31	1.320	2.297	668.6	−117.7	5.82

5 State-of-the-Art Comparison

In Table 2, the performances of the proposed DCO are summarized and compared with state-of-the-art LC-tank DCO solutions in 28 nm. Compared with [15], the proposed DCO is able to operate with lower power consumption but takes a larger area. This could be explained by the lower oscillation frequency and wider tuning range of the proposed DCO, requiring a more area-hungry LC-tank. Considering [16] and [17], both exhibits very low power consumption but with a limited tuning frequency range. The [18] shows the best performances in terms of phase noise at 1 MHz compared to all other DCOs, but this is paid for with a power consumption an order higher than the other. Considering the tuning range, the proposed DCO shows the widest tuning range, 38.02% wider than [15], 67.73% wider than [16] and [17], and 17.73% wider than [18].

Table 2. The state-of-the-art comparison

	[15]*	[16]$^{\pm}$	[17]$^{\lambda}$	[18]°	This work
Technology	28 nm	28 nm FDSOI	28 nm	28 nm	28 nm
Typology	NMOS	NMOS with P-tail	Class-D	NMOS	CMOS
Frequency range [GHz]	2.78–3.75	2.4788–2.4793	2.35	2.4–4	1.24–2.51
Percentual Frequency range	29.71%	0.04%	–	50%	67.73%
Frequency resolution [MHz]	5	0.008	n.a	n.a	1.4–9.7
Supply voltage [V]	0.9	1	0.225–0.5	1.8	0.9
Temp. Range [°C]	n.a	0–100	n.a	n.a	0–125
Power consumption [mW]	6.6	1.5	0.38	58.6	2.77–5.82
Phase Noise @ 1 MHz [dBc/Hz]	−127.5	−113.0	−115.9	−132.0	−117.7
Area [mm^2]	0.15	No layout	0.17	0.15	0.19

* Common Mode Resonance architecture, $^{\pm}$Pre-Layout results, $^{\lambda}$Dynamic supply voltage control, °Analog and Digital tuning range control.

6 Conclusions and Future Developments

In this work, we presented the design in 28 nm TSMC technology of an integrated LC-DCO operating in the frequency range between 1.24 GHz and 2.51 GHz with a resolution lower than 10 MHz.

The oscillation frequency is digitally controlled by an 11-bit word through the use of coarse and fine capacitors banks. The former implements a thermometric code thanks to the use of couples of capacitors and MOSFETs as switches. The second instead uses the natural binary code implemented with properly sized MOSFET varactors.

Considering that the quality factor of the LC-tank decreases at low frequency for the two main reasons reported in Sects. 2.2 and 2.3, in the proposed DCO design, an increment of the tail current in the cross-coupled devices is adopted. Specifically, three MOSFETs driven by the bits 10, 18, and 28 of the coarse word are activated for low oscillation frequencies, increasing the master current in Fig. 4. Thanks to this solution the loop gain remains above 4.3 dB even in the worst PVT condition (slow-slow process, 0.81 V supply voltage, and 0 °C operating temperature).

In conclusion, post-layout simulations show compliance with the required frequency range and resolution with performances that stand if compared with the state-of-the-art solutions.

However, future improvements of the designed LC-DCO could be realized to increase its robustness to PVT variations. In particular, the tail current generated with the MOSFET M0 shown in Fig. 4 is sensitive to the device performance in different PVT corners. Therefore, it is necessary to implement a more efficient solution for generating the tail current, replacing the simple diode-connected device. The proposed solution to increase the gain of the LC-DCO for low oscillation frequencies, using MOSFETs M10, M18, and M28, is still valid, with appropriate modifications, even with other tail current generator architectures.

In addition to improving the performance of the proposed architecture designed in 28 nm CMOS technology, we planned to improve the LC-DCO performance by using advanced technology such as 12 nm CMOS technology. Indeed, this technology, having a higher cutoff frequency than 28 nm technology, could improve the performances (e.g., phase noise) of the proposed LC-DCO. In addition, a further push toward migration to a more advanced technology node is that the LC-DCO, if used for a DVFS, will have to be integrated into a processor and thus follow the current trend in technology development.

Acknowledgments. Work partially supported by MIUR with the *Dipartimenti di Eccellenza CrossLab-ForeLab* projects and by European Union's Horizon 2020 research and innovation program under grant agreement No 101036168 (*European Process Initiative SGA2*).

References

1. Chun, K.B., et al.:A frequency scaling model for energy efficient DVFS designs based on circuit delay optimization. In: 2015 International Symposium on Consumer Electronics (ISCE), pp. 1–2 (2015)
2. Park, C., et al.: A new DVFS algorithm to minimize energy consumption on system-on-chip architecture and electrical characteristics. In: 2022 IEEE International Conference on Consumer Electronics (ICCE), pp. 1–4 (2022)
3. Jawad, H., et al.: SysScale: exploiting multi-domain dynamic voltage and frequency scaling for energy efficient mobile processors. In: 2020 ACM/IEEE 47th Annual International Symposium on Computer Architecture (ISCA). IEEE (2020)

4. Albea, C., et al.: Architecture and robust control of a digital frequency-locked loop for fine-grain dynamic voltage and frequency scaling in globally asynchronous locally synchronous structures. J. Low Power Electron. **7**(3), 328–340 (2011)
5. Mestice, M., et al.: Analysis and design of integrated blocks for a 6.25 GHz spacefibre PLL. Sensors **20**, 4013 (2020)
6. Zhou, P., et al.: Distributed on-chip switched-capacitor DC-DC converters supporting DVFS in multicore systems. IEEE Trans. Very Large Scale Integr. (VLSI) Syst. **22**(9), 1954–1967 (2014)
7. Ciarpi, G., et al.: Inductorless DC/DC converter for aerospace applications with insulation features. IEEE Trans. Circ. Syst. II Exp. Bri. **67**(9), 1659–1663 (2020)
8. Hsieh, M., et al.: Comparison of LC and ring VCOs for PLLs in a 90 nm digital CMOS process. In: ISOCC, pp. 19–22 (2006)
9. Monda, D., et al.: Analysis and comparison of rad-hard ring and LC-tank controlled oscillators in 65 nm for SpaceFibre applications. Sensors **20**, 4612 (2020)
10. Mačaitis, V., et al.: Comparison of LC-VCO and LC-DCO parameters in 65 nm CMOS technology. In: 2017 5th IEEE Workshop on Advances in Information, Electronic and Electrical Engineering (AIEEE). IEEE (2017)
11. Bhat, M., et al.: Comparison of phase noise in cross-coupled voltage controlled oscillator topologies. In: 2018 3rd International Conference on Inventive Computation Technologies (ICICT). IEEE (2018)
12. Bisiaux, P., et al.: Design of a 1.5 GHz low jitter DCO ring in 28 nm CMOS Process. In: 2020 European Conference on Circuit Theory and Design (ECCTD), pp. 1–5 (2020)
13. Elhajjami, I. et al.: Optimal design of RF integrated inductors via differential evolution algorithm. In: 2020 1st International Conference on Innovative Research in Applied Science, Engineering and Technology (IRASET), pp. 1–6 (2020)
14. Ciarpi, G., et al.: Design and characterization of 10 Gb/s and 1 Grad TID-tolerant optical modulator driver. IEEE Trans. Circ. Syst. I: Reg. Papers **69**, 1–13 (2022)
15. Murphy, D., Darabi, H., Wu, H.: Implicit common-mode resonance in LC oscillators. IEEE J. Solid-State Circ. **52**(3), 812–821 (2017)
16. Panetas-Felouris, O., Vlassis, S., et al.: A bluetooth low energy DCO in 28 nm FDSOI. In: 2020 43rd International Conference on Telecommunications and Signal Processing (TSP), pp. 321–324 (2020)
17. Yoshihara, Y., Majima, H., Fujimoto, R.: A 0.171-mW, 2.4-GHz Class-D VCO with dynamic supply voltage control. In: ESSCIRC 2014 - 40th European Solid State Circuits Conference (ESSCIRC), pp. 339–342 (2014)
18. Zhang, Y., et al.: A 2.4-to-4 GHz VCO achieving −137 dB c/Hz phase noise at 1 MHz offset in 28 nm CMOS. In: 2019 IEEE International Conference on Integrated Circuits, Technologies and Applications (ICTA), pp. 15–16 (2019)

Machine Learning

Affordance Segmentation Using RGB-D Sensors for Application in Portable Embedded Systems

Edoardo Ragusa(✉), Matteo Pastorino Ghezzi, Rodolfo Zunino, and Paolo Gastaldo

Department of Electrical, Electronic, Telecommunication Engineering and Naval Architecture DITEN, University of Genoa, Genova, Italy
edoardo.ragusa@edu.unige.it, 4358664@studenti.unige.it, {rodolfo.zunino,paolo.gastaldo}@unige.it

Abstract. Depth sensors play a major role in the control pipeline of semi-autonomous prostheses. The feed from these sensors can complement, or even substitute, the information retrieved by RGB cameras. The paper explores the application of depth sensors to affordance detection, to recognize the graspable part of an object in foreground images. The experiments confirm that the depth ("D") components can boost the accuracy in predicting the affordable parts of an object, when they are matched with state-of-the-art computer vision architectures. A prototype of a portable inference system with real-time performances has been assembled using a Jetson Tx2 and a Realsense D435i camera, confirming the proposal's suitability for semi-autonomous prostheses.

Keywords: Affordance prediction · Grasp selection · CNNs · prosthetics · embedded systems

1 Introduction

Deep learning (DL) architectures can achieve state-of-the-art results on challenging problems such as natural language processing [1], robotics [2], and molecular science [3]. DL is indeed becoming ubiquitous in computer vision applications [4,5]. This in turn means that such applications must rely on computing systems that can address the computational requirements of deep learning architectures [6].

Eye-on-hands paradigm is becoming predominant in modern prosthesis control [7]: modern high-end prosthesis often embed cameras [8,9], that can support cognitive-like processing and therefore enable semi-autonomous actions (e.g., automatic preshaping).

Semi-autonomous movements require depth (D) sensors because one cannot grasp an object without knowing its relative distance [10,11]. The environment representation provide by depth sensors somehow overlap with the information sensed from RGB cameras (as illustrated in Figs. 1(a) and (b)). This paper explores the interaction between RGB and D data, and provides design criteria to get more efficient contact-less sensing pipelines in semi-autonomous control applications.

R. Berta and A. De Gloria (Eds.): ApplePies 2022, LNEE 1036, pp. 109–116, 2023.
https://doi.org/10.1007/978-3-031-30333-3_14

Semi-autonomous control, by means of traded autonomy, could substantially improve prosthesis performance while alleviating a user from the cognitive burden of control [12]. Toward that end, a prosthesis should recognize the graspable portion of an object [13]. Affordance segmentation (AS) first identifies the potential interactions that an object can afford, then segments the object itself into parts according to these functions [14] (Fig. 1(c)). Previous research proved that AS solutions developed for robotics [15] can suitably isolate the graspable parts of the object [16,17]. The proposed approach can be hosted on portable embedded devices, thus mitigating the limitations imposed by cloud computing-based solutions [18,19]. Power and memory consumptions become major constraints when one thinks that the computing unit support simultaneously other tasks to achieve semi-autonomous control [11,20–22]. The paper pursues a pair of alternative deep neural-networks approaches to tackle the reformulated problem [16], involving RGB data and D information when the observed target object is in the foreground. The first solution implies a configuration with depth data only, thus envisioning the possibility to remove RGB cameras. The second solution mixes the entire information from RGB-D cameras.

Generalization performance were compared against a recent baseline [16] using a pair of popular benchmarks, namely, the UMD dataset [23] and the Multi-View dataset (MV) [24]. The results confirmed the effectiveness of the proposal. Eventually, a prototype of a portable inference system composed of a NVIDIA Jetson TX2 interfaced with an Intel RealSense camera was developed and attained real-time inference timings.

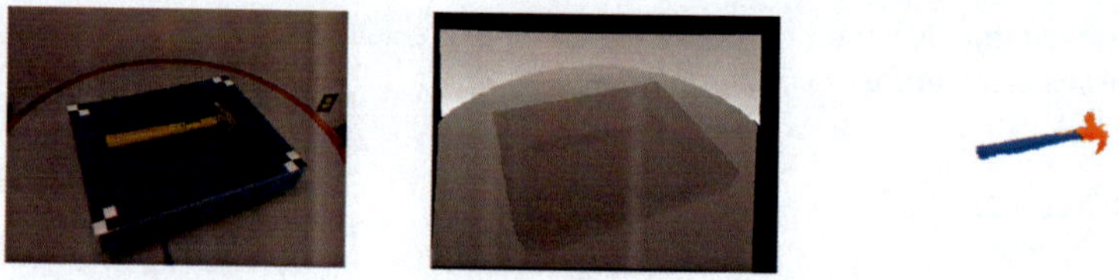

(a) (b) (c)

Fig. 1. Example of objects sensed by RGB camera (a), D camera (b) and the corresponding affordance map (c) extracted from the UMD dataset.

2 Materials and Methods

A straightforward approach to exploit depth data is to only use D information and drop RGB components altogether, hence a prosthesis would only carry D sensors. Depth is represented using a single channel, i.e. a grey-scale image, which is amenable for deep-learning processing architectures. In this research, we tested an EfficientNetB0-Unet architecture similar to the one adopted in [16]. The architecture accounts for 10.84 M parameters and 4.09 GFLOPs for a single inference using input size 224 × 224. This network was selected because it suitably balanced generalization performance and the number of parameters [16].

A second option to exploit depth data is a two-tier architecture. One branch might processes the RGB inputs, whereas the other branch would handle depth data. In this paper, these branches coincide and the information is merged according to a late fusion approach, in which the upper layers of both architectures are concatenated. Eventually, a 2D convolutional layer aggregates the information carried by the two components. Thanks to that late fusion, the network can learn appropriate descriptors for the two input sources. In fact, D and RGB data relate to different physical information, therefore merging these sources in the input (initial) stages of the architecture would basically prove sub-optimal. The two-tier architecture involves 21.68 M parameters and 8.18 GFLOPs when input size is 224 × 224. As expected, this solution doubles hardware requirements with respect to the first solution but it is expected to improve generalization performance.

Reaching real-time performance in portable devices is a task of primary importance if one thinks that stable, fast, and unlimited connections cannot be given for granted in many real-life scenarios. This research presents a prototype of the portable system. The system senses RGB-D data from the Realsense camera and uses a Jetson TX2 to support the inference procedure. The TX2 is a powerful System On Module (SOM) that hosts two processors, namely a dual-core NVIDIA Denver2 and a quad-core ARM Cortex-A57; it equips an 8 GB RAM and an integrated 256-core Pascal GPU. The entire Jetson TX2 module measures 5 cm × 9 cm and provides a large amount of power that supports the inference phase of deep neural networks.

The proposed prototype requires a significant amount of energy. In practice, using lithium batteries, a system can guarantee a few hours of operations. However, traded autonomy requires a trigger for an operation that limits the active time of the system significantly. Accordingly, this device can be hosted in a prosthesis or appropriate support carried by the user.

3 Experiments and Results

The experiments were divided into two sets. First, a detailed analysis compared the performance of the architecture feed with different input sources. Then the performances of the embedded prototype were measured.

The present paper targets the same application presented in [16] using a different/additional input source, i.e. the D information. Accordingly, the models presented in [16] are the natural baseline. In [16] the authors tested architectures based on MobileNets Version 1 and 3, EfficentNet B0, and Bisenet. EfficentNet B0 will represent the baseline comparison because together with MobileNets Version 1 obtained the best trade-off among computation cost and generalization performance. In addition, the performance of EfficentNet B0 provides a

Table 1. Effect of distortions on the generalization performance through augmentation techniques.

	Test aug	Test no-aug
Train aug	0.36, 0.56	0.42, 0.70
Train no-aug	0.26, 0.50	0.64, 0.79

valid reference for two other works. In [17] the affordance detection task was extended considering the localization of the object inside the framework. In this scenario too EfficientNetB0 with RGB input represents a valid option for the affordance segmentation part of the pipeline. Vice-versa, [25] tackles the "robotic-like" affordance detection problem. The experiments presented in [16] confirmed that this architecture suits the more general setup with a larger number of classes. Accordingly, EfficentNetB0 other than being the selected target architecture of the present work represents a concise way to compare the presented approach with three strictly related works in a direct or indirect fashion.

3.1 Generalization Performance

Generalization performance was measured using two datasets: UMD is a well-known benchmark for affordance segmentation that contains 28.843 RGB-D images of 7 categories of objects. 5135 images served as a test set. Similarly, the MW dataset [24] accounts 23605 RGB-D images collected using 37 classes of objects. Following the setup proposed in [16] all the grasping affordances were grouped into a unique class and the other affordances were grouped into a single class, i.e. don't grasp. The two datasets proposes very different sets of objects, providing de-facto two different benchmarks. The learning problem consists of a three-class pixel-wise classification where the predictor should discriminate between grasp, don't grasp, and background.

Depth images only carry geometrical information, and lose color information. A preliminary analysis assessed the impact of color in the learning phase by applying color and geometrical augmentation to the data. The UMD dataset, in particular, represented an interesting benchmark because the objects were placed over a blue background. In Table 1, the rows mark the training setup configurations, including both the original images and the augmented training set, whereas the columns refer to the same configurations in the testing set. Each cell of the table gives the average Intersection over Union (IoU) for classes "grasp" and "no-grasp", respectively. The background was excluded because, due to the large number of pixels belonging to that class, the average result always was about 99%.

Empirical results confirmed that colors affected RGB-based approaches heavily. This in practice might become a drawback in class assignment. The peculiar features of the UMD dataset emphasized that phenomenon.

Table 2 reports on the generalization performance measured on the UMD and MW datasets. By using the format adopted in Table 1, the Table gives the IoU indexes for both classes ("grasp" and "no-grasp"). The rows mark the input source options, the row associated with the RGB input sets the baseline comparison.

The experiments on the UMD dataset confirmed that, when targeting foreground framing, the depth component can substitute effectively the RGB camera in a well-posed scenario. In the case of MW, results proved that valuable information could be yet extracted, but RGB represented the most valuable option, probably due to noise introduced by the depth sensor. A possible explanation is that MW, as opposed to UMD, covered a variety of objects, such as books and boxes, that were characterized by flat surfaces. Both tests on UMD and MW confirmed that the D component in combination with RGB provided marginal improvements when analyzing foreground images. This seems to confirm that D information is more likely to be a substitute than an integration RGB data.

4 Deployment

Table 3 reports the performance of the real-time portable system. It is divided into two parts: the upper one is marked as "Single" and gives the performances of the architecture with one input source (RGB or D) identified as single branch; the lower section refers to thc architecture with two input sources and is marked as "Double". All the tests involved models that were optimized in FP16 precision with the tensorRT optimization tool. The columns give the average memory usage, the power required by the SOM, the power required by the GPU alone, and the frame rate, respectively. The measurements were averaged over 2 min of inference. Each row covers different O.S. configurations to control hardware usage: for the sake of readability, the settings were sorted in ascending order based on the number of resources available. In practice, MAXQ indicates the setting with the smallest amount of available resources, while MAXN implies that all the available hardware resources were used. The MAXQ configuration did not support the inference phase of the double branch architecture.

Memory-usage results highlighted the high-performance nature of the Jetson device. In fact, a considerable amount of memory had been allocated to ensure low latency. The FPS score confirmed that the architecture could run on embedded devices in real-time. As anticipated, power requirements proved quite strict. Interestingly enough, the performances for the two-branch scheme almost compare with those for the Single branch architecture. In addition, TX2 represents

Table 2. Generalization performance of the proposed networks

Input	UMD		MW	
	grasp	no-grasp	grasp	no-grasp
RGB [16]	0.64	0.79	0.74	0.60
Depth	0.63	0.78	0.71	0.46
RGB-D	0.65	0.80	0.69	0.59

Table 3. Analysis of the performance of the portable real time embedded device.

Network	Conf.	Mem.	Pow. SOM (W)	Pow. GPU (W)	Frame (fps)
Single	MAXQ	9.5	5.50	1.90	8
	MAXP CORE ALL	9.6	8.90	4.16	10
	MAXN	9.6	11.80	5.50	15
Double	MAXP CORE ALL	11.6	10.33	5.45	7
	MAXN	11.6	12.52	6.72	11

a useful development board that allows fast prototyping using the same SDK employed by Jetson Nano. As a result, one might reduce cluttering and power requirements by switching to the low-performing version of the device.

5 Conclusion

This paper presented an analysis of the usage of depth sensors to approach complex computer vision for a prosthetic application. Two deep learning solutions exploiting D data to discriminate affordable parts of objects in the foreground offered a valuable baseline to explore the effects of D component in foreground objects. The deployment on high-end portable embedded devices proved the suitability of the proposed solution for prosthetic scenarios.

References

1. Ni, J., Young, T., Pandelea, V., Xue, F., Cambria, E.: Recent advances in deep learning based dialogue systems: a systematic survey. Artif. Intell. Rev. 1–101 (2022)
2. Gianoglio, C., Ragusa, E., Gastaldo, P., Valle, M.: A novel learning strategy for the trade-off between accuracy and computational cost: a touch modalities classification case study. IEEE Sens. J. **22**(1), 659–670 (2022)
3. Atz, K., Grisoni, F., Schneider, G.: Geometric deep learning on molecular representations. Nat. Mach. Intell. **3**, 1023–1032 (2021)
4. O'Mahony, N., et al.: Deep learning vs. traditional computer vision. In: Arai, K., Kapoor, S. (eds.) CVC 2019. AISC, vol. 943, pp. 128–144. Springer, Cham (2020). https://doi.org/10.1007/978-3-030-17795-9_10
5. Ragusa, E., Apicella, T., Gianoglio, C., Zunino, R., Gastaldo, P.: Design and deployment of an image polarity detector with visual attention. Cogn. Comput. **14**(1), 261–273 (2022)
6. Wang, X., Han, Y., Leung, V.C., Niyato, D., Yan, X., Chen, X.: Convergence of edge computing and deep learning: a comprehensive survey. IEEE Commun. Surv. Tutor. **22**(2), 869–904 (2020)
7. Vasile, F., Maiettini, E., Pasquale, G., Florio, A., Boccardo, N., Natale, L.: Grasp pre-shape selection by synthetic training: Eye-in-hand shared control on the Hannes prosthesis. arXiv preprint arXiv:2203.09812 (2022)

8. Hundhausen, F., Grimm, R., Stieber, L., Asfour, T.: Fast reactive grasping with in-finger vision and in-hand FPGA-accelerated CNNs. In: 2021 IEEE/RSJ International Conference on Intelligent Robots and Systems (IROS), pp. 6825–6832. IEEE (2021)
9. Hundhausen, F., Megerle, D., Asfour, T.: Resource-aware object classification and segmentation for semi-autonomous grasping with prosthetic hands. In: 2019 IEEE-RAS 19th International Conference on Humanoid Robots (Humanoids), pp. 215–221. IEEE (2019)
10. Markovic, M., Dosen, S., Popovic, D., Graimann, B., Farina, D.: Sensor fusion and computer vision for context-aware control of a multi degree-of-freedom prosthesis. J. Neural Eng. **12**(6), 066022 (2015)
11. Castro, M.N., Dosen, S.: Continuous semi-autonomous prosthesis control using a depth sensor on the hand. Frontiers in Neurorobotics **16** (2022)
12. Mouchoux, J., Carisi, S., Dosen, S., Farina, D., Schilling, A.F., Markovic, M.: Artificial perception and semiautonomous control in myoelectric hand prostheses increases performance and decreases effort. IEEE Trans. Rob. **37**(4), 1298–1312 (2021)
13. Li, Y., Wang, P., Li, R., Tao, M., Liu, Z., Qiao, H.: A survey of multifingered robotic manipulation: biological results, structural evolvements and learning methods. Front. Neurorobot. **53** (2022)
14. Hassanin, M., Khan, S., Tahtali, M.: Visual affordance and function understanding: a survey. ACM Comput. Surv. (CSUR) **54**(3), 1–35 (2021)
15. Do, T.T., Nguyen, A., Reid, I.: Affordancenet: an end-to-end deep learning approach for object affordance detection. In: 2018 IEEE international conference on robotics and automation (ICRA), pp. 1–5. IEEE (2018)
16. Ragusa, E., Gianoglio, C., Dosen, S., Gastaldo, P.: Hardware-aware affordance detection for application in portable embedded systems. IEEE Access **9**, 123178–123193 (2021)
17. Apicella, T., Cavallaro, A., Berta, R., Gastaldo, P., Bellotti, F., Ragusa, E.: An affordance detection pipeline for resource-constrained devices. In: 2021 28th IEEE International Conference on Electronics, Circuits, and Systems (ICECS), pp. 1–6. IEEE (2021)
18. Nguyen, A.T., et al.: A portable, self-contained neuroprosthetic hand with deep learning-based finger control. J. Neural Eng. **18**(5), 056051 (2021)
19. Weiner, P., Starke, J., Rader, S., Hundhausen, F., Asfour, T.: Designing prosthetic hands with embodied intelligence: the kit prosthetic hands. Front. Neurorobot. **16** (2022)
20. Parajuli, N., et al.: Real-time EMG based pattern recognition control for hand prostheses: a review on existing methods, challenges and future implementation. Sensors **19**(20), 4596 (2019)
21. Ragusa, E., Gianoglio, C., Zunino, R., Gastaldo, P.: Data-driven video grasping classification for low-power embedded system. In: 2019 26th IEEE International Conference on Electronics, Circuits and Systems (ICECS), pp. 871–874. IEEE (2019)
22. Seminara, L., Capurro, M., Valle, M.: Tactile data processing method for the reconstruction of contact force distributions. Mechatronics **27**, 28–37 (2015)
23. Myers, A., Teo, C.L., Fermüller, C., Aloimonos, Y.: Affordance detection of tool parts from geometric features. In: 2015 IEEE International Conference on Robotics and Automation (ICRA), pp. 1374–1381. IEEE (2015)

24. Khalifa, Z.O., Shah, S.A.A.: Towards visual affordance learning: a benchmark for affordance segmentation and recognition. arXiv preprint arXiv:2203.14092 (2022)
25. Nguyen, A., Kanoulas, D., Caldwell, D.G., Tsagarakis, N.G.: Object-based affordances detection with convolutional neural networks and dense conditional random fields. In: 2017 IEEE/RSJ International Conference on Intelligent Robots and Systems (IROS), pp. 5908–5915. IEEE (2017)

ML-Based Classifier for Precision Agriculture on Embedded Systems

Romina Soledad Molina[1,2,3](✉), Valentina Carrer[1], Maynor Ballina[2], Maria Liz Crespo[2], Luciana Bollati[4], Daniel Sequeiro[4], Stefano Marsi[1], and Giovanni Ramponi[1]

[1] IPL, Dipartimento di Ingegneria e Architettura, Università degli Studi di Trieste, Piazzale Europa, 1, 34127 Trieste, TS, Italy
rominasoledad.molina@phd.units.it, valentina.carrer@studenti.units.it
{marsi,ramponi}@units.it

[2] MLAB, The Abdus Salam International Centre for Theoretical Physics, Via Beirut, 31, 34151 Grignano, TS, Italy
{rmolina,mballina,mcrespo}@ictp.it

[3] LEIS, Departamento de electrónica, Universidad Nacional de San Luis, Av. Ejército de los Andes 950, D5700 San Luis, Argentina

[4] Nectras, Av Belgrano 758, Sunchales, Santa Fe, Argentina

Abstract. In precision agriculture, effective pest control helps to reduce yield loss and pesticide application. In this research, the pest to be detected and controlled is the moth lobesia botrana, which mainly attacks the vineyard. We present an automatic pest classifier based on machine learning, considering resource-constrained devices in IoT systems. Transfer learning and an ensemble of compression techniques are used to reduce the size of the classifier with a good trade-off between efficiency, effectiveness, and resource utilization. This procedure allows the achievement of a fully on-chip deployment in two technologies: esp32 and SoC-based FPGA Xilinx PYNQ-Z1 and KRIA.

Keywords: machine learning · compression · embedded systems · precision agriculture · pest control

1 Introduction

A system based on the Internet of Things (IoT) is mainly composed of sensors, actuators, wireless connections, an energy module, and an embedded system. Precision agriculture can benefit from this type of system because of its features such as energy efficiency, reliability, mobility, durability, computation ability, and communication [1]. In this context, IoT traps have been developed to aid pest control and detection [2,3], improve food safety, and enable new understanding of pest trends and their characteristics. A further advantage of the IoT approach is its scalability to fields of diverse size.

R. Berta and A. De Gloria (Eds.): ApplePies 2022, LNEE 1036, pp. 117–124, 2023.
https://doi.org/10.1007/978-3-031-30333-3_15

In agriculture, fruit crops are mainly affected by various diseases and pests during their growth process. If the control is not timely, this will lead to a reduction in the soil or even a loss of harvest. The accurate and effective control of insect pests is important to help fruit growers improve fruit yield. One of the methods to solve this task is a manual approach, which is time-consuming and susceptible to errors. Trends in data processing have provided different ways to solve this problem, using automatic and efficient image recognition methods [4–7].

The use of embedded systems based on FPGA for IoT enables different measurements to be performed in parallel. Thus, logic design can collect and process information from the environment using different types of sensors (air temperature, solar radiation, soil temperature, etc.) while processing complex algorithms based on machine learning (ML) and reducing power consumption. Owing to its flexibility in reconfiguration, this type of device has begun to be used in agriculture [8].

In this study, the contribution of [9] is extended as follows:

- an automatic pest classification based on ML, considering resource-constrained devices, is presented. The actual front-end data-acquisition system is based on an esp32 microcontroller;
- we exploit compression techniques to demonstrate that the implementation is feasible in two types of embedded systems (SoC-based FPGA and esp32) in terms of memory footprint. The application we focus on requires implementation in a low power consumption system, where data processing is not necessary in real-time. To this end, transfer learning and compression techniques are employed to obtain the final model to be deployed in both architectures, allowing the achievement of a fully on-chip deployment, which is a desirable feature for a precision agriculture application implemented on the edge;
- a standard training set available in the literature is used for training, granting a stable and effective performance.

2 Experimental Setting Under Study

The pest to be detected and controlled is the moth *lobesia botrana*, which mainly attacks the vineyard. The moth is captured using intelligent IoT traps designed by Nectras, as depicted in the left image of Fig. 1.

The capture surface (16×18 cm) is gummed with pheromones and an Arduino OV2640 2M camera is placed at the top of the IoT trap. Periodically, images are sent to the cloud through a custom hardware developed by Nectras and stored in .jpg format with resolutions of 800×600 and 1600×1200 pixels in the RGB color space. An example of a captured image is shown in Fig. 1 (center image), where it is noticed the presence of different type of insects, the pheromone lure, and other objects.

The original application developed by Nectras has a detection phase that is responsible for detecting and cropping elements present in the image. To simplify the classification problem for identifying *lobesia botrana*, the pheromone lure is

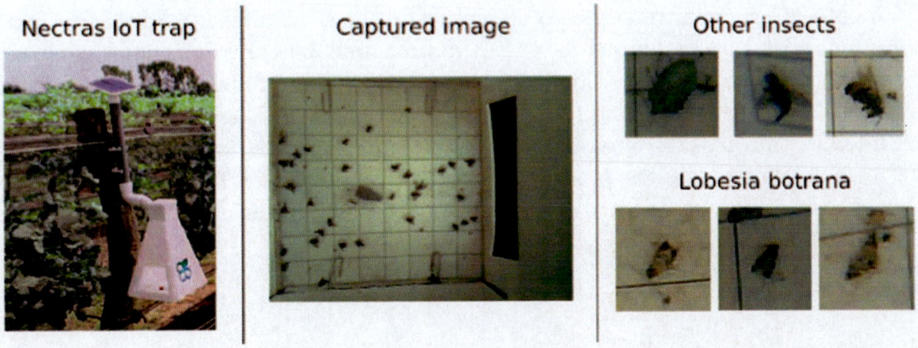

Fig. 1. IoT trap (left). Capture image (center). Classes to be detected by the classifier (right).

discarded during the detection phase based on its size. Binary classification is performed using machine learning techniques to classify two types of classes, presented on the right side of Fig. 1: other insects (class 0) and the *lobesia botrana* (class 1).

3 Pest Classifier Based on Machine Learning

When implementing ML-based models in embedded systems, the complexity and performance of the design are related to the number of required resources. Compression techniques are essential for deploying ML models on resource-constrained devices while maintaining efficiency and effectiveness [10]. Knowledge distillation [11] is based on the premise that a larger model (teacher) is trained to distill knowledge to a smaller and faster target network (distilled or student), which is computationally less expensive.

In this study, the teacher architecture is a binary classifier based on VGG-16, and it is obtained through the **transfer learning** method. The learning process allows the VGG-16 model to be loaded with weights obtained from the previous training with ImageNet. The last layers of the model are replaced by a classifier built for the new problem, based on [9]: first, a layer that flattens the output of the VGG-16 network, then three fully connected layers with 50, 25, and 10 neurons, respectively, and finally, a softmax layer with two outputs. The first four layers of this network are frozen, so they are not trainable and the rest of the model will be fine-tuned. The final architecture is composed of 14,818,706 parameters.

The knowledge of the teacher (32-bits floating point) is distilled to a student network, previously defined with quantization and pruning strategies. The overall student network architecture is presented in Fig. 2 and has the same number of layers as the teacher, but with fewer kernels and neurons. Two student networks are obtained, considering the target device: QStudentFPGA with 1,677 parameters (to be implemented on the SoC-based FPGA) and QStudentESP32 composed of 2,524 parameters (to be ported on the esp32 device). For the former, the number of bits is defined as 8 for the first four 2D-convolutional layers and 4 for the remaining layers. Concerning the latter, the number of bits is defined as 8-bits for all layers. Finally, a target sparsity of 50% is applied to remove redundant parameters. Quantization- and pruning-aware training (QP-AT) is performed, which implies pruning, quantizing, and retraining the model such that it can adjust and learn according to the newly quantized values and the distribution of neurons and connections.

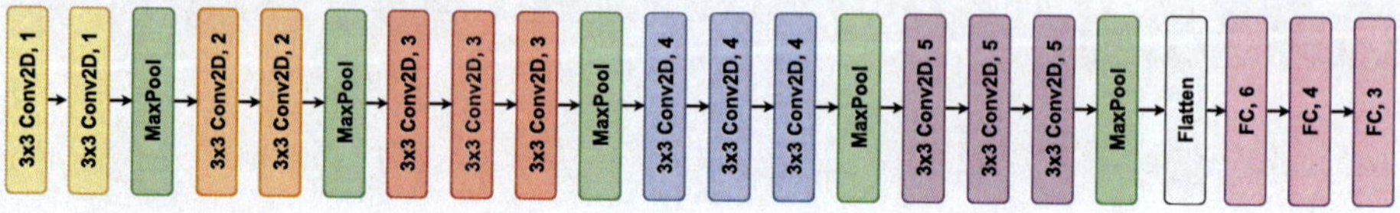

Fig. 2. Distilled architecture.

3.1 Dataset

Because of the reduced size of the dataset provided by the original application through Nectras front-end data-acquisition system, and to avoid augmentation techniques, *Pest24* [12] is selected for fine-tuning the teacher and training the student networks. This is a large-scale dataset composed of 25 thousand images and almost 200 thousand instances of pests.

Several datasets for pest analysis in precision agriculture are available in the open literature. They differ in the source of the images and the captured insects: (i) *internet*, such as IP102 [13], which is based on the collection of different images obtained from internet, (ii) *crop fields*, such as AgriPest [14], where the insects are captured directly in their habitat in different parts of the plant; and (iii) *traps*, such as Pest24, which is composed of images where the insects get stuck after being attracted, usually by pheromone lures. The selection of Pest24 for this research is motivated by the need of a special purpose dataset for sticky paper traps.

For binary classification, the dataset is divided into two classes: Coleoptera (class 0) and Lepidoptera (class 1). Some samples are shown in Fig. 3. Regarding the inference process for the student networks, the test is performed with two different datasets composed of images from in-field traps: *Pest24* and a test set from the original application in Argentina (ARG dataset) provided by Nectras.

4 Experiments and Results

Experimental Setup: We performed the experiments on a CPU Core i7 3.4 GHz 64 GB RAM GeForce GTX 1070, using Python 3.6.7, TensorFlow 1.12, and QKeras for quantization-aware training when targeting FPGA devices. SoC-based FPGA and esp32 TTGO LoRa32 SX1276 OLED boards are selected. For the SoC-based FPGA, the network is translated into an HLS project using

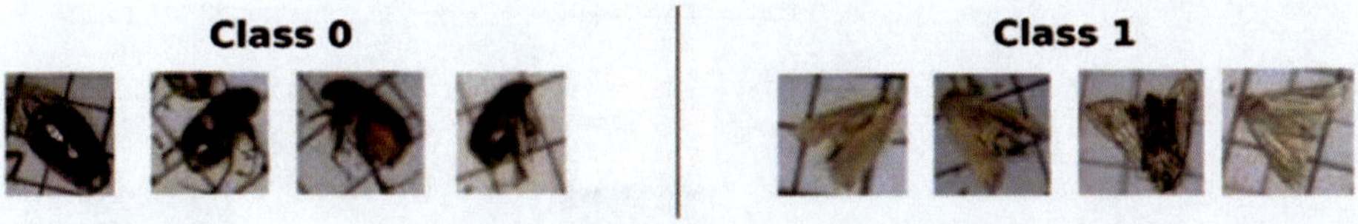

Fig. 3. Samples of Pest24 dataset.

hls4ml [15] and Vitis 2021.1. For esp32 implementation, Arduino IDE 1.8.19 is used with a modified version of TensorFlowLite (TFLite) library called TensorFlowLite_ESP32 V1.0.0, optimized for the ESP32 modules.

The teacher network is obtained through **transfer learning** technique. The framework to implement the learning process is Keras, which allows the VGG-16 model to be loaded with the weights obtained from previous training with ImageNet. Regarding the hyperparameters, the Adam optimizer is used with a learning rate of 0.0001, and the regularizer in each layer is L2 with a factor of 0.0001. The batch size is 32, as is the number of epochs, although an early stopping mechanism is employed, which stops the training when the validation loss does not improve for five epochs (where a change of 0.005 counts as no improvement), then the weights of the model are restored to when the loss stopped decreasing. The confusion matrix for the teacher network is shown in Fig. 4. The size of the final model is **177.6Mb**, making it difficult to implement in an embedded system. Thus, a compressed student network is generated through knowledge distillation technique [11].

To obtain the student networks (QStudentFPGA, QStudentESP32), the distillation loss is configured as KLDivergence, Adam optimizer with 0.0001, batch size of 32, 128 epochs, an early stopping mechanism, and a callback monitor to reduce the learning rate when a specific metric has stopped improving after five epochs. The L2 regularizer is configured as in the teacher training. Pruning and quantization-aware training is performed, which implies pruning [16], quantizing [17], and retraining the model such that it can adjust and learn according to the newly quantized values and the distribution of neurons and connections. The confusion matrix for each student network is shown in Fig. 4, where it is noticed a high accuracy despite the compressed network and the impact of the different datasets. Nevertheless, we can observe the potential of transfer learning when working with datasets directly related to an application. This could help in the early stages of implementation of this system. Indeed, in-field fine-tuning is not necessary if the reduced performances shown in the matrices 4 and 5 in Fig. 4 are deemed acceptable.

For the embedded implementation assessments, to deploy the 32FPStudent on esp32 device, the model is translated to TFLite and converted to a data structure in .h file imported into the source code. An interpreter is created and assigned to physical memory and contains the model in .h file, the memory space that holds the TFLite operations provided to the interpreter to reproduce the model, and a tensor area (limited by the amount of RAM of the microcontroller)

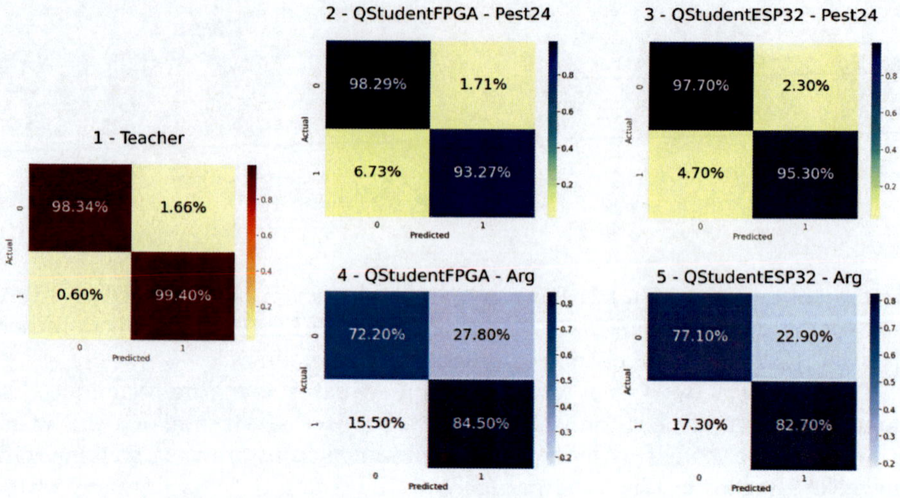

Fig. 4. Confusion matrix: VGG-16-based teacher (1), QStudentFPGA *Pest24* dataset (2), QStudentESP32 *Pest24* dataset (3), QStudentFPGA ARG dataset (4), and QStudentESP32 ARG dataset (5).

for the intermediate results of the model and the sizes of the inputs and outputs. For this implementation, the model is fully mapped in the internal RAM of 512KB, with a total occupancy of 26%.

For the SoC-based FPGA implementation, the memory footprint is measured in terms of the hardware utilization (DSP, RAM, FF, and LUT). Therefore, the hls4ml package is employed to generate a high-level synthesis project to map the inference to the programmable logic. Once the IP core is created and integrated with the framework hardware, the place and route reports provide the final resource utilization. For this purpose, we employed architectures based on FPGA: Xilinx PYNQ-Z1 and KRIA, considering a clock of 5 ns. Table 1 presents the final resource utilization after place and route implementation.

Table 1. Utilization from place and route reports (post-implementation).

SoC-FPGA Fabric	BRAM	DSP	FF	LUT
KRIA	62 (22%)	26 (2%)	19797 (8%)	23644 (20%)
PYNQ-Z1	88 (63%)	26 (12%)	29441 (28%)	21543 (41%)

As the original application was based on Nectras' hardware, which includes a power supply based on a solar panel, the benefit of implementing the system on an esp32 platform relies on a completely autonomous pest detection system on the existent platform. Thus, replication of the nodes in the field is affordable, due to the cost of this type of devices.

Regarding the FPGA-based solution, the selected platforms are low- and mid-end FPGA, considering a possible implementation in the field that implies a cheaper solution for replication using this technology.

5 Conclusions

In this paper, we presented an ML-based classifier for precision agriculture. We exploited compression techniques to demonstrate that the implementation is feasible in resource-constrained devices such as SoC-based FPGA and esp32, in terms of memory footprint, allowing the achievement of a fully-on-chip deployment. We verified the potential of transfer learning when working with a dataset directly related to the application, thereby facilitating the introduction of ML-based classifiers during the early stages of implementing this type of system in the field. Nevertheless, changes in the external conditions can impact the input images, leading to accuracy degradation over time. Thus, monitoring data drift has to be considered when deploying machine learning models for production.

References

1. Quy, V.K., et al.: IoT-enabled smart agriculture: architecture, applications, and challenges. MDPI Appl. Sci. **12**, 3396 (2022). https://doi.org/10.3390/app12073396
2. Brunelli, D., Polonelli, T., Benini, L.: Ultra-low energy pest detection for smart agriculture. IEEE Sens. 1–4 (2020). https://doi.org/10.1109/SENSORS47125.2020.9278587
3. Albanese, A., Nardello, M., Brunelli, D.: Automated pest detection with DNN on the edge for precision agriculture. IEEE J. Emerg. Sel. Top. Circuits Syst. **11**(3), 458–467 (2021). https://doi.org/10.1109/JETCAS.2021.3101740
4. Ding, W., taylor, G.: Automatic moth detection from trap images for pest management. Comput. Electron. Agric. **123**, 17–28 (2016). https://doi.org/10.1016/j.compag.2016.02.003
5. Liu, L., et al.: Pestnet: an end-to-end deep learning approach for large-scale multi-class pest detection and classification. IEEE Access **7**, 45:301–45:312 (2019). https://doi.org/10.1109/ACCESS.2019.2909522
6. Lu, C.-Y.L., Rustia, D.J., Lin, T.-T.: Generative adversarial network based image augmentation for insect pest classification enhancement. In: IFAC-PapersOnLine, vol. 52, pp. 1–5 (2019). https://doi.org/10.1016/j.ifacol.2019.12.406
7. Barbedo, J.: Detecting and classifying pests in crops using proximal images and machine learning: a review. In: MDPI AI, vol. 1, pp. 312–328 (2020). https://doi.org/10.3390/ai1020021
8. Ayaz, M., Ammad-Uddin, M., Sharif, Z., Mansour, A., Aggoune, E.-H.M.: Internet-of-things (IoT)-based smart agriculture: toward making the fields talk. IEEE Access **7**, 129551–129583 (2019). https://doi.org/10.1109/ACCESS.2019.2932609
9. Suárez, A., Molina, R.S., Ramponi, G., Petrino, R., Bollati, L., Sequeiros, D.: Pest detection and classification to reduce pesticide use in fruit crops based on deep neural networks and image processing. In: 2021 XIX Workshop on Information Processing and Control (RPIC), pp. 1–6 (2021). https://doi.org/10.1109/RPIC53795.2021.9648485

10. Choudhary, T., Mishra, V., Goswami, A., Sarangapani, J.: A comprehensive survey on model compression and acceleration. Artif. Intell. Rev. **53**(7), 5113–5155 (2020). https://doi.org/10.1007/s10462-020-09816-7
11. Hinton G., Vinyals O., Dean J.: Distilling the knowledge in a neural network. arXiv preprint arXiv:1503.02531 **2**(7) (2015). https://doi.org/10.48550/arXiv.1503.02531
12. Wang, Q.-J., et al.: Pest24: a large-scale very small object dataset of agricultural pests for multi-target detection. Comput. Electron. Agric. **175** (2020). https://doi.org/10.1016/j.compag.2020.105585
13. Wu, X., Zhan, C., Lai, Y.K., Cheng, M.M., Yang, J.: Ip102: a large-scale benchmark dataset for insect pest recognition. In: Proceedings of the IEEE/CVF Conference on Computer Vision and Pattern Recognition, pp. 8787–8796 (2019). https://doi.org/10.1109/CVPR.2019.00899
14. Wang, R., Liu, L., Xie, C., Yang, P., Li, R., Zhou, M.: AgriPest: a large-scale domain-specific benchmark dataset for practical agricultural pest detection in the wild. Sensors **21**(5), 1601 (2021). https://doi.org/10.3390/s21051601
15. Duarte, J., et al.: Fast inference of deep neural networks in FPGAs for particle physics. J. Instrum. **13**(07) (2018). https://doi.org/10.1088/1748-0221/13/07/P07027
16. Blalock, D., Gonzalez Ortiz, J.J., Frankle, J., Guttag, J.: What is the state of neural network pruning?. In: Proceedings of Machine Learning and Systems, vol, 2, pp. 129–146 (2020)
17. Liang T., Glossner J., Wang L., Shi S., Zhang X.: Pruning and quantization for deep neural network acceleration: a survey. Neurocomputing **461**, 370–403 (2021). https://doi.org/10.1016/j.neucom.2021.07.045

Deep Reinforcement Learning for Automated Car Parking

Luca Lazzaroni(✉), Francesco Bellotti, Alessio Capello, Marianna Cossu, Alessandro De Gloria, and Riccardo Berta

Department of Electrical, Electronic and Telecommunication Engineering (DITEN), University of Genoa, Via Opera Pia 11a, 16145 Genoa, Italy
{luca.lazzaroni,alessio.capello,marianna.cossu}@edu.unige.it,
{francesco.bellotti,alessandro.degloria,riccardo.berta}@unige.it

Abstract. This article explores the development of a Deep Reinforcement Learning (DRL) -based agent able to perform both path planning and trajectory execution, processing sensor perception information and directly controlling the steering wheel and the acceleration, like a normal driver. As a preliminary investigation, we limit our research to low-speed manoeuvers, in a challenging narrow drivable area. The vehicle's agent completely relies on the real-time information from the sensors, thus avoiding the need of a map. We show the validity of the proposed system in a simulated car parking test, in which the agent has been able to achieve high target reach rates, with a limited number of manoeuvers (gear inversion rate), outperforming the well-established Hybrid A-Star path planning algorithm in both the metrics. Further research is needed for improving the generalization ability of the agent and its application in more dynamic driving environments.

Keywords: Automated driving · autonomous agents · proximal policy optimization · reinforcement learning · low-speed manoeuvering · valet parking · unity ML-Agents

1 Introduction

Motion and path planning is a fundamental early step of the automated driving (AD) pipeline (e.g., see [1] for a detailed and comparative method overview). An emerging trend in the area is represented by the use of agents (i.e., artificial brains) that learn policies through reinforcement learning (RL), thus taking advantage of trial and error in safe operating environments (e.g., simulators), that provide feedback (rewards and/or penalties) to the agent's actions. This approach is expected to lead to systems that learn incrementally and, given the high flexibility of the underlying neural model, possibly developing new behaviors, not constrained by a-priori beliefs (apart from the definition of the rewards, that determine the learning of the agent). The goal of an RL agent, in fact, is to maximize the sum of the expected rewards in each episode.

While these two steps are usually distinct in a typical AD pipeline, we intend to explore the development of an RL agent able to perform both path planning and trajectory

R. Berta and A. De Gloria (Eds.): ApplePies 2022, LNEE 1036, pp. 125–130, 2023.
https://doi.org/10.1007/978-3-031-30333-3_16

execution, processing sensor perception information and directly controlling the steering wheel and the acceleration, like a normal driver. As a preliminary investigation, we limit our research to low-speed maneuvers, in a challenging narrow drivable area, thus taking the use case of a vehicle approaching to and entering a parking slot in a public garage. The vehicle's agent completely relies on the real-time information from the sensors, thus avoiding the need of a map.

The remainder of the article is organized as follows. Section 2 presents the simulation engine and the tools used, while Sect. 3 the models and the environment setting along with the training setup. Results are provided in Sect. 4 and finally, Sect. 5 draws conclusions.

2 Deep Reinforcement Learning in Unity 3D

In order to train our parking system, we chose Unity as our simulation engine [2] particularly because of its versatility and usability. In addition, Unity supports the training and inference of neural models through the open-source ML-Agents toolkit [3]. ML-Agents enables the development of simulated environments inside the Unity Editor and the interaction with them through a Python API. On the Unity side, the ML-Agents C# library allows defining environments to train on. It also includes a set of state-of-the-art RL algorithms, such as Soft Actor-Critic (SAC) [4] and Proximal Policy Optimization (PPO) [5] etc. An ML-Agents learning environment consists of three main components:

- Agent: collects observations from the surrounding environment, executes actions, and obtains rewards based on the new observation collected after an action has been accomplished.
- Brain: specifies the dimension of both the observations and the actions spaces. These can be discrete or continuous. The brain is in charge of choosing the next action, given the current observation, according to the learnt policy and the current state of the environment.
- Academy: responsible for global coordination of the environment simulation since it orchestrates all the Agent and Brain objects in a Unity scene.

3 Modelling and Training

For the development of our environment, we started from an autonomous parking scenario baseline, described in [6]. In this project, the agent (i.e., the controlled car) is trained to reach a target parking lot among all the available ones. The maneuvering area is about 80 m^2 wide. The agent can choose whether to park forward or backward, in which case it gets the maximum reward.

Starting from this scenario, we developed a simple public garage environment (Fig. 1). The lighting of the scene has been designed to reduce as much as possible the resources dedicated to rendering, in order to minimize the impact on the training time. Compared to [6], the maneuvering area is way more restricted (i.e., 5.3 m $\times$ 3.5 m), and the lines delimiting the parking lots have been replaced by walls, which further strongly limits the drivable area, requiring the vehicle to perform careful maneuvers. At each

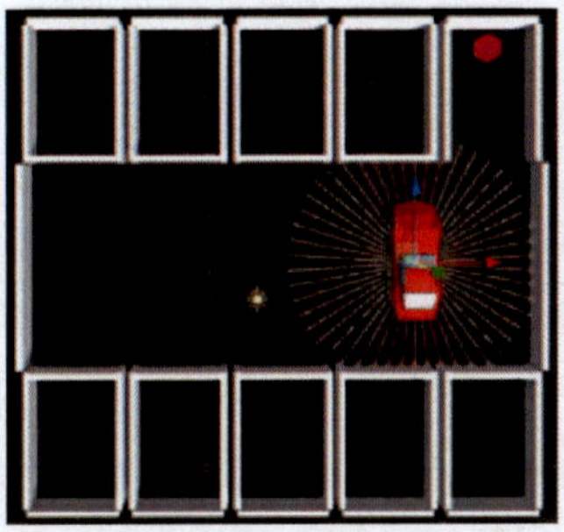

Fig. 1. The parking environment developed in Unity. The desired point of arrival is shown as a purple sphere. LiDAR sensor's rays visible, red-colored indicate an object hit. Car 3D model provided by [7].

episode, the vehicle's starting point and orientation are random. The target to reach is represented by the only one open parking lot, which is chosen randomly at each episode.

Regarding the characteristics of the vehicle, the car's dimensions are 4.90 m in length and 1.80 m in width, the total mass is 1500 kg, the maximum motor torque is 400 Nm, the maximum brake torque is 300 Nm, and the maximum steering angle is 33°. A path is considered valid only if the vehicle reaches and overlaps the target purple sphere, without collisions. The vehicle is modeled through a simple kinematic bicycle model, which was considered suited to describe low-speed maneuvers. The observations that the agent collects from the environment at each step are taken with a Raycast Unity object, which casts a total of 50 rays of fixed length (i.e., 5 m), resulting in a 1D array representing the surrounding space at 360°. The Raycast models a LiDAR sensor.

3.1 Rewards

We distinguish two main kinds of reward: dense (provided at each simulation step) and sparse (provided only when some events occur). We defined three reward contributions, two sparse and one dense (1).

$$reward = 0.01 * \left(-c_1 d^2 + \frac{-c_2(1-\cos\theta)}{d+1}\right) + \begin{cases} 0\ if\ goal \\ -1\ if\ collision \end{cases}. \quad (1)$$

θ is the angle between the vehicle's heading and the distance to the goal vectors. The scalar distance between agent and goal, d, is calculated through the Manhattan formula (2), which is better suited to the environment, while $c1$ and $c2$ are scaling factors, being set to 0.01 and 1.5, respectively.

$$d = \left|agent_x - target_x\right| + \left|agent_y - target_y\right|. \quad (2)$$

In (1), the dense contribution aims to minimize the distance between the agent and the goal as well as to arrive at destination with the correct orientation. The square of the distance from the goal is penalized, in order to incentivize the agent to get closer to the target. The orientation penalty is reduced with the distance since the vertical alignment constraint for the vehicle is relevant only close to the target.

3.2 Training

We trained our RL model using the PPO algorithm [5]. PPO is a stochastic policy gradient method for RL which clips policy changes so that the policy does not update too rapidly, which would hinder the training by creating instability. PPO uses a neural network to approximate the ideal function that maps an agent's observations to the action decision. We chose a simple Multi-Layer Perceptron (MLP) network, composed of 2 hidden layers of 256 neurons each (Fig. 2). The input size is 416; 408 input values originate from the LiDAR sensor, and the remaining 8 are the 2D values of distance between agent and goal, agent's speed, agent's heading, and goal's heading, respectively. The output size is 3, representing the throttle, the steering angle, and the brake values to be imparted to the vehicle, respectively. Batch size is set to 512 and initial learning rate 10^{-4}, with a linear schedule reaching 0 at the end of the training.

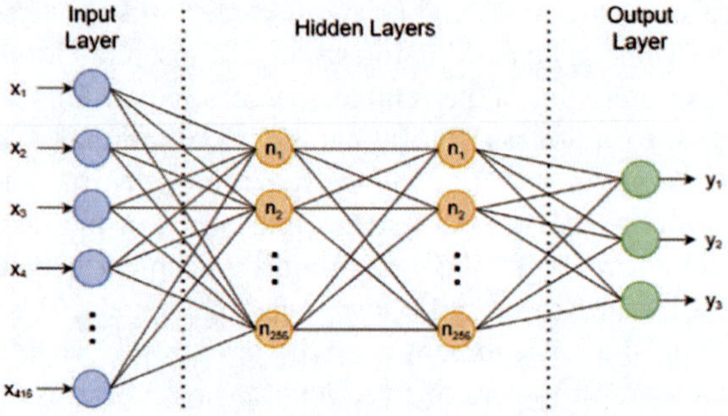

Fig. 2. The PPO decision network: an MLP with 2 hidden layers of 256 neurons each.

A key factor for the training was the choice of a curriculum learning strategy [8]. Curriculum learning is a training approach in which the agent starts learning the basic tasks, then moves on to ever more complex ones. This well fits our use case, since we can first train the agent to reach the goal (thus exploring a wide set of situations and settings), then to also avoid collisions. In the first phase, we set collisions as non-terminal events, i.e., when they occur, we do not terminate the episode, so to let the agent explore the environment more extensively. Therefore, an episode was terminated only if the agent reached the goal or after a fixed number of steps. The collision penalty was also relatively low, −0.01. When the agent reached a 100% goal rate, we started the second phase, in which collisions were considered as terminal events and their penalty set to −1. Training took around 25 M steps for the first phase, and 57M for the second one, for a total of 82 M steps.

4 Results

After 82 M training steps, the model is able to reach a 94% goal-reach rate. The training trend is shown in Fig. 3. As visible from Fig. 3a and Fig. 3b, at 25 M steps there is a sharp drop in both the reward and the reached goal rate, followed by a marked decrease in the collisions rate. This discontinuity is due to the mentioned increase of the collision penalty by a 100 × factor, and by enforcing a collision as a terminal event. As a consequence, the agent, after learning how to reach the goal in the first part of training (100% goals reached at 25 M steps (Fig. 3b)), started learning how to reach it without collisions, totaling a final rate of 94% of goals reached without collisions.

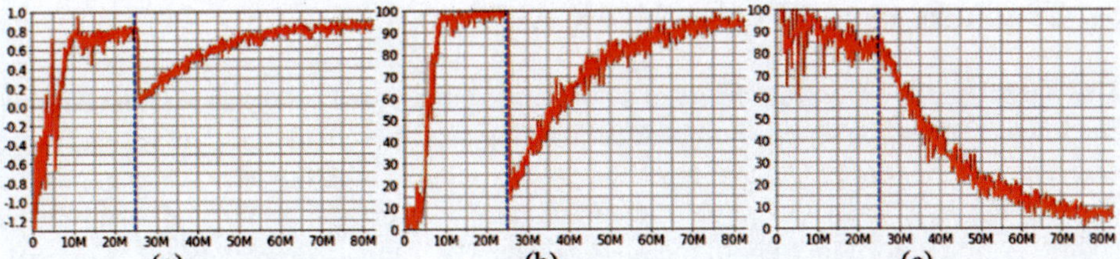

Fig. 3. The PPO network training over 80M steps. (a) is the reward, as described in (1), whereas (b) and (c) are the ratio of reached goals and collisions, respectively. The dashed blue line indicates the collision management changing.

For the sake of performance assessment, we compared our agent with the popular Hybrid A-Star path planning algorithm, implemented in the same car and environment we used, starting from the Unity package provided by [9]. The comparison metrics we chose are the success (i.e., goal reach) rate and the average number of maneuvers performed (measured in terms of number of gear inversions) per episode. Results are presented in Table 1.

Table 1. Comparison between PPO and Hybrid A-Star over 100 episodes

Algorithm	Avg. number of manoeuvers	Success rate (%)
PPO	3.36	94
Hybrid A-Star	4.48	85

The table shows that our model achieves a higher success rate, with a lower average number of gear inversions. We argue that the lower success rate by Hybrid A-Star may be imputed to the discretization level of the map. We set the resolution value to 25 cm, since, with lower values, the map pre-processing time was around one minute. As a consequence, while the algorithm was always able to output a path to the goal, its execution may lead to collisions with the walls, being the parking lot very narrow. A higher resolution value may increase the Hybrid A-Star's success rate, but at the cost of a huge map processing time, which makes the approach unsuited to real-world scenarios. Regarding the gear inversions, we argue that this difference between the two approaches is related to the Hybrid A-Star's main focus being on minimizing the road traveled, which results in more complex trajectories compared to our DRL approach, which instead tends to have a more human-like behavior (i.e., smoother trajectories and fewer maneuvers). A key step forward made by our model compared to the Hybrid A-Star is given by the lack of the need for the map of the area, that the Hybrid A-Star algorithm employs to compute a path to the goal. Our model, on the other hand, only relies on the real-time LiDAR data and the knowledge of the goal's coordinates.

5 Conclusions and Future Work

This paper has shown the feasibility of a DRL-based integrated approach to path planning and low-speed maneuver execution. A car parking test, in a challenging narrow drivable area, showed that the PPO agent has been able to achieve high target reach rates, with a limited number of maneuvers (gear inversion rate), outperforming the well-established Hybrid A-Star algorithm in both the metrics. Our DRL agent completely relies on real-time LiDAR information, without needing a map. These preliminary results open significant new research opportunities. A key goal before starting tests in real-world vehicles will be to increase the generalization ability of the agent, that will need to be trained and tested in more complex and dynamic environments. Also, different neural network architectures will be tested and compared with the currently used PPO. Another important step could be in the direction of implementing other types of maneuvers (i.e., not only at low speed), employing a more complex vehicle model.

References

1. Claussmann, L., Revilloud, M., Gruyer, D., Glaser, S.: A review of motion planning for highway autonomous driving. IEEE Trans. Intell. Transp. Syst. **21**, 1826–1848 (2020). https://doi.org/10.1109/TITS.2019.2913998
2. Unity Technologies: Unity Real-Time Development Platform | 3D, 2D VR & AR Engine. https://unity.com/. Accessed 30 May 2022
3. Juliani, A., et al.: Unity: a general platform for intelligent agents (2020). http://arxiv.org/abs/1809.02627
4. Haarnoja, T., Zhou, A., Abbeel, P., Levine, S.: Soft actor-critic: off-policy maximum entropy deep reinforcement learning with a stochastic actor. In: Proceedings of the 35th International Conference on Machine Learning, pp. 1861–1870. PMLR (2018)
5. Schulman, J., Wolski, F., Dhariwal, P., Radford, A., Klimov, O.: Proximal policy optimization algorithms (2017). http://arxiv.org/abs/1707.06347
6. K, R.: Autonomous Car Parking using ML-Agents. https://medium.com/xrpractices/autonomous-car-parking-using-ml-agents-d780a366fe46. Accessed 27 May 2022
7. ShrineToLoud: Chevrolet Corvette 1980 Different colours. https://sketchfab.com/3d-models/chevrolet-corvette-1980-different-colours-7e428bdb3ab54b4e9ac610e545fd9d03. Accessed 30 May 2022
8. Bengio, Y., Louradour, J., Collobert, R., Weston, J.: Curriculum learning. In: Proceedings of the 26th Annual International Conference on Machine Learning, pp. 41–48. Association for Computing Machinery, New York (2009). https://doi.org/10.1145/1553374.1553380
9. Nordeus, E.: Self Driving Vehicle. https://github.com/Habrador/Self-driving-vehicle. Accessed 10 June 2022

Machine Learning Techniques for Anomaly-Based Detection System on CSE-CIC-IDS2018 Dataset

Abdussalam Elhanashi[1](✉), Kaouther Gasmi[2], Andrea Begni[1], Pierpaolo Dini[1], Qinghe Zheng[3], and Sergio Saponara[1]

[1] Department of Information Engineering, University of Pisa, Pisa, Italy
aelhanashi@student.unipi.it

[2] Department of the Computer Science, University of Tunis, Tunis, Tunisia

[3] School of Intelligence Engineering, Shandong Management University, Dingxiang Road, Jinan 250357, China

Abstract. Anomaly-based detection is a novel form of an intrusion detection system, which has become the focus of many researchers for cybersecurity systems. Data manages most business decisions. With more access to data, it is necessary to interrupt and analyze them correctly. When it comes to security, the first step is to determine the outliers as a security threat. Machine learning and deep learning techniques have proven to recognize anomalous attack patterns that deviate from normal network behavior. Machine learning can be utilized to learn the characteristic of data and help to improve the speed of detection. In this research, we present our approach to implementing an algorithm for the anomaly detection framework in complex and unbalanced data. The proposed method has been applied to a CSE-CIC-IDS2018 dataset. It is the most recent dataset that is publicly available, an extensive dataset that includes a wide range of attack types. This data has been pre-processed and cleaned to find helpful information for classification by the proposed models. We performed a correlation methodology to filter irrelevant anomalies and grouped the correlated anomalies into a single feature to minimize detection time. A stacked autoencoder has been used to reduce the dimensionality of the dataset. We exploited different machine learning algorithms such as (Random Forest, GaussianNB, and multilayer perceptron) to classify the streamed data. Our experimental results outperformed the superiority of the proposed approach to identify anomalous components and manage threat detection in cybersecurity applications.

Keywords: Anomaly-based detection · cybersecurity · machine learning · CSE-CIC- IDS2018 · stacked autoencoder

1 Introduction

The recent improvements in the IoT ecosystem enabled IoT networks to be vulnerable to cyber-attacks [1]. Vast data are generated from IoT devices and offloaded to distributed data centers. Fog and clouds may pave the way to abnormal behaviors. For example,

R. Berta and A. De Gloria (Eds.): ApplePies 2022, LNEE 1036, pp. 131–140, 2023.
https://doi.org/10.1007/978-3-031-30333-3_17

an end-user can claim to become an alternative user and acquire illegitimate benefits and accomplish man-in-the-middle attacks, and denial of service attacks may occur [2]. In many situations, an intrusion tries to violate one of the security objectives, which are confidentiality, integrity, and data availability. Hence many tools and methods are developed to secure wireless networks from intrusions. An intrusion detection system (IDS) based on anomaly detection is widely used for security evaluation. [3–5] This detects malicious traffic and interrupts intrusions. However, the IDS suffers from several limitations, such as real-time analysis and detection, generated alarm, and data quality, that can decrease the detection rate and accuracy performances [6]. Recently, Machine Learning (ML) techniques have been integrated to enhance intrusion detection systems and reinforce computer security. Many research contributions explore how to incorporate ML approaches in the intrusion detection system to achieve reliable IDS with an accurate performance by improving the data quality [7, 8]. The recent cybersecurity datasets CSE-CIC-IDS 2017 and CSE-CIC- IDS 2018 were created by the Canadian Institute of Cybersecurity. Both have used the notion of profiles for generating cybersecurity datasets [9]. These datasets present an in-depth knowledge of attacks for IDS evaluation. The CSE-CIC-IDS 2017 dataset has attracted many researchers to suggest an intrusion detection model based on ML [10, 11]. CSE-CIC-IDS 2018 is the latest dataset that includes different attack scenarios: Brute Force, Heartbleed, DoS, DDoS, Web Attack, Botnet, and Inside Infiltration. CSE-CIC-IDS 2018 was created to train ML models-based intrusion detection for massive network traffic [12]. In this paper, we propose a novel network intrusion detection approach to filter irrelevant anomalies and group the correlated anomalies into a single feature to minimize detection time. Experimental results on the CICIDS2018 dataset demonstrate that our proposed method performs well in terms of accuracy recall and precision for machine learning classifiers. The main contributions of this research are: -

- We propose a solution for fast detection for different types of attacks to identify possible intruders and eliminate damage effects.
- We build a novel approach for feature selection by utilizing an entropy decision approach to enhance the data quality.
- Machine learning algorithms classify unforeseen data by measuring the neighbor points. This is a discrete categorical target which is binary data that includes regular traffic and attacks.

Hereafter, the paper is organized as follows: Sect. 1 presents the introduction. Section 2 explores the methodology of this research. Section 3 and 4 offers the Experiment setup and results. Finally, conclusions are drawn in Section 5.

There are very few works in the literature that present innovative IDS with application on the CSE-CIC-IDS-2018 dataset, especially regarding Stacked Autoencoder-based methods. In the following, we report the most recent work on AI-based IDS to highlight the contribution of our work concerning the state of the art. The authors of one study [13] presented an IDS based on the Convolutional Neural Network (CNN) model, focusing on the Denial-of-Service attack type, testing the model on the KDD99 and CSE-CIC-IDS-2018 datasets. The work presented by the authors has a limitation that focuses on binary classification. At the same time, our study proposes a multi-class model with comparable detection performance. In another study [14], the authors proposed the comparison of different ML models to the CIC-IDS-2018 dataset, obtaining an excellent result in the reduced problem of anomaly detection, in which each of the attack types is grouped in one class of anomalous traffic. Also, in this case, our contribution is to apply an innovative paradigm respect classic ML, and to manage the multi-classification problem, obtaining excellent detection accuracy. The authors of another study [3] compared RNN and CNN models with application to the CIC- IDS-2018 dataset, managing the multi-class detection problem. Concerning these authors, we propose the application of the most recent learning paradigm, and we reach better accuracy in the detection of the single attack type. Authors in a study done in 2018[15] proposed an IDS based on the Fast Learning Network (FLN) model applied to the KDD99 dataset. A recent study from 2021 [16] suggested a novel feature selection technique and a comparison of various ML models for IDS implementation, testing the obtained results on the KDD99 dataset. Concerning such articles, we propose a more innovative IDS model with the added value of applying our proposed method to a more complex dataset for multi-class problems instead of binary detection. Authors in two different studies done in 2019 [17, 18] applied ML models with application to the UNSW-NB15 dataset and also the management of multiclass problems. Concerning these works, we propose the application of innovative IDs on a more complex dataset, still obtaining comparable detection performance. What we want to highlight in our work is that we present a very novel approach to IDs implementation respective to state of the art, exploiting advanced feature selection and preliminary dataset management using Stacked Autoencoder as the core structure of the proposed method, obtaining very high performance within the application on one of the most complex datasets for multiclass detection problems.

2 Methodology

In this section, we will discuss our methodology and the proposed approaches to validate and implement the proposed solution by improving the feature extraction and classification techniques to obtain reliable IDS with accurate performance. The main goal of the proposed approach is to collect and preprocess the data. Hence, the algorithm executes the process to gather and accumulate the necessary data from network traffic. Once the data is collected, a specific preprocessing is performed on the gathered data. The preprocessing phase evaluates the introduced data and eliminates the incompatible data types. Once the data are preprocessed, it is fed to the machine learning classifiers to distinguish between normal and attack data types in the network traffic. The implementation of components that represent our approach is shown in Figure 1.

In this work, we will propose a novel approach by applying various machine learning algorithms versus the other methodologies to detect anomaly detection precisely and initiate an immediate response for monitoring the state of software and hardware running in the network.

2.1 Dataset Description

Dataset CSE-CIC-IDS2018 is designed to assess predictive models on anomaly- based intrusion detection on network traffic. The Communication Security Establishment

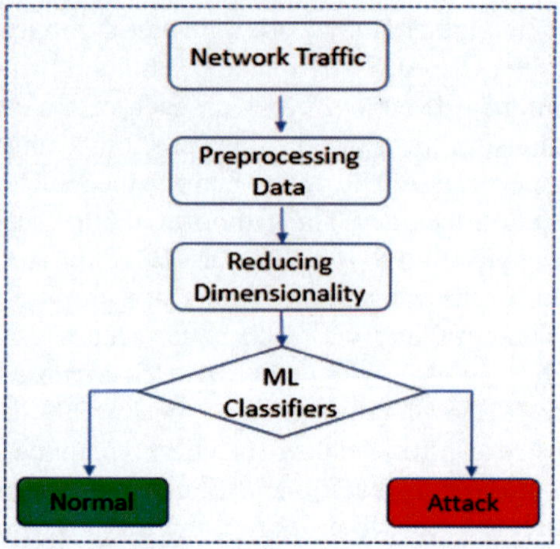

Fig. 1. System overview for anomaly-based network intrusion detection

(CSE) and the Canadian Institute for Cybersecurity (CIC) released the iteration of the intrusion detection dataset. CSE-CIC-IDS2018 has imbalanced data, which is structurally similar to CSE-CIC-IDS2017. However, CSE-CIC-IDS2018 has been prepared from a more extensive network of simulated attack machines and a simulated target. This dataset contains 16,233,002 instances collected from 10 days of network traffic. 18.2% of these instances are attack traffic. 50 machines are included in the assaulting foundation, and the causality association has 5 offices with a total of 30 servers and 420 machines. Table 1 presents the distribution in percentages of types of network traffic for CSE-CIC-IDS2018. CSE-CIC-IDS2018 is distributed into 10 CSV files, which are available in an open-source cloud. 10 CSV files consist of 80 independent features. These traffic highlights have been extracted by CICFlowMeter- V3.

2.2 Data Preprocessing

The issue of an imbalanced dataset in intrusion detection system modeling is a critical concern. In machine learning, especially in classification modeling, introducing machine learning models to the balanced dataset in the training phase has a significant effect on the classifier's performance. The CSE-CIC-IDS2018 dataset comes from a large number of normal and abnormal features in the system log. To overcome the problems of imbalanced data, we introduced different techniques to normalize the dataset. The first step is to eliminate any flows that contain invalid data. Due to the rounding factor, flows with a Flow Duration of 0 may have various characteristics on average of becoming undetermined (NaN and infinity). It is essential to replace the NaN values with the maximum values of the columns which contain these values. The NaN values will be eliminated, aiming to avoid the effect of these values on the model performance during the training process. In the second step, we convert cytological variables into numeric variables to string the data, and it can be easily understood for machine learning models for further processing. We have proposed an approach to enhancing the performance of the models by further preprocessing the dataset considering the ratio of benign (Normal

Table 1. CSE-CIC-IDS2018 Network traffic distribution

Traffic type	Distribution %
Benign (Normal traffic)	81.8
DDOS attacks-HOIC	4.54
DDoS attacks-LOIC-HTTP	3.81
DoS attacks-Hulk	3.05
BoT	1.89
FTP-BruteForce	1.28
SSH-BruteForce	1.26
Infilteration	1.06
DoS attacks-SlowHTTPTest	0.89
DoS attacks-GoldenEye	0.24
DoS attacks-SlowLoris	0.09
DDoS attacks-LOIC-UDP	0.06
Brute Force -Web	0.02
Brute Force -XSS	0.011
SQL Injection	0.003

traffic) and attacked labeled data. We applied Pearson Correlation for feature selection of the introduced data, see Figure 2. The feature selection approach plays a vital role in the model's performance in the given problem due to the massive amount of measured data. The model algorithm should be capable of fast prediction and effective in real-time applications. The purpose of the feature selection is to locate the most practical features that will result in more accurate models being trained and less computational time required. In machine learning classification, correlation criteria are commonly used in the filter technique. Correlation is a measure of the linear relationship of two or more features. The most correlated data with the target will be selected to establish the model. The Pearson correlation approach is one of the best feature selection criteria, which is described in equation (1):

$$Corr(i)\frac{cov(a_i, b)}{\sqrt{var(a_i) * var(b_i)}} \tag{1}$$

where *Corr*(*i*) is the Pearson correlation ai is the *i*th feature, *b* is the targeted label, and *cov* () is the covariance function, and *var* () is the variance function.

To choose the features with a high correlated target, we set the threshold to >0.85 for relevant components. Assuming that the selected features are related to one another, in that instance, the one with the lowest correlation to the target can be eliminated. Furthermore, the parts that indicate a high correlation can be unified.

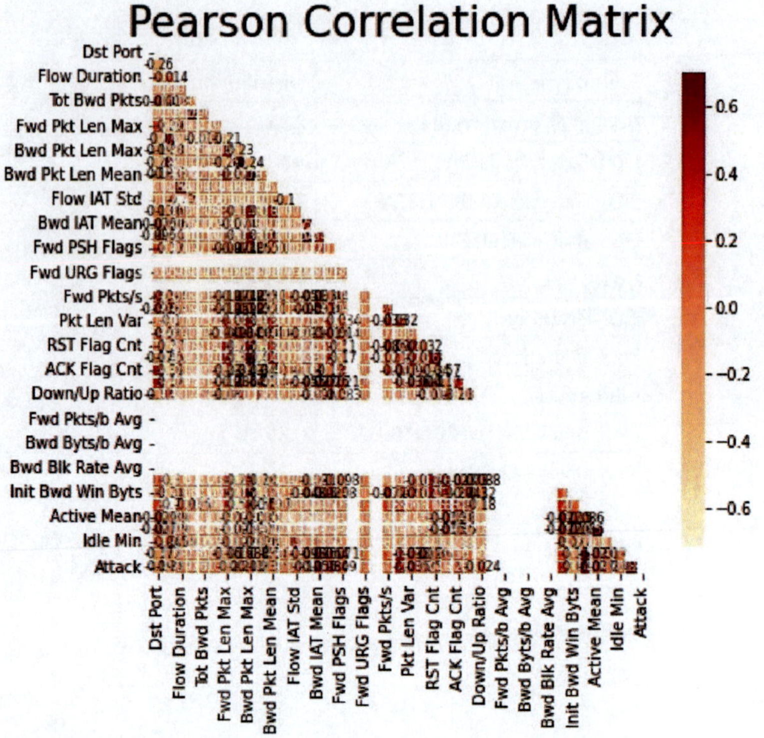

Fig. 2. A graphical representation of the Pearson correlation matrix that shows pairwise feature correlation computed over the entire dataset

3 Experiment Setup

We used a stacked autoencoder in our experiments to reduce the dimensionality for the preprocessed data. This is to enhance the training time and improve computational efficiency. We split the data into 80% for training and 20% for validating data. We set the number of epochs at 30, and we adjusted the batch size to 5000. The loose curve shows promising results between the training and validation sets for stack autoencoder with this dataset. See Fig. 3. After the dimensionality reduction process, the data is ready to be trained with machine learning models. The supervised classification models used in this research are Random Forest classifier, GaussianNB Classifier, and Multi-layer perceptron model. To assess the performance of each model for this dataset, these classifiers were chosen based on their effectiveness in solving the classification problems. For all experiments in this work, we utilized NVIDIA Tesla K80 GPU platform on Google colab cloud service, which has been created for classification tasks of machine learning models.

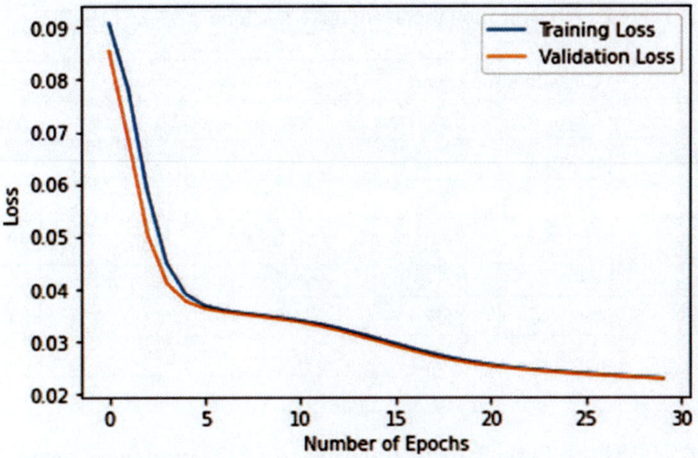

Fig. 3. Training and validation loss curve for stack autoencoder.

4 Results and Discussion

In this section, we summarize our experimental results for detection intrusions by using selected machine learning classifiers over the CSE-CIC-IDS2018 dataset. With the optimized parameters, we trained the CSE-CIC-IDS2018 dataset with three classifiers. We carried out the experiments by feeding the dataset into the ML classifiers. The dataset assessment is an important rule to validate the intrusion detection approach. To evaluate the performance of deployed models, we used confusion matrix criteria (precision, recall, f-score, and accuracy) to assess the commission for the proposed machine learning classifiers.

We started by comparing the detection assessment among the three classifiers. Based on the results of our experiments, the performance of the Random Forest classifier achieved the highest accuracy, with 99.6%, and the multi-layer perceptron model obtained an accuracy of 99.3%. In contrast, the GaussianNB classifier was less accurate, with only 88% using the dataset of CSE-CIC-IDS2018. Random forest & multi-layer perceptron algorithms can perform better on a selected dataset while the false positive rate is very low. These classifiers reduced the disparity in the training set and synthesized with high proximity with the actual data. The major role of these algorithms is to extract and improve complex information from disproportionate features in the CSE-CIC-IDS2018 dataset. However, the GaussianNB classifier has a zero-frequency problem. In the case where the class does not appear during the training phase, it will be assigned to the probability of zero, thus will result in overfitting and it will affect the performance of this classifier. The demonstrated results are summarized in Table 2.

The results show that the proposed classifiers can achieve a high detection accuracy while producing a low false positive rate. It indicated that Random Forest and Multi-layer perceptron algorithms have the highest performance detecting abnormalities while the GaussianNB model has the most deficient performance. In addition to that, we compared the proposed approach to the other methodologies [19–21], which were experimented on the CSE-CIC-IDS2018 dataset with the best accuracy of 99.6% concerning 96%,

Table 2. Classification report for three machine learning classifiers

Method	Classification report				
	Class	Precision	Recall	F1-score	Accuracy
Random forest	Normal	99.42%	99.1%	99.25%	99.6%
	Attack	99.3%	99.05%	99.17%	
Multi-layer	Normal	99.01%	99.05%	99.02%	99.3%
	Attack	99.15%	99%	99.09%	
Gaussian NB	Normal	85%	81%	82.9%	88%
	Attack	49%	85%	62%	

90.25%, and 96.2%, see Table 3. As mentioned earlier, the introduced dataset has a class imbalance. The imbalance has reduced the performance of machine learning models to classify attacks and standard features in the intrusion system. The high imbalance can make identification of the minority feature more inconvenient for learners, especially for big data, and manages to introduce a bias in the benefits of majority features. IDS need an informative characteristic of traffic data, so it is important to identify one in order to get better performance. This study proposed a Pearson correlation-based trait selection method to determine whether a trait is important. Pearson correlation-based feature selection is a module that uses selected metrics to identify unrelated attributes and exclude redundant columns from the data set. The columns are returned ordered by their feature scores. The important features has improved the performance of the classification in comparison to the other methodologies. We applied the techniques for the gathered data. We obtained transformed data by implementing dataset quality and increased the accuracy of our approach. In addition to that, the proposed system achieved the classification task for anomaly-based detection, and it was effective in distinguishing the incoming traffic as normal or attack type.

Table 3. Comparison of our approach vs. state of the art by using the CSE-CIC-IDS2018 dataset

Methods	Algorithm	Accuracy
This work	**Random forest**	**99.6%**
Kim et al. [19]	CNN IDS	96%
Farhan et al. [20]	DNN	90.25%
Lin et al. [21]	LSTM	96.2%

5 Conclusions

Intrusion detection is an improved approach to monitoring networks and data for malicious activities and policy violations. In this paper, we propose a reliable anomaly intrusion detection technique based on machine learning classifiers. Three different machine learning models were employed in our experiments which include Random Forest, GaussianNB, and Multi-layer perceptron classifiers. The pre-processing stage increased the detection rate and the accuracy of the model's entire CSE-CIC-IDS2018 dataset. Feature selection based on Pearson's correlation process enhanced the imbalanced dataset with high-quality data considering the ratio of benign (Normal traffic) and attack-labeled data. From the obtained results, Random Forest and Multi-layer perceptron showed better performance than the GaussianNB classifier. Indeed, the proposed approach achieved high accuracy compared with the other methodologies. In the future, we will extend our exploration to utilize convolutional neural network algorithms to find out the optimal model along with CNN parameters for the CSE-CIC-IDS2018 dataset.

References

1. Cerullo, G., et al.: Iot and sensor networks security. In: Security and Resilience in Intelligent Data-Centric Systems and Communication Networks, pp. 77–101. Academic Press, Cambridge (2018)
2. Tu, S., et al.: Security in fog computing: A novel technique to tackle an impersonation attack. IEEE Access **6**, 74993–75001 (2018)
3. Kim, J., Shin, Y., Choi, E.: An intrusion detection model based on a convolutional neural network. J. Multimedia Inf. Syst. **6**(4), 165–172 (2019)
4. D'hooge, L., et al.: Classification hardness for supervised learners on 20 years of intrusion detection data. IEEE Access **7**, 167455–167469 (2019)
5. Stiawan, D., et al.: CICIDS-2017 dataset feature analysis with information gain for anomaly detection. IEEE Access **8**, 132911–132921 (2020)
6. Guezzaz, A., et al.: A global intrusion detection system using pcapsocks sniffer and multilayer perceptron classifier. Int. J. Netw. Security. **21**(3), 438–450 (2019)
7. Thaseen, I.S., Poorva, B., Ushasree, P.S.: Network intrusion detection using machine learning techniques. In: 2020 International Conference on Emerging Trends in Information Technology and Engineering (ic-ETITE). IEEE (2020)
8. Beghdad, R.: Training all the KDD data set to classify and detect attacks. Neural Netw. World **17**(2), 81 (2007)
9. Thakkar, A., Lohiya, R.: A review of the advancement in intrusion detection datasets. Procedia Comput. Sci. **167**, 636–645 (2020)
10. Yulianto, A., Sukarno, P., Suwastika, N.A.: Improving adaboost- based intrusion detection system (IDS) performance on CIC IDS 2017 dataset. J. Phys.: Conf. Ser. **1192**(1) (2019)
11. Kurniabudi, D.S., Darmawijoyo, M.Y.B., Bamhdi, A.M., Budiarto, R.: CICIDS-2017 dataset feature analysis with information gain for anomaly detection. IEEE Access **8**, 132911–132921 (2020). https://doi.org/10.1109/ACCESS.2020.3009843
12. Kanimozhi, V., Jacob, T.P.: Artificial intelligence-based network intrusion detection with hyper-parameter optimization tuning on the realistic cyber dataset CSE-CIC- IDS2018 using cloud computing. In: 2019 international conference on communication and signal processing (ICCSP). IEEE (2019)

13. Kim, J., et al.: CNN-based network intrusion detection against denial-of-service attacks. Electronics **9**(6), 916 (2020)
14. Kilincer, I.F., Ertam, F., Sengur, A.: Machine learning methods for cyber security intrusion detection: datasets and comparative study. Comput. Netw. **188**, 10784 (2021)
15. Ali, M.H., et al.: A new intrusion detection system based on fast learning network and particle swarm optimization. IEEE Access **6**, 20255–20261 (2018)
16. Dini, P., Saponara, S.: Analysis, design, and comparison of machine-learning techniques for networking intrusion detection. Designs **5**(1), 9 (2021)
17. Bagui, S., et al.: Using machine learning techniques to identify rare cyber-attacks on the UNSW-NB15 dataset. Security and Privacy **2**(6), e91 (2019)
18. Aboueata, N., et al.: Supervised machine learning techniques for efficient network intrusion detection. In: 2019 28th International Conference on Computer Communication and Networks (ICCCN). IEEE (2019)
19. Kim, J., Shin, Y., Choi, E.: An intrusion detection model based on a convolutional neural network. J. Multimed. Inf. Syst. **6**, 165–172 (2019)
20. Farhan, R.I., Abeer, T.M., Nidaa, F.H.: Optimized deep learning with binary PSO for intrusion detection on CSE-CIC-IDS2018 dataset. J. Al Qadisiyah Comput. Sci. Math. **12**, 16 (2020)
21. Lin, P., Ye, K., Cheng-Zhong, X.: Dynamic network anomaly detection system by using deep learning techniques. In: Da Silva, D., Wang, Q., Zhang, L.-J. (eds.) Cloud Computing – CLOUD 2019: 12th International Conference, Held as Part of the Services Conference Federation, SCF 2019, San Diego, CA, USA, June 25–30, 2019, Proceedings, pp. 161–176. Springer International Publishing, Cham (2019). https://doi.org/10.1007/978-3-030-23502-4_12

AI-Based Sound Event Detection on IoT Nodes: Requirements Evaluation

D. Errico[1], M. Re[2], V. Colombo[2], G. C. Cardarilli[2], M. Martina[1], and M. Ruo Roch[1](✉)

[1] Dipartimento di Elettronica e Telecomunicazioni, Politecnico di Torino, Turin, Italy
massimo.ruoroch@polito.it

[2] Dipartimento di Ingegneria Elettronica, Università di Roma, Tor Vergata, Rome, Italy

Abstract. Sensors capable to detect specific audio events can be deployed inside smart cities, to improve citizens safety. A sensor cloud would allow to georeference relevant incidents, like screams, car crashes, or gunshots. An IoT based approach requires the development and deployment of smart nodes combining minimal power consumption and reasonable preprocessing capabilities, to minimize both power supply requirements and the amount of transmitted data. In this work, a possible system architecture is presented, and a detailed analysis of IA approaches to sound event detection is carried-out. Optimizations for IoT nodes deployments are then applied, and a performance comparison to current algorithms is presented.

Keywords: Smart city · Safety · Ubiquitous computing · Audio event · Artificial intelligence

1 Introduction

The Internet of Things (IoT) paradigma is well suited to the deployment of smart sensors in a urban environment, contributing to the development of the so-called Smart Cities in a near future [1].

The initial idea of bare sensors collecting data from the environment and sending them to a remote server infrastructure, possibly in the cloud, has been evolved, due to the possibility to have significant computing power, at low cost, inside each single node. From this point of view, a layered architecture has been derived, allowing to move typical server-based functionalities on the node itself. A possible classification of the operations carried-out by an IoT complete system, even if not standardized yet, is described in [2,3]. Five layers can be identified:

- Perception layer: Raw informations are collected from the real world, and processed, to discard not interesting data (e.g. noise), or recognize specific events, locally.
- Transport layer: Preprocessed data are sent to remote servers for further elaboration. It is based on LAN or WAN technologies, as LoRa, LTE, WiFi, Bluetooth, etc.

R. Berta and A. De Gloria (Eds.): ApplePies 2022, LNEE 1036, pp. 141–148, 2023.
https://doi.org/10.1007/978-3-031-30333-3_18

- Processing layer: Further process, classify, and store data collected by the sensors, seen as a whole. Here correlations and comparisons among data collected by different nodes can be applied, too, to assess or improve information quality.
- Application layer: Gathers collected informations and convert them in a form suitable for user visualization and eventually decision making. Data are distributed on a service based model (e.g. visualization of pollutant density maps, or traffic jam maps)
- Business layer: Manages the system as a whole.

In this context, we are interested in developing an IoT based system, able to collect audio data inside a urban environment, and to detect specific events. Thinking to a safety related application, we want not only to detect specific sounds, like screams, car crashes, or gunshots, but we want to be able to find the position of the event source, too, and to build a reliable system, resilient to the malfunctioning of a reasonable number of nodes. Last, the system must be pervasive, to be able to fully cover a city, with negligible deployment cost.

A possible solution to satisfy these requirements is to design IoT sensors deployed on transportation systems, both public and/or private ones. In fact, in a city, the amount of vehicles and their spatial distribution, is more than enough to guarantee a wide distribution of sensors, even if only a small percentage of cars were so equipped.

It must be noted that a vehicle can work as sensor base when traveling, but also when it is parked, as movement is not mandatory, and only the knowledge of the position is needed to apply sensor fusion techniques in the above described processing layer.

A first choice is then to deploy IoT nodes on private cars. As a consequence, this requires very low power consumption, as it is not possible to draw too much energy from the vehicle battery when it is stationary. For the same reason, it is mandatory to collect audio data through low power microphones, and a continuous transmission of raw data samples is not affordable, as RF circuitry is typically power hungry.

The idea is then to design a node capable to continuously acquire audio samples, eventually waking up a local microcontroller for further data processing when the sound level overtakes a predefined threshold. At the end of the processing, an event is possibly detected, and, in this case, compressed information about it is sent, together with timing and positioning data, to the cloud.

To perform an audio event classification, artificial intelligence techniques are widely used in literature [4–9], but these works cannot be applied on an IoT node without significant modifications.

This paper aims to evaluate currently adopted deep learning algorithms for sound event detection, and to optimize them for continuous execution on a low power IoT node.

2 Methodology

To have a fair comparison of different algorithms and architectures, we have decided to select an input data set already generated from the DCASE

[10,11]community. In fact, it is the base of challenges, specifically dedicated to "Detection and Classification of Acoustic Scenes and Events". In these challenges, competitors demonstrate AI algorithms and architectures for the recognition of specific acoustic events from real word audio sequences. It is very important to evaluate performances with real, noisy and complex sequences, including a mix of sounds coming from different sources.

The systems used to perform this kind of computation are typically based on GPUs, or on complex custom accelerators, which are not able to satisfy the low power requirements of our solution.

To overcome this problem, a suitable, and available, hardware platform has been identified and used to evaluate the event detection algorithm. This platform is a development kit built around a 32 bit, ARM based low power microcontroller (STM32L486) [12].

Due to the possible need to integrate computing capabilities with hardware accelerators based on FPGA, a special purpose board, based on the same microcontroller, and including a low power Altera FPGA has been identified [13]. This system is capable to accurately measure, in real-time, power consumption related to the FPGA core and the I/O circuitry.

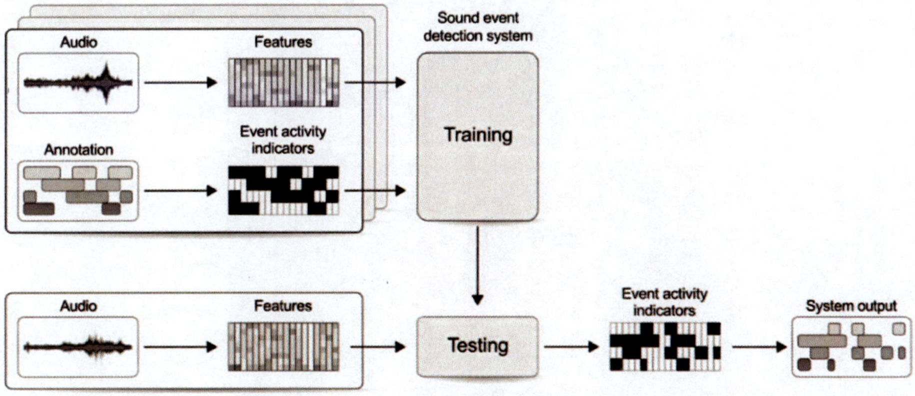

Fig. 1. Data flow for the training and classification phases

The detection algorithm is split in two phases:

- Feature extraction. The audio signal is a monodimensional and redundant sequence of samples. To better couple it to a neural engine, a pre-processing is needed, to extract so-called features, sent to the AI algorithm. Typically, the output of the pre-processing is a bidimensional map. Spectrograms, i.e.time-frequency transformations of subsequences of the signals are mostly used, with linear (STFT) or logarithmic scale (Mel spectrogram).
- Classification. The generated image is sent to a convolutional network, trained with a subset of the DCASE set. After the training, the other signals in the set are applied, and the effectiveness of the recognition is evaluated.

For each phase, hardware resources consumption, execution time, and power are analyzed, and they are used to make comparison of the different solutions. Figure 1 shows the described approach.

The pre-processing phase must be described in C-language and run on the target platform, to measure power consumption and execution times. A significant advantage of the choosen microcontroller is the availability of a tool which automatically maps neural engine architectures to C-language firmware, estimating performances, and the memory requirements.

3 Implementation and Results

The starting point of our evaluation is the implementation of the first phase, i.e. the generation of the Mel spectrogram for an arbitrary input subsequence. The DCASE samples are sequences with a length of 10 s, subdivided in 240 windows. Each window is overlapped to the adjacent one by 360 samples. The resulting output is depicted in Fig. 2.

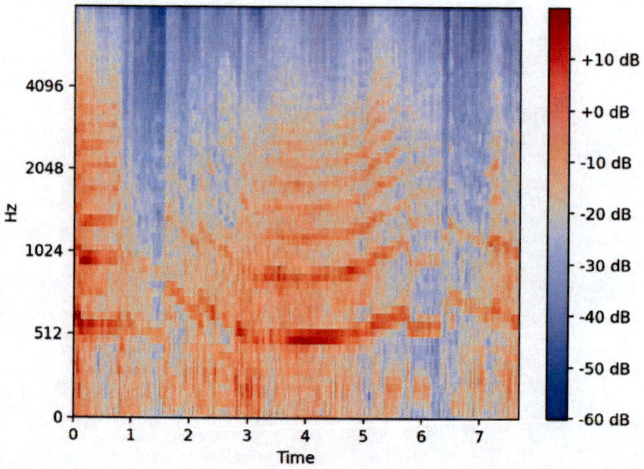

Fig. 2. Graphical representation of the Mel spectrogram for an input data window

Concerning the second phase of the algorithm, the implemented neural network is the one used by the winners of the DCASE challenge 2017.

The original structure is built up of several layers with GLU (gated linear units), followed by RNN (recurrent neural network) layers, which have as input the image generated by the previous phase.

Authors used GLUs in place of the activation function ReLU, to control the information flowing to subsequent levels. In fact, useful informations in the spectrogram have the property to be concentrated in reduced areas, and these units eliminate portions of the image without significant data. This is different from the distribution of relevant data in standard images, where the whole field of view must be considered.

Recurrent layers, instead, are used to introduce a feedback, enhancing time correlation between adjacent segments in the stream of images.

The output of these layers is sent to fully connected ones, where sigmoid and softmax activation functions act in parallel. The softmax activation enhances temporal localization of the event, being multiplied by the sigmoid output. Last, the overall classification is obtained through a weighted average of the values collected for all segments of the sequence.

The detailed architecture of the neural engine is visible in Fig. 3, where fully connected layers are on the right, and convolutional ones are on the left.

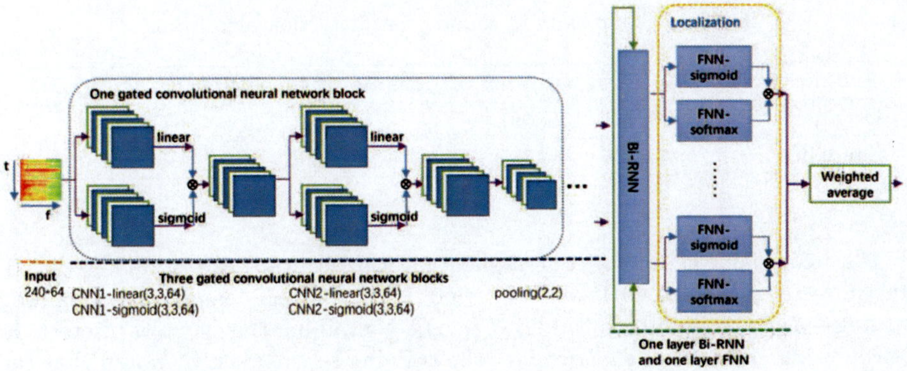

Fig. 3. Architecture of the CNN used for sound event detection

The model has been implemented through Tensorflow, trained, and then converted to C language. The resulting complexity is unfeasible for a low power embedded system, as shown in Table 1 (first row).

As a consequence, a complexity reduction must be applied to the model, to allow its implementation on the selected hardware platform. The basic idea, according to the desired application, is to not design a general sound event classifier, but a more specialized circuit, which is optimized to detect only a reduced set of events.

The first simplification step is the elimination of the GLU and RNN levels, followed by a reduction in dimensionality of the pooling level, only along the frequency axis. This approach allows to directly obtain the probabilities of event presence in each audio segment. Then, the probabilities are filtered by the sigmoid function, and the event detection is obtained through a threshold, i.e. the event is detected if it appears in a sufficiently high number of audio segments. The resulting architecture is shown in Fig. 4.

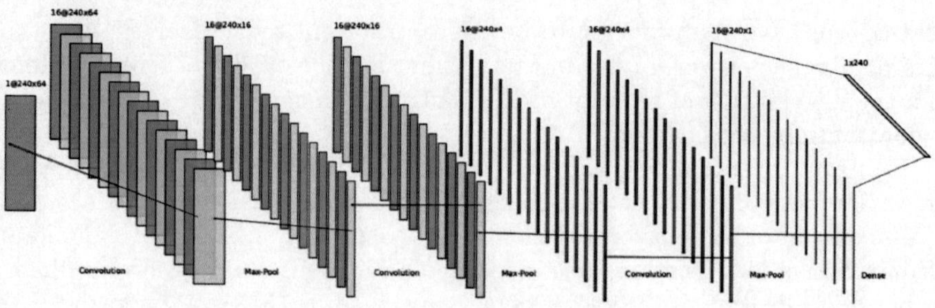

Fig. 4. Architecture of the CCN used for sound event detection

The complexity reduction steps led to the results in Table 1, second row.

Table 1. Complexity of sound event detection algorithms

Architecture	MACC	Number of params	RAM(KB)	FLASH(KB)
Original	790062480	1758097	7560	6710
Simplified	13922656	4785	305	19

As a consequence of model simplification, the accuracy of results is reduced, but it can be restored to its initial value, if not better, changing the training strategy, using as input only audio segments containing the event of interest. In other words, using *strong labeling* for the training set. It must be noted that the original DCASE dataset is *weakly labeled*, as only the presence of a specific event is attached to each audio sequence, without specifying precise time references. Table 2 shows the resulting performances.

Table 2. Performance of the original and modified model, with different training strategies

Metrics	Original model		Modifiedmodel	
	Original training	Modified training	Original training	Modified training
Precision (%)	25.5	37.8	22.0	55.50
Recall (%)	79.2	61.1	45.1	64.2
F1 (%)	38.6	46.7	29.6	59.5
Acc (%)	64.8	80.5	73.1	87.8

A further optimization step is introduced, converting the original floating point model to a fixed point one. In particular, parameters are represented as 8 bit signed integers. The reduction in precision is negligible, as accuracy passes from 80.6% to 80.3%, while the execution time is reduced by 33% passing from 1.085 to 0.717 s.

The obtained execution times with the choosen microcontroller are anyway not sufficient, as each audio segment must be processed in a time less than the segment duration itself, i.e. 40 ms. It means that a 20x reduction factor in the execution time is till required, to allow a real-time audio signal processing. This result cannot be obtained just with further firmware optimizations. As a consequence, a low power FPGA accelerator, used to off-load portions of the neural network algorithm is required. The MCU will remain in charge of acquiring audio signal from digital microphones, and to perform the first stage of the event detection algorithm (Mel spectrogram generation). Moreover, the microcontroller will be used to manage RF data transmission to the cloud.

4 Conclusions and Future Work

In this paper, a methodology to adapt GPU-based algorithms for sound event recognition to IoT nodes has been presented. This allows to optimize and run classification on hardware platforms with reduced computing power. As a direct consequence, power consumption minimization is achieved, too.

Even after optimizations, timing results are anyway not satisfactory if the target platform includes only a standard low cost microcontroller, and a real-time processing of audio streams is required. Thus, a low power FPGA must be integrated into the node, to speed up execution of time critical neural network algorithms.

The implementation of the FPGA accelerator is currently under development and it will be accomplished taking advantage of the low power Altera FPGA already available on the selected platform [13].

Once trained and tested with the DCASE data set, the node will be deployed on the field, to assess real world performances.

References

1. Zanella, A., Bui, N., Castellani, A., Vangelista, L., Zorzi, M.: Internet of Things for smart cities. IEEE Internet Things J. **1**(1), 22–32 (2014). https://doi.org/10.1109/JIOT.2014.2306328
2. Sethi, P., Sarangi, S.: Internet of Things: architectures, protocols, and applications. J. Electr. Comput. Eng. **2017**, 1–25 (2017). https://doi.org/10.1155/2017/9324035
3. Vashi, S., Ram, J., Modi, J., Verma, S., Prakash, C.: Internet of Things (IoT): a vision, architectural elements, and security issues. In: 2017 International Conference on I-SMAC (IoT in Social, Mobile, Analytics and Cloud) (I-SMAC), pp. 492–496 (2017). https://doi.org/10.1109/I-SMAC.2017.8058399
4. Cakir, E., Heittola, T., Huttunen, H., Virtanen, T.: Polyphonic sound event detection using multi label deep neural networks. In: 2015 International Joint Conference on Neural Networks (IJCNN), pp. 1–7 (2015). https://doi.org/10.1109/IJCNN.2015.7280624

5. Liu, M., Wan, C.: Feature selection for automatic classification of musical instrument sounds. In: Proceedings of the 1st ACM/IEEE-CS Joint Conference on Digital Libraries. JCDL '01, pp. 247–248. Association for Computing Machinery, Roanoke (2001). ISBN 1581133456. https://doi.org/10.1145/379437.379663
6. Vij, D., Aggarwal, N., Raman, B., Ramakrishnan, K.K., Bansal, D.: Acoustic scene classification based on spectral analysis and feature-level channel combination. Rapp. tecn. DCASE2016 Challenge, September 2016
7. Foleiss, J.H., Tavares, T.F.: Mel-band features for DCASE 2016 acoustic scene classification task. In: DCASE, September 2016
8. Battaglino, D., Lepauloux, L., Evans, N.: Acoustic scene classification using convolutional neural networks. In: DCASE 2016, Workshop on Detection and Classification of Acoustic Scenes and Events, 3 September 2016, Budapest, Hungary (2016)
9. Valenti, M., Diment, A., Parascandolo, G., Squartini, S., Virtanen, T.: A convolutional neural network approach for acoustic scene classification. In: MAG (2017). https://doi.org/10.1109/IJCNN.2017.7966035
10. Mesaros, A., Heittola, T., Virtanen, T.: TUT database for acoustic scene classification and sound event detection. In: 2016 24th European Signal Processing Conference (EUSIPCO), pp. 1128–1132 (2016)
11. DCASE community web site. https://dcase.community
12. LTE Cellular to Cloud Pack with STM32L496AG MCU. STMicroelectronics. https://www.st.com/en/evaluation-tools/p-l496g-cell02.html#overview
13. Roch, M.R., Martina, M.: VirtLAB: a low-cost platform for electronics lab experiments. Sensors **22**, 4840 (2022). https://doi.org/10.3390/s22134840

Contextual Bandits Algorithms for Reconfigurable Hardware Accelerators

Marco Angioli(✉), Marcello Barbirotta(✉), Abdallah Cheikh(✉), Antonio Mastrandrea(✉), Francesco Menichelli(✉), Saeid Jamili(✉), and Mauro Olivieri(✉)

Sapienza Università di Roma, Via Eudossiana 18, Rome, Italy
{marco.angioli,marcello.barbirotta,abdallah.cheikh,antonio.mastrandrea, francesco.menichelli,saeid.jamili,mauro.olivieri}@uniroma1.it

Abstract. Reconfigurable processing cores for IoT and edge computing applications are emerging topics to calibrate costs, energy consumption and area occupation with performance and reliability on Commercial Off the Shelf (COTS) devices. This work analyzes how to take advantage of Machine Learning to potentially automate the reconfiguration process of a hardware accelerator inside the Klessydra Vector Coprocessor Unit (VCU), choosing the best configuration according to the workload. The problem is modeled with a contextual bandits approach using the Linear UCB algorithms and validated with offline Python simulations.

Keywords: Contextual Bandits · Linear UCB · Disjoint · Hybrid · Reconfigurable hardware accelerators · Klessydra T13

1 Introduction

Contextual multi-armed Bandits (CB) are sequential decision-making problems in which the algorithm must choose the best action based on the context. These algorithms work online using a learning procedure based on a rewarding mechanism. A branch of CB algorithms is the Linear UCB presented in [1]. The algorithms use the Upper Confidence Bound (UCB) [2] strategy assuming a linear relationship between the context and the reward. At every time instant, the algorithm observes a context, chooses an action and receives a reward related to the chosen option. The reward is then used to update the model parameters so the algorithm can learn in an online way as the number of iterations increases, collecting enough information to predict the relation between the observed context and the reward associated with every action.

This work is organized as follows: in Sect. 2, related CB works are reviewed and the Linear UCB algorithms are presented. The reconfiguration is modeled analytically as a CB problem in Sect. 3 and the reward computation procedure as well as the case of study is provided. The algorithm performance is evaluated in Sect. 4 using offline simulations on Python. Finally, in Sect. 5 the obtained results are summarized and future developments are shown.

R. Berta and A. De Gloria (Eds.): ApplePies 2022, LNEE 1036, pp. 149–154, 2023.
https://doi.org/10.1007/978-3-031-30333-3_19

2 Related Works

Contextual bandits algorithms are widely used in sequential decision problems. Authors in [3] modeled the analysis of mouse skin cancer as a CB problem carrying out a Gaussian linear regression, showing an extension of more than 50% on the subjects' life duration compared with other strategies. In [1], CB algorithms model the choice of the highlighted article on the Yahoo home page solving the problem using Linear UCB. In the same way [4] treats the personalized movie choice on the Netflix homepage, modeling that as a CB problem.

2.1 Linear UCB Contextual Bandits Algorithms

In a contextual bandits problem, the context observed by the system at time t can be modeled through a feature vector x_t, whose size depends on the problem. Under this hypothesis, the reward can be expressed as a linear combination of the context features and some unknown coefficients θ, and for each iteration, a ridge regression problem is carried out. Authors in [1] present two different algorithms based on this hypothesis, namely Disjoint and Hybrid, using the UCB technique as an exploration strategy called Linear UCB.

The *Disjoint algorithm* is the simplest version, in which all actions are considered independent. Each action is analytically modeled through three matrices without shared elements. At each iteration, the algorithm computes the UCB for each action and selects the one with the lowest value. The UCB is computed adding the expected reward for the configuration a to the uncertainty contribution weighted by the confidence factor α.

The *Hybrid Linear UCB* is the second algorithm in which the actions are modelled through feature vectors. Two shared matrices are presents, while for each action there are four matrices. In this case, the UCB is computed using a term that does not depend on the selected action and allows the estimation of the rewards for not tested options but similar to others already observed. For both algorithms, after choosing the operation, the received reward is used to update the model.

3 Methodology

The target of the proposed work is the optimal accelerator configuration choice for the Klessydra Vector Coprocessor Unit (VCU) [5–7] according to the workload, taking advantage of the CB algorithms. The requested operation is the context, while the chosen action is the configuration (Fig. 1). The reward is computed from the performance data, produced by the execution of the requested operation with the selected accelerator configuration.

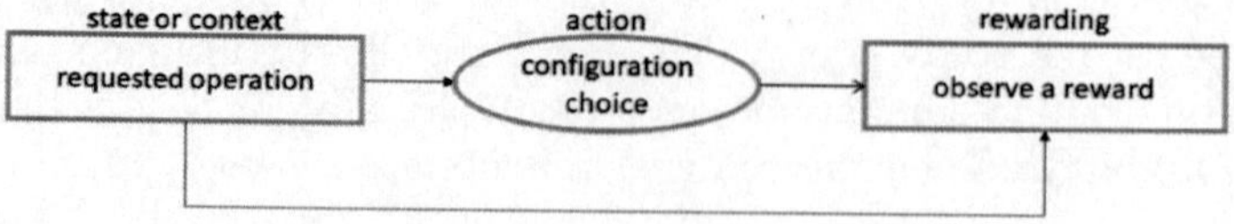

Fig. 1. Reconfiguration problem modeled as a contextual bandits problem

3.1 Case Study

The Klessydra project [5–8] is a processing core family based on the RISC-V instruction set architecture and fully compatible with the PULPino open-source System-on-Chip platform. The Klessydra-T13 version is a parametric design that allows the execution of three hardware threads (harts) in a four-stage pipeline thanks to the Interleaved Multi-threading (IMT) paradigm. This architecture includes highly reconfigurable hardware accelerators for optimized and super-scalar vector operations in a dedicated VCU. By changing the parameters M, F and D, the accelerator configuration can be adjusted, evaluating the balance between thread-level parallelism (TLP) and data-level parallelism (DLP). In [7], it is observed how the performance changes varying the accelerator configuration for different operations. This problem can be generalized and solved by CB algorithms.

To describe every operation reported in [7], the vector size is equal to 14. One example is shown below.

MatMul 64 × 64 matrices →

Conv.	FFT	MatMul	compos	filt_0	filt_1	0x0	4x4	8x8	16x16	32x32	64x64	samples_0	samples_1
0	0	1	0	1	0	1	0	0	0	0	1	1	0

The configuration in [7] are SISD, Symmetric MIMD and Heterogenous MIMD, each tested for different DLP, $D = 1, 2, 4, 8$, for a total of 12 possible choices. The action feature vector length has to be equal to 8. One example below.

$Sym.\ MIMD\ +\ SIMD$ 8 →

M_0	M_3	F_0	F_1	D_1	D_2	D_4	D_8
0	1	0	1	0	0	0	1

Both feature vectors are based on 0 and 1, to ensure that no action or operation has greater weight both in the model choice and in the update process.

3.2 Reward Computation Starting from Performance Data

CB algorithms are based on rewarding mechanism. In our study, the reward computation must consider the operation execution time and the mean energy consumption, and it is computed as a linear combination of them, like shown in Eq. 1, where the weight of that contributions can be chosen changing the value of the coefficients c_{time} and c_{energy}.

$$r_{t,a} = c_{time}\ x_{time} + c_{energy}\ x_{energy} \tag{1}$$

Before computing the reward, each performance data is normalized with a max-min technique to avoid that one prevails over the other.

3.3 Offline Simulation

Offline analysis [9] are important for algorithms that work in real-time, where at each iteration only the reward related to the selected action is known. These simulations allows the evaluation of important metrics such as the regret [10]. In this paper we use the unbiased simulation procedure [9] to test Linear UCB algorithm performances with a complete dataset extracted from the data reported in [7]. Every element is composed by the following items: the context x_t, the action number a and the reward $r_{a,t}$. At each iteration, one dataset element is read and, if the action is the same selected by the algorithm, the iteration is considered valid, it is added to the simulation history, and the reward is used to update the model.

4 Experimental Results

Using the offline analysis shown in Sect. 3.3, we verified the Linear UCB performances on the hardware accelerator reconfiguration problem. For every algorithm, we carried out some Python simulations varying the exploration parameter α, plotting the results in Fig. 2 and Fig. 3 as a function of the mean reward and the regret. For the Disjoint algorithm (Fig. 2), the α parameter determines the uncertainty weight in the UCB computation. For small α values, the reward and the regret do not improve in time, remaining almost constant at the initial value, while for high α values the algorithm explores with an high frequency even after finding the best action.

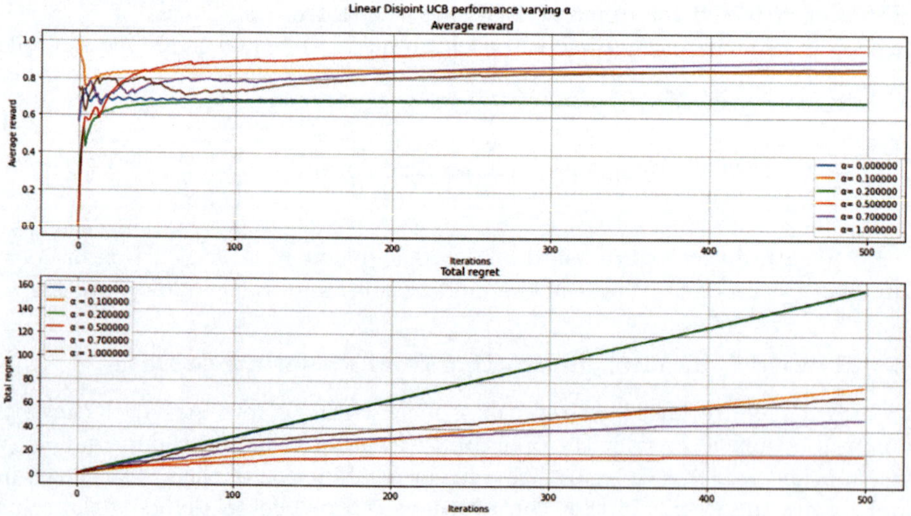

Fig. 2. Disjoint Linear UCB algorithm performance varying alpha

In the Hybrid simulations reported in Fig. 3, the algorithm knows the relationship between each action and it converges even for low α values. The good performance for both algorithm can be appreciated from the trend of the average

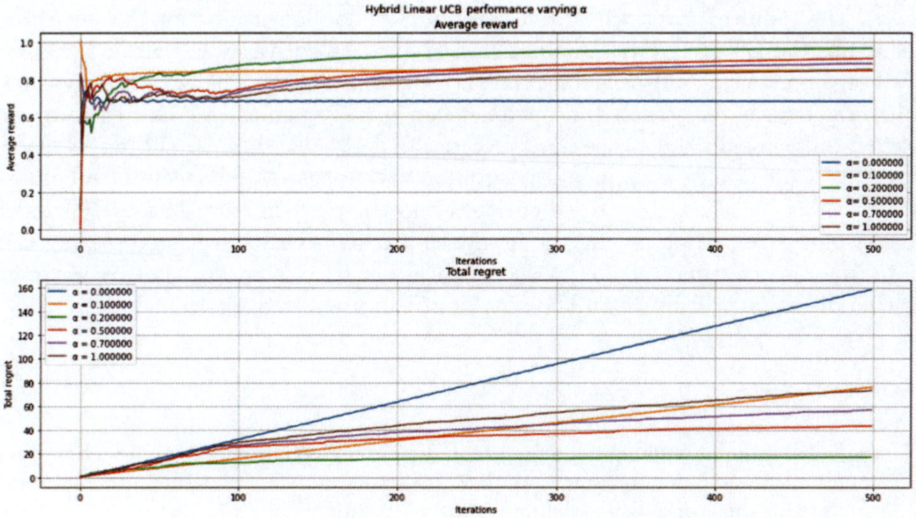

Fig. 3. Hybrid Linear UCB algorithm performance varying alpha

reward and regret. In this case the reward tends to the unit value, and it means that the algorithm converges to optimal solutions. In Fig. 4 is possible to see the actions chosen by the Hybrid algorithm and how the trained model selects the optimal configuration for the requested operation.

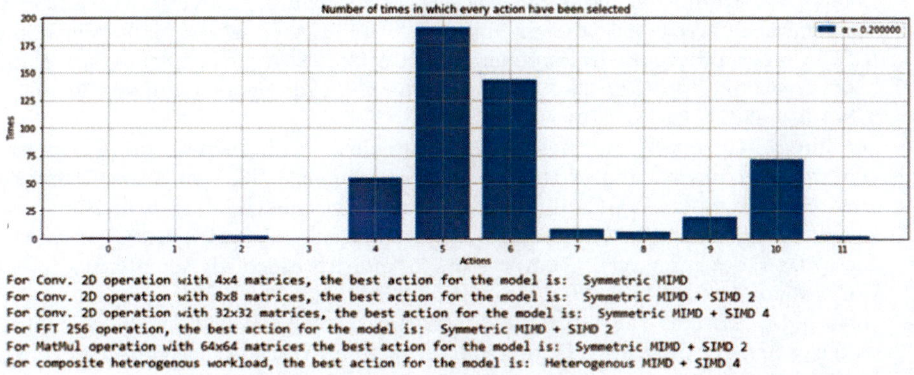

Fig. 4. Selected actions by Hybrid algorithm with $\alpha = 0.2$

5 Conclusions

We presented a study on modeling a hardware reconfiguration problem as a contextual bandits case, taking the Klessydra-T13 processor with its VCU as a case study. The approach has been demonstrated by showing how to analytically model the required operations and the possible configurations for the accelerator using the Linear UCB (Disjoint and Hybrid) algorithms [1] through offline Python simulations, varying the exploration parameter α. The results show that both algorithms can reach the optimization goal, opening the way to a direct use of these inside the processor, in a parallel execution with the application tasks. Whenever a new application workload is executed, the accelerator configuration may be selected by the algorithm and the performance data can be used to compute a reward, allowing the model to learn at every iteration, creating the potential of a runtime reconfiguration process in FPGA boards. Future activity in this direction will build on the results of the presented work.

References

1. Li, L., Chu, W., Langford, J., Schapire, R.E.: A contextual-bandit approach to personalized news article recommendation. In: Proceedings of the 19th International Conference on World Wide Web, pp. 661–670 (2010)
2. Auer, P., Cesa-Bianchi, N., Fischer, P.: Finite-time analysis of the multiarmed bandit problem. Mach. Learn. **47**(2), 235–256 (2002)
3. Durand, A., Achilleos, C., Iacovides, D., Strati, K., Mitsis, G.D., Pineau, J.: Contextual bandits for adapting treatment in a mouse model of de novo carcinogenesis. In: Machine Learning for Healthcare Conference, pp. 67–82. PMLR (2018)
4. Amat, F., Chandrashekar, A., Jebara, T., Basilico, J.: Artwork personalization at netflix. In: Proceedings of the 12th ACM Conference on Recommender Systems, RecSys '18, New York, NY, USA, pp. 487–488. Association for Computing Machinery (2018)
5. Cheikh, A., Sordillo, S., Mastrandrea, A., Menichelli, F., Olivieri, M.: Efficient mathematical accelerator design coupled with an interleaved multi-threading RISC-V microprocessor. In: Saponara, S., De Gloria, A. (eds.) ApplePies 2019. LNEE, vol. 627, pp. 529–539. Springer, Cham (2020). https://doi.org/10.1007/978-3-030-37277-4_62
6. Sordillo, S., Cheikh, A., Mastrandrea, A., Menichelli, F., Olivieri, M.: Customizable vector acceleration in extreme-edge computing: a RISC-V software/hardware architecture study on VGG-16 implementation. Electronics **10**(4), 518 (2021)
7. Cheikh, A., Sordillo, S., Mastrandrea, A., Menichelli, F., Scotti, G., Olivieri, M.: Klessydra-t: designing vector coprocessors for multithreaded edge-computing cores. IEEE Micro **41**(2), 64–71 (2021)
8. Olivieri, M., Cheikh, A., Cerutti, G., Mastrandrea, A., Menichelli, F.: Investigation on the optimal pipeline organization in RISC-V multi-threaded soft processor cores. In: 2017 New Generation of CAS (NGCAS), pp. 45–48. IEEE (2017)
9. Li, L., Chu, W., Langford, J., Wang, X.: Unbiased offline evaluation of contextual-bandit-based news article recommendation algorithms. In: Proceedings of the Fourth ACM International Conference on Web Search and Data Mining, pp. 297–306 (2011)
10. Lattimore, T., Szepesvári, C.: Bandit Algorithms. Cambridge University Press, Cambridge (2020)

Transport, Energy, Security, Health

Low-Cost Lithium-Ion Battery Characterization Setup Based on Auxiliary Batteries

Niccolò Nicodemo(✉), Roberto Di Rienzo, Alessandro Verani,
Federico Baronti, Roberto Roncella, and Roberto Saletti

Dipartimento di Ingegneria dell'Informazione, Università di Pisa, Via Caruso 16, 56122 Pisa, Italy
{niccolo.nicodemo,alessandro.verani}@phd.unipi.it,
{roberto.dirienzo,federico.baronti,roberto.roncella,
roberto.saletti}@unipi.it

Abstract. Characterization tests are key to improve the Lithium-Ion batteries performance. Unfortunately, they require expensive equipment that small companies and laboratories can hardly afford. In this paper a novel low-cost battery characterization setup is proposed. It uses an auxiliary battery to strongly reduce the required maximum instrument power and cost. A simulation framework of the proposed setup is developed in Matlab/Simulink environment and is applied to the characterization of a 48 V mild hybrid battery. This case study is used to both explore setup advantages and guide through its sizing. The obtained results show a strong reduction of the power required to the test equipment and a reduction up to about 70% of the energy drawn from the grid.

Keywords: Lithium Ion Batteries · Battery Characterization · Low Cost Test Setup

1 Introduction

Lithium-Ion Batteries (LIBs) are constantly improved to cover an increasing number of applications in very different fields, such as stationary applications, mobility and medium power tools. A good characterization of LIBs allows us to develop and tune accurate models to predict the battery behaviors and optimize its use [1]. Unfortunately, commercial characterization instruments are very expensive and small companies and laboratory can rarely afford them. Many custom characterization experimental setups have been presented in literature [2–4]. They are usually based on general purpose power supplies and electronic loads that charge and discharge the battery, while a measurement system acquires battery voltage, current and temperature [2]. The use of general purpose instruments reduces the new investments required to enter the battery development field. Auto-producing some parts of the setup is another possible solution to shrink the setup cost, as proved by several projects such as the battery emulator presented in [4], the cell characterization platform of [2] and the electronic

R. Berta and A. De Gloria (Eds.): ApplePies 2022, LNEE 1036, pp. 157–162, 2023.
https://doi.org/10.1007/978-3-031-30333-3_20

load presented in [3]. A new characterization setup is presented in this paper. Its novelty lies in the usage of an auxiliary battery to strongly reduce the power requirements of the setup instruments. The general block diagram of the proposed solution is reported in Fig. 1a.

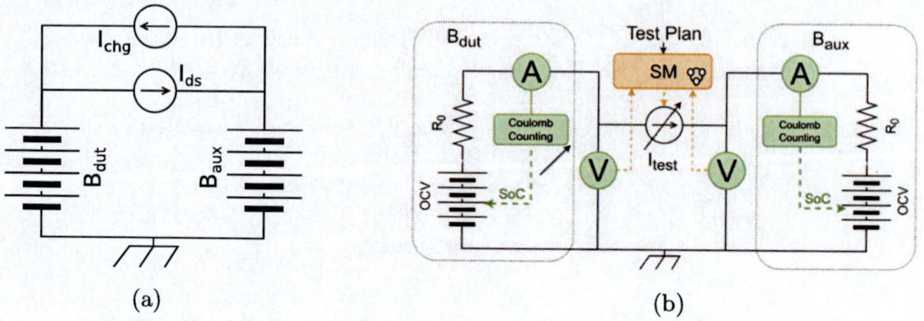

Fig. 1. Setup block diagram (a) and simulation platform model (b)

The battery under test (B_{dut}) can be charged and discharged using the power supply I_{chg} and the electronic load I_{ds}, respectively. I_{chg} and I_{ds} are series connected with the auxiliary battery B_{aux}, that strongly reduces the power required from the instruments. The lower the voltage difference between B_{dut} and B_{aux}, the lower the power required from I_{chg} and I_{ds} and so their cost. Moreover, the auxiliary battery strongly reduces the energy wasted in the characterization tests. In fact, part of the energy moved from B_{dut} during the discharge process is stored in B_{aux} and is then used to charge B_{dut} in the next charging phases. However, the introduction of the auxiliary battery increases the complexity of the setup and could represent a further cost. Luckily, laboratories working with LIBs usually have several unused batteries from their projects that can be used for the purpose without adding new costs. Sizing properly B_{aux} becomes essential to the correct functioning of the proposed characterization setup. A Matlab/Simulink platform has been developed to simulate a characterization test campaign and help the users to correctly size the setup. The simulation platform is described in Sect. 2 while Sect. 3 provides the main theoretical constraints of the proposed setup. A case-study based on a 48 V mild hybrid battery characterization is reported in Sect. 4. It also shows the sizing procedure of the novel characterization setup and discusses the simulation results. Finally, the main conclusions are summarized in the last section.

2 Simulation Platform

A simulation platform has been developed to study the benefits and the application limits of the novel characterization setup. The platform uses the Simscape library of Simulik/Matlab environment and can be freely downloaded from [5]. Its general block diagram is reported in Fig. 1b. The battery behavior is modeled with the Equivalent Series Resistance model [6], which consists of a resistor series-connected with a voltage generator. The generator models the relationship between the battery State of Charge (SoC) and its Open Circuit Voltage (OCV), while the resistor R_0 models the dynamic variation of the battery voltage with the battery current. Battery SoC can be estimated using Coulomb Counting algorithm. In particular, the algorithm models the battery as a charge reservoir with a capacity equal to Q_B, an initial stored charge of $Q(t_0)$, and a coulomb efficiency equal to μ. Battery SoC at t time is calculated using Eq. 1.

$$SoC(t) = \frac{Q(t_0) - \mu \int_{t_0}^{t} i(\tau)\, d\tau}{Q_B} \tag{1}$$

The OCV-SoC relation of the batteries is provided to the platform as n-point Look Up Table (LUT). I_{chg} and I_{ds} are both modelled with a single current generator (I_{test}) which sets the desired test current. I_{test} is controlled by a state machine (SM) developed in the StateFlow environment. The SM manages the test procedure. It consists of a step array that specifies test current, maximum step duration and exit condition on B_{dut} voltage for each test step.

3 Theoretical Setup Constrains

The majority of general purpose power supply and electronic load instruments require a minimum dropout voltage ($V_{I_{test}}^{min}$) to work properly. Considering the proposed setup, this can be expressed in the following constraint:

$$OCV_{B_{dut}} > OCV_{B_{aux}} + V_{I_{test}}^{min} + (R_{0\ aux} + R_{0\ dut}) \cdot I \tag{2}$$

where I is positive if B_{dut} is discharged. This condition must be guaranteed during the whole test campaign and with all the SoCs assumed by the batteries. Less cells compose B_{aux}, easier is satisfying this constraint, but higher is the $V_{I_{test}}$ and thus I_{test} power. The developed simulation platform is very useful in evaluating Eq. 2 and finding a proper B_{aux} composition that minimizes $V_{I_{test}}$ and the instrument power. Another key point is the sizing of B_{aux} capacity. Since the two batteries exchange charge, $Q_{B_{aux}}$ must be equal or greater than $Q_{B_{dut}}$ to allow the complete charge and discharge of $Q_{B_{dut}}$. Moreover, LIBs age slower if they cycle in the SoC range from 20 to 80% [7,8]. $Q_{B_{aux}}$ should thus be sized to work only in this range preventing B_{aux} premature aging. Using Eq. 1 and neglecting μ, the $Q_{B_{aux}}$ the following constrain is obtained:

$$Q_{B_{aux}} = \frac{Q_{B_{dut}}}{SoC_{aux}^{max} - SoC_{aux}^{min}} \cdot 100 \tag{3}$$

where SoC_{aux}^{max} and SoC_{aux}^{min} are the higher and lower SoC range limits of B_{aux}, respectively. The last setup constrain is the initial SoC of the auxiliary battery. It can easily be retrieved from Eq. 1 given the initial SoC of B_{dut}, which is a test-defined parameter. As an example, if the initial SoC of B_{dut} is set to 50%, the B_{aux} one is equal to the center of the $[SoC_{aux}^{max}, SoC_{aux}^{min}]$ range. This condition is used in the rest of the paper as starting point of the characterization test campaign.

4 Characterization of a 48 V Mild Hybrid Battery

A 48 V mild hybrid battery is used as case-study to analyze the novel setup. The B_{dut} parameters are obtained from [9]. This battery is composed of 14 series-connected lithium iron phosphate (LPF) cells with a capacity of 8 Ah and an overall series resistance of 10 mΩ. The coulomb efficiency of the battery is set to 0.95 that is a worst-case value for a LPF batteries. Constant Current (CC) and Hybrid Pulse Power Characterization (HPPC) [10] tests are considered to cover various aspects of battery characterization. Table 1 reports all the test steps including the auxiliary ones to bring the B_{dut} at the starting conditions of the characterization steps. Current values are expressed using C-rate (C). 1C is the current value that moves in one hour a quantity of charge equal to the nominal battery capacity.

Table 1. Characterization test description

#1-Auxiliary	#2-1C Capacity	#3-Fast charge [1]	#4-Hybrid Pulse Power
Chrg. @1C, (V_{end}: 48.94V) Pause 3600 s	Disch. @1C, (V_{end}: 35.64 V) Pause 3600 s	Chrg. @4.8C, (SoC_{end}: 20%) Chrg. @5.2C, (SoC_{end}: 60%) Chrg. @4.16C, (SoC_{end}: 80%) Chrg. @1C, (V_{end}: 48.94 V) Pause 3600s	10 repetitions: Disch. @5C for 10 s Pause 40 s Chrg. @5C for 10 s Pause 40 s Disch. @1C for 360 s Pause 3600 s Disch. @1C, (V_{end} of 35.64 V)

The simulation platform is used to obtain the optimum sizing of B_{aux} in terms of series-connected cell number and cell capacity. The characteristics of the B_{aux} cells are obtained starting from the B_{dut} ones keeping constant the product of the cell capacity and its series resistance [11]. In this way, the cell capacity can be changed keeping realistic B_{aux} model. The simulation platform is applied to the case study for a B_{aux} made of 0 to 13 cells. The upper limit is chosen equal to the number of B_{dut} cells minus one. Instead, the simulation with 0 cells represents, instead, the classic experimental setup without the auxiliary battery. B_{aux} SoC range is set to [80%, 20%] while $Q_{B_{aux}}$ to 13.4 Ah using Eq. 3. Simulations with 11, 12, and 13 series-connected cells do not satisfy the theoretical setup constraint reported in Eq. 2 considering a $V_{I_{test}}^{min}$ of 1 V. For this reason, B_{aux} can be composed of up to 10 series-connected cells.

Figure 2 shows the test current profile, the SoC of the batteries and the B_{dut} and I_{test} generator power signals in a test with a B_{aux} composed of 10 series-connected cells. As we can note, the current of B_{dut} follows the test plan in Table 1 highlighting the correct operation of the proposed simulation platform. The comparison between the maximum power of B_{dut} and I_{test} shows the main

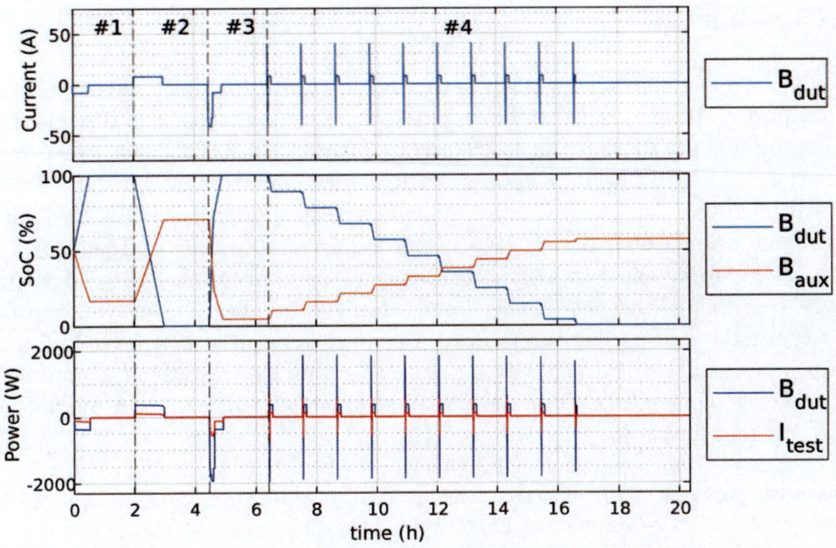

Fig. 2. Results of the simulated test plan for a B_{aux} with 10 cells

advantage of the novel proposed setup. In particular, I_{test} average power is about 29% of the B_{dut} one. Moreover, the maximum charge and discharge I_{test} power are 747 W and 715 W, with respect to 1957 W and 1938 W of the B_{dut} maximum power. As we can note, the proposed setup is able to control very high power tests with lower power and lower cost instruments. The maximum instrument power values for all the simulations are reported in Fig. 3a, which shows how I_{test} power decreases when the number of the B_{aux} cells increases

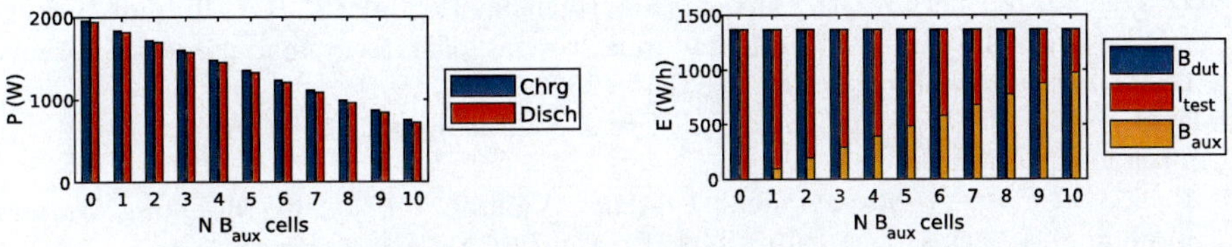

Fig. 3. Maximum I_{test} Power (a) and Transferred Energy (b) at various B_{aux}

Finally, the absolute power value of B_{dut}, I_{test}, and B_{aux} is integrated and reported in Fig. 3b to obtain the energy moved from/to each setup component. Since I_{test} generator energy is directly related to the energy drawn from the grid to execute the test, the introduction of the auxiliary battery strongly reduces this energy with a positive impact for the characterization cost and environmental pollution. For 10 cells, I_{test} energy is only 29% of the energy moved from/to B_{dut} while the remaining 71% is moved to/from B_{aux}. In this way, the energy extracted from the battery under test is harvested into B_{aux}, ready to be re-used to perform further steps.

5 Conclusion

A novel low-cost characterization setup for Lithium-Ion batteries is proposed in this paper. It uses an auxiliary battery to strongly reduce the maximum test equipment power and the energy drawn from grid during test execution. A simulation platform of the proposed setup is developed using Matlab/Simulink and is applied to a typical 48 V mild hybrid battery characterization. Simulations show a reduction up to about 70% of the equipment power and the used grid energy. Therefore, the proposed setup allows us to perform very high power battery test using low power and low-cost instruments. This advantage could help the smaller companies and laboratories in the characterization of high power lithium-ion battery without investing a large amount of resources in commercial high power testing setups. The developed simulation platform is provided freely to the community on [5].

Acknowledgments. This research was partially supported by CrossLab project, funded by MIUR "Department of Excellence" program.

References

1. Attia, P., et al.: Closed-loop optimization of fast-charging protocols for batteries with machine learning. Nature **578**, 397–402 (2020)
2. Carloni, A., Baronti, F., Di Rienzo, R., Roncella, R., Saletti, R.: Open and flexible Li-ion battery tester based on Python language and Raspberry Pi. Electronics **7**, 454 (2018)
3. Weßkamp, P., Haußmann, P., Melbert, J.: 600-A test system for aging analysis of automotive Li-ion cells with high resolution and wide bandwidth. IEEE Trans. Instrum. Measure. **65**, 1651–1660 (2016)
4. Di Rienzo, R., Verani, A., Baronti, F., Roncella, R., Saletti, R.: Modular battery emulator for development and functional testing of battery management systems: the cell emulator. Electronics **11**, 1215 (2022)
5. http://github.com/batterylabunipi/Low-Cost_Lithium-Ion_Battery_Characterization_Setup
6. Plett, G.: Battery Management Systems, Volume 1: Battery Modeling Battery Modeling. Artech House Publishers, Boston (2015)
7. Jiang, J., et al.: Optimized operating range for large-format $LiFePO_4$/graphite batteries. J. Electrochem. Soc. **161**, A336 (2013)
8. Ecker, M., et al.: Calendar and cycle life study of Li(NiMnCo)O_2-based 18650 lithium-ion batteries. J. Power Sources **248**, 839–851 (2014)
9. Lee, S., Cherry, J., Safoutin, M., McDonald, J., Olechiw, M.: Modeling and validation of 48V mild hybrid lithium-ion battery pack. SAE Int. J. Altern. Powertrains. **7**, 273–288 (2018)
10. Walker, L.: Battery Test Manual For 48 Volt Mild Hybrid Electric Vehicles. (Idaho National Lab. (INL), Idaho Falls, ID (United States) (2017)
11. Baronti, F., Di Rienzo, R., Papazafiropulos, N., Roncella, R., Saletti, R.: Investigation of series-parallel connections of multi-module batteries for electrified vehicles. In: 2014 IEEE International Electric Vehicle Conference (IEVC), pp. 1–7 (2014)

Model-Based Vital Control Architecture for Highly Automated Train Operations

Giovanni Mezzina[1(✉)], Cataldo L. Saragaglia[1], Mario Barbareschi[2], Diana Serra[2], Salvatore De Simone[2], Alberto Moriconi[2], and Daniela De Venuto[1]

[1] Department of Electrical and Information Engineering, Politecnico di Bari, 70125 Bari, Italy
{giovanni.mezzina,cataldo.saragaglia}@poliba.it
[2] Rete Ferroviaria Italiana S.p.A, 80021 Afragola, NA, Italy
{m.barbareschi,d.serra,sa.desimone}@rfi.it

Abstract. The railway panorama is experiencing notable development due to the introduction of always new infrastructures with high grade of automation, realizing the so-called ATO over ETCS (AoE) framework on mainlines. Despite AoE achieved optimal results for supervised operations, unattended train operations management is still in an embryonal stage. In this context, this paper proposes a model-based architecture, Vital Control Module (VCM), that improves the safety of unattended convoys operating on infrastructures with highly automated AoE. The model, developed in Matlab/Simulink, supervises the framework operativity, estimates train speed and communicates with the trackside-connected operator. Exploiting these functions, VCM can substitute ETCS when faulty or unpowered, and detect hazardous situations intervening in a negligible time. To test the model, a custom and interactive emulator has been realized. Experimental results showed that the VCM can detect hazards and mitigate them in 451 ms–533 ms, halving the time required by the related standard constraints.

1 Introduction

The introduction of the European Rail Traffic Management System/European Train Control System (ERTMS/ETCS) on High Speed/High Capacity (HS/HC) mainlines, helped to solve the problems deriving from the constant increase in users and train traffic [1]. In this context, the European railway industry is prompting new research and development directions to address the needs of the HC/HS mainline served by ERTMS/ETCS infrastructures. Among these objectives, the use of unattended rail vehicles for maintenance and inspection gathered the principal interest. A high grade of automation for unattended rail vehicles is generally possible through the extensive use of the Automatic Train Operation equipment, which can achieve a Grade of Automation of level 4 (ATO GoA 4) according to the classification in [2]. Nevertheless, to date, ERTMS/ETCS ecosystems for mainline, which ensure a high GoA (GoA 4) are still in an embryonal stage, and the first definitive outcomes are expected until 2025[3]. The above-mentioned ecosystem, also called ATO over ETCS (AoE), already approached the railway panorama, but it has been demonstrated only for low GoAs (GoA2 and in rare cases GoA3). Nevertheless, it was not demonstrated for unmanned and unattended scenarios. For this reason, the new

R. Berta and A. De Gloria (Eds.): ApplePies 2022, LNEE 1036, pp. 163–170, 2023.
https://doi.org/10.1007/978-3-031-30333-3_21

frontier of research and developments is starting to consider the definition of new Automatic Train Protection (ATP) systems that can be incorporated in the AoE framework at the on-board level of a rail vehicle ensuring the GoA3/4 function as per the European guidelines in [3].

In this context, we present the model-based application architecture of an ATO Vital Control Module (VCM). This architecture is designed to be integrated as ATP equipment into an AoE GoA4 onboard system, ensuring supervision and monitoring functions. It also provides a set of safety-related procedures that can partially replace the ATP and emergency brake functionalities of ETCS when it is faulty or isolated. The VCM application logic has been designed to enhance convoy safety and its availability, including a dedicated procedure for the recovery of operations without needing in-loco human interventions. The proposed VCM has been implemented via Simulink/Stateflow and integrated into a simulator environment for use case assessments and the analysis of the system behavior in response to external interactions.

The paper is organized as follows. Section 2 outlines the model-based architecture. Section 3 presents and discusses the simulator environment, two use-case scenarios, and their intervention time.

2 Model-Based Architecture

2.1 System Overview

Figure 1 shows a simplified schematic overview of the ATO onboard equipment for the AoE integration. The main block of the framework is the ATO-OB, which manages the overall automatic train operation according to the journey profile and drives the convoy. ATO-OB is also interfaced with the ERTMS/ETCS system which provides functions and speed supervision according to [4].

In the proposed AoE framework, the onboard equipment is supplied with an additional module, the VCM (red block - Fig. 1). The VCM global architecture has been expanded in the box on the right side of Fig. 1. The VCM logic is supplied with several interfaces: (i) ATO-OB for the supervision of the operations (ii) the ATO odometry system (ATO-ODO) that provides VCM with data for speed estimation; (iii) the Balise Transmission Module (BTM) that transmit the intercepted telegrams. The application logic of VCM is intended to run on a custom PCB realized and described in our previous works [5–8], here named Vital Control Board (VCB).

VCB interfaces the ETCS, and the Emergency Brake (EB) via dedicated functional blocks [8–13]. The first block, i.e., ETCS Check one, provides VCM with monitor and control functions on the ETCS module power supply. While, the EB Check has the same role, but concerning the EB system. Through these interfaces, VCM can manage the isolation mode of the ETCS system and directly activate and reset the emergency brake.

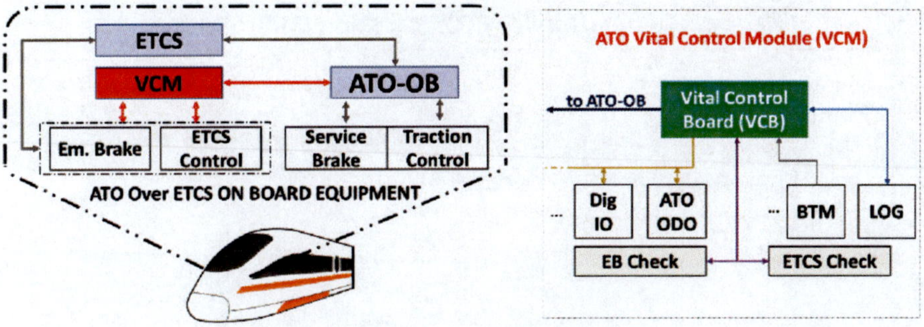

Fig. 1. AoE architecture with VCM integration

2.2 VCM Logic

The VCM logic consists of three main tasks detailed in the following.

Periodic VCM Communication. This task oversees the generation of all the periodic messages constituting the communication with the ATO-OB, the trackside equipment, and the remote-connected operator. The exchanged message is composed of requests, notifications, and flags useful to monitor the vitality of the onboard system. The message fields are used to notify train status and events and provide the high-level interface for the operator command. Figure 2 shows the model-based/Simulink implementation of one of the above-mentioned requests: the ETCS isolation requests. The request is generated by VCM only when the train is in a standstill position, the odometric functions are reliable and the ETCS system is online. This request allows the remote operator to command the ETCS isolation.

Vital Control and Odometry. This task includes the main functions to elaborate on the information received from onboard and track-side equipment (including operator interface). It is also in charge to monitor the ETCS isolation mode (through power supply readings), as well as the status of the EB (via solenoid valves check). This task oversees the elaboration of tachometric sensors data from a dedicated board in the rack (ATO-ODO – Fig. 1), through a dedicated odometric algorithm for speed and position estimation. This estimation is continuously supervised by VCM and provided to ATO-OB only if ETCS is isolated. In the latter case, VCM can request EB intervention if a non-permitted speed is detected, ensuring the convoy safety. Moreover, the task periodically checks the ATO-OB operational mode and its vital signals. It monitors the remote operator's vigilance and emergency brake requests. Finally, this block exploits all the above-mentioned monitoring functions to detect hazardous events requesting EB actions.

EB and ETCS Procedures Management. This task embeds all the procedures concerning commands toward ETCS and the EB system. VCM manages the procedure for changing the isolation mode of ETCS when requested by the remote-connected operator. In this case, VCM can turn on (inserting) or turn off (isolating) ETCS, commanding a

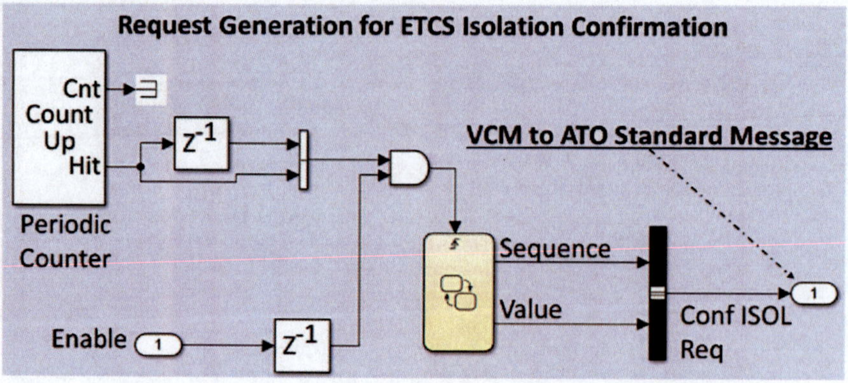

Fig. 2. Simulink/Stateflow for ETCS isolation request generation

radio relay that acts on the ETCS power supply. Then, the task checks if the appropriate state change occurred by reading the relay status. VCM can also control and activate the EB system for both actuating a braking procedure or for recovery of operations purposes. The latter operations allow to safely recover the train operations without in-loco human intervention. For this purpose, VCM must check if the ATO ODO system is fully operative. Alternatively, VMC can enable and manage the ETCS switch-on procedure, if requested by the remote operator. This is necessary because ATO-OB needs, at least, one odometry source (from ATO-ODO as a subsystem of VCM or ETCS) to work properly.

3 Experimental Results

3.1 Simulation Environment

The proposed model-based architecture has been realized in MATLAB®/Simulink R2020b. Specifically, the whole architecture has been realized as mixed Simulink/Stateflow blocks. To realize a testing environment for the VCM model, placing it in a loop, an emulator with an integrated GUI has been developed. It has the role of replacing and emulating most of the train dynamics (both on-board and track-side systems), generating the appropriate signals for VCM. Exploiting the here-proposed simulator, it is possible to assess both automatically and manually, several use-case scenarios and test the behaviors of VCM in response to different stimulations. Figure 3 shows the GUI of the simulator realized in Simulink. GUI includes different sections with buttons, light indicators, and a slider in the bottom section. Specifically, buttons can be used to manually set up the status of the on-board simulated equipment. The GUI's light indicators are, instead, used for signaling and notifications purposes. Finally, the slider is used to set the current speed of the train.

3.2 VCM Intervention Timing: Use-Cases

To provide a complete overview of the here proposed model-based architecture, two explanatory use-cases are proposed in the following. The first analyzed scenario concerns the longest timing of intervention of VCM (worst case), while the second one outlines one of the fastest interventions of the architecture. All the below-presented scenarios consider a computation cycle of 1 ms [14].

The first scenario is composed as follows: (1) with the train in the standstill position on the track, the operator commands VCM to isolate the ETCS system (see Sect. 2.2 - Periodic VCM Communication); (2) with ETCS successfully isolated, the operator increases the train speed via the simulator GUI; (iii) VCM receives and elaborates the data from ATO-ODO every 80 ms. When a non-permitted speed is detected, the VCM model reacts by issuing the emergency brake command. Figure 4 shows the reaction time of VCM when the speed limit is exceeded. In the proposed example, the intervention time of the VCM system, from the onset of the hazardous event to the generation of the brake signal, takes about 53 ms. This time mostly depends on the periodic communication between VCM and ATO-ODO (i.e. 80 ms). The reaction time contribution of VCM is the 3 ms in the final part of the simulation. Ultimately, the considered intervention can take, in the worst case, up to 83 ms.

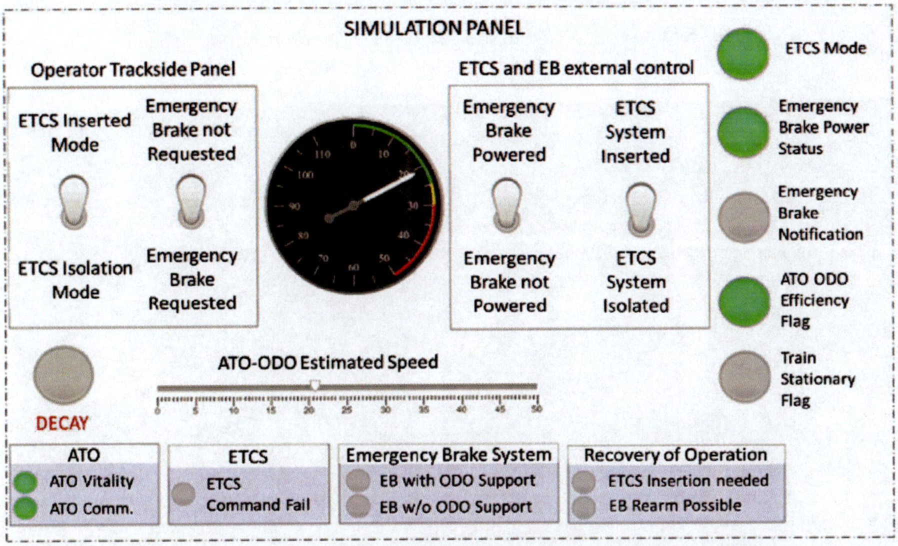

Fig. 3. GUI of the Simulation Environment used to test the VMC logic

The second scenario (i.e., fastest VCM intervention) concerns the detection of a failure in the ATO-OB system. VCM constantly supervises the vitality signal and operational mode of the ATO-OB equipment (Sec. 2.2 - Vital Control and Odometry). The detection of ATO-OB in failure mode leads VCM to command the emergency brakes. Figure 5 reports the main outputs of this scenario.

In this case, the VCM system reaction takes about 7 ms. A theoretical bias, for this best-case, depends on the data exchange periodicity between the ATO-OB and VCM systems: 20 ms. The reaction time contribution of VCM is the 1 ms in the final part of the simulation. Ultimately, the considered intervention can take, in the worst case, up

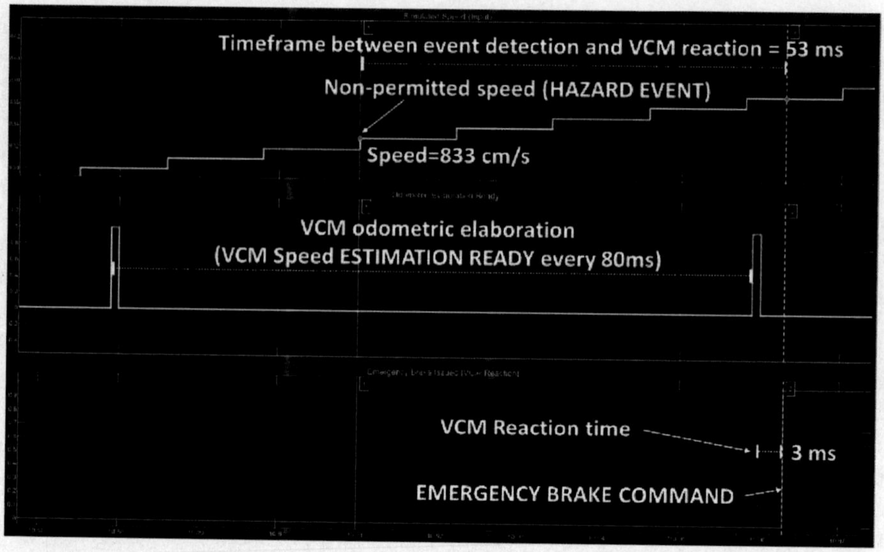

Fig. 4. Reaction of VCM to the detection of a non-permitted speed

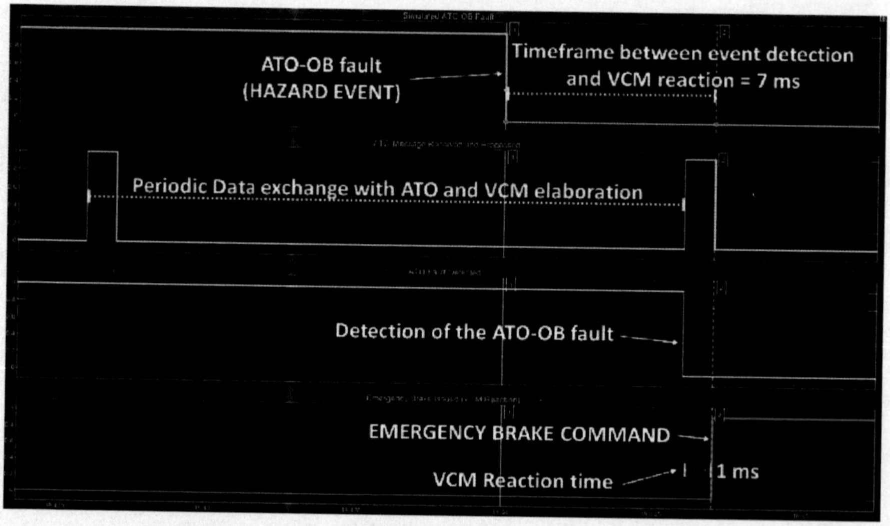

Fig. 5. Reaction of VCM to the detection of the ATO fail

to 21 ms. In both these cases, the major component of delay can be attributed to the data exchange between VCM with ATO-OB or other peripherals. In any case, when a hazardous event is detected, the model is always capable to react in a maximum of 6 cycles (6 ms). Considering a standard pneumatic plates system, which takes ~ 450 ms to actuate the emergency braking, the proposed architecture should take a maximum of 550 ms to command the braking. Thus, the architecture must intervene in a period shorter than this limit to be compliant with the specifications provided by the related subset [15].

Considering the worst VCM reaction time, the overall process, from detection to braking actuation, takes about 533 ms at most. This is largely below the ETCS requirements, which specifies a response time of seconds.

4 Conclusions

New research trends in the railway sector are prompting the design of additional ATP functionalities for ATO frameworks with GoA3/4 to permit unattended train operations. For this aim, an architecture capable of expanding the vital control capabilities of ETCS, as well as supporting and temporarily substituting this system for safe unattended train operations has been presented. The proposed architecture, designed to minimize the impact on existing ETCS technology, oversees the monitoring of various equipment of the ATO framework, evaluating the overall operativity, and manages mitigating actions when hazardous situations are detected during unattended train operations. The proposed architecture has been designed and implemented on Simulink/Stateflow. For test purposes, a testbench, composed of an emulator to put the VCM in the loop, has been realized. Experimental results showed that considering the worst scenario, the proposed architecture can complete the emergency braking in 533 ms. This intervention timing is below the *current* related standard specification constraints of about 467 ms. Ultimately, the proposed architecture presents manifold advantages that range among safe unattended train operations in railway lines not served by the ERTMS/ETCS infrastructure, new and complementary vital control capabilities for the existing ATP systems, and the possibility of a quick recovery of train forced in standstill position with faulty ETCS.

References

1. Deutsch, P.: Overview ERTMS/ETCS baseline 3 and Beyond. In: Collart-Dutilleul, S. (ed.) Operating Rules and Interoperability in Trans-National High-Speed Rail, pp. 29–94. Springer, Cham (2022). https://doi.org/10.1007/978-3-030-72003-2_3
2. Lagay, R., Adell, G.M.: The autonomous train: a game changer for the railways industry. In: 2018 16th International Conference on Intelligent Transportation Systems Telecommunications (ITST), pp. 1–5 (2018). https://doi.org/10.1109/ITST.2018.8566728
3. Digital & Automated up to Autonomous Train Operations TOPIC ID: HORIZON-ER-JU-022-FA2–01. https://tinyurl.com/38r4ah2e. Accessed 08 June 2022
4. ERA, UNISIG, EEIG, ERTMS USERS GROUP - Subset-026. ERTMS/ETCS System Requirements Specifications (2016)
5. Mezzina, G., Barbareschi, M., De Simone, S., Di Benedetto, A., Narracci, G., Saragaglia, C.L., Serra, D., De Venuto, D.: Smart on-board surveillance module for safe autonomous train operations. In: Saponara, S., De Gloria, A. (eds.) ApplePies 2021. LNEE, vol. 866, pp. 317–325. Springer, Cham (2022). https://doi.org/10.1007/978-3-030-95498-7_44
6. De Venuto, D., Annese, V.F., Mezzina, G., Defazio, G.: FPGA-based embedded cyber-physical platform to assess gait and postural stability in parkinson's disease. IEEE Trans. Compon. Packa. Manuf. Technol. **8**(7), 1167–1179 (2018). https://doi.org/10.1109/TCPMT.2018.2810103
7. De Venuto, D., Annese, V.F., Mezzina, G., Ruta, M., Di Sciascio, E.: Brain-computer interface using P300: a gaming approach for neurocognitive impairment diagnosis. In: 2016 IEEE International High Level Design Validation and Test Workshop (HLDVT), pp. 93–99 (2016). https://doi.org/10.1109/HLDVT.2016.7748261

8. De Venuto, D., Annese, V.F., Defazio, G., Gallo, V.L., Mezzina, G.: Gait analysis and quantitative drug effect evaluation in Parkinson disease by jointly EEG-EMG monitoring. In: 2017 12th International Conference on Design & Technology of Integrated Systems in Nanoscale Era (DTIS), pp. 1–6 (2017). https://doi.org/10.1109/DTIS.2017.7930171
9. De Venuto, D., Mezzina, G.: Spatio-temporal optimization of perishable goods' shelf life by a pro-active WSN-based architecture. Sensors **18**, 2126 (2018). https://doi.org/10.3390/s18072126
10. Blagojevic, M., Kayal, M., Gervais, M., De Venuto, D.: "SOI hall-sensor front end for energy measurement. IEEE Sens. J. **6**(4), 1016–1021 (2006). https://doi.org/10.1109/JSEN.2006.877996
11. De Venuto, D., Castro, D.T., Ponomarev, Y., Stikvoort, E.: Low power 12-bit SAR ADC for autonomous wireless sensors network interface. In: 2009 3rd International Workshop on Advances in sensors and Interfaces, pp. 115–120 (2009). https://doi.org/10.1109/IWASI.2009.5184780
12. De Venuto, D., Stikvoort, E., Tio Castro, D., Ponomarev, Y.: Ultra low-power 12-bit SAR ADC for RFID applications. In: 2010 Design, Automation & Test in Europe Conference & Exhibition (DATE 2010), pp. 1071–1075 (2010). https://doi.org/10.1109/DATE.2010.5456968
13. Biccario, G.E., Annese, V.F., Cipriani, S., De Venuto, D.: WSN-based near real-time environmental monitoring for shelf life prediction through data processing to improve food safety and certification. In: 2014 11th International Conference on Informatics in Control, Automation and Robotics (ICINCO), pp. 777–782 (2014). https://doi.org/10.5220/0005102407770782
14. Annese, V.F., De Venuto, D.: Fall-risk assessment by combined movement related potentials and co-contraction index monitoring. In: 2015 IEEE Biomedical Circuits and Systems Conference (BioCAS), pp. 1–4 (2015). https://doi.org/10.1109/BioCAS.2015.7348366
15. ERA, UNISIG, EEIG, ERTMS USERS GROUP - Subset-041. ERTMS/ETCS Performance Requirements for Interoperability (2016)

Exposure of the Human Head to 5G Electromagnetic Radiations: Modeling and Analysis

Sara Alameddine[1], Dina Al-Houmsy[1], Ali Mohsen[1], Houssein Hajj Hassan[2], Ali Ibrahim[1], and Mohamad Hajj-Hassan[1](✉)

[1] Lebanese International University, Bekaa, Lebanon
mohamad.hajjhassan@liu.edu.lb

[2] The Interantional University of Beirut, Beirut, Lebanon

Abstract. 5G, the 5th generation cellular network provides new utilities to different kind of business and industry sectors such as healthcare, electronics, and communications. In addition, it provides numerous features that distinguish it from previous generations such as high speed of internet reaching 10 Gbps, increase in the amount of data transmitted "wide bandwidth", and advanced antennas. Although 5G has promising future, it still has a blurry side that disquiet most of the scientists, since there are no final studies ensuring that the frequency of 5G has no effects on human health especially on the brain, additionally that the people usage of the cellphone will reach its peak with 72.6% by 2025. In this paper, numerical simulations of different ranges of frequencies are conducted on a new designed model according to different international standards representing the biological, physical, and chemical characteristics of the sub-layers of the human head, to investigate the effect of 5G radiations on the head. Here, we investigate the effects of Specific Absorption Rate (SAR) averaging mass, the heat that results from other sources, and the propagation of the wave on the correlation with temperature elevation under the effect of 5G exposure over time. The obtained results are employed to check the safety of the simulated scenario.

1 Introduction

Since the beginning of the telecommunication revolution, the communication network passed through several evolutions, starting from 1st generation up to 5th generation and currently the 6th generation is under research. Each generation has its own specific frequency bandwidth and features. To begin with, the 1st generation was the first wireless network offered to the world in 1980 with an operating frequency of 800 MHz. It was based on analog systems and provided a voice of poor quality. Later, the 2nd generation have been shown, this generation was launched in 1991 with 1.8 GHz frequency used. It was based on digital radio waves signals and provided more efficient network than the 1st generation as it tended to reduce the noise in the line. The 2nd generation was followed by 3rd generation, which was launched in 2000 with used frequency ranging between 1.6 GHz and 2.5 GHz. It provided same features of the that of 2nd generation

R. Berta and A. De Gloria (Eds.): ApplePies 2022, LNEE 1036, pp. 171–178, 2023.
https://doi.org/10.1007/978-3-031-30333-3_22

with an increase in the data capacity transmitted at low cost. With the presence of 3rd generation many applications have emerged such as emails, video downloading, and picture sharing. Then the 4th generation was brought to light in 2010 with a frequency range bounded between 2 GHz and 8 GHz. It offered high speed internet, in addition to the high quality and high capacity for the users with an improvement in the security of the data services. Finally, the latest generation started to be spread world widely is the 5th generation due to its reliable features that opens new opportunities with unlimited benefits to most of the careers with different backgrounds. 5th generation was launched in 2020 but it still under research, the frequencies utilized in this generation are divided in to 3 parts low band and it is bounded between 600 MHz–850 MHz, mid band and it is bounded between 2.5 GHz–3.7 GHz and high band and it bounded between 25 GHz–39 GHz. The 5th generation provides a fast speed and low latency, it solved the drawbacks of the previous generations such as lack of coverage, dropped calls and low performance at cell edges, and it can handle much more connected smart devices at the same time with an increase in the data transmitted rate.

To study the effect of EMF radiations emitted by the antennas of the phone, the biological structure of the human head should be expounded well. The human head is a complex structure consists of several layers to protect one of the most important organs in the human body, the brain. The layers that cover the brain are pia mater, arachnoid mater, dura mater, skull, muscle, fat, skin, and pinna as illustrated in Fig. 1.

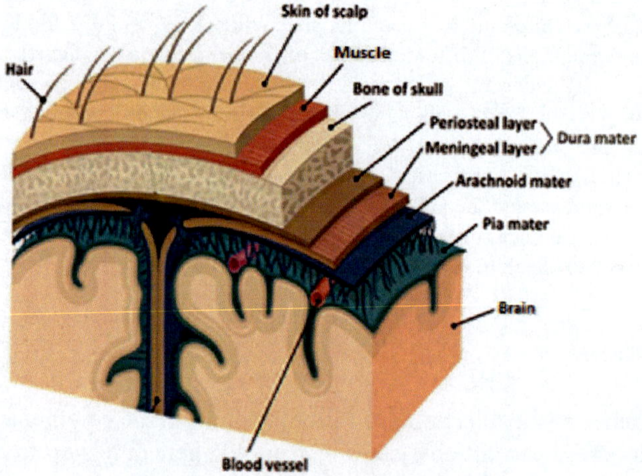

Fig. 1. Layers of the head [1].

One of the most sophisticated organs in the human body is the brain. It consists of millions of wired neurons communicating with each other to accomplish and control all the body functions such as controlling the voluntary and involuntary movements, controlling the other organs of the body like heart, lungs, kidneys, liver, etc. and much more other functions. The brain is made up of three main parts including the cerebrum that considered to be the largest part of the brain and it is composted of the right and the left hemispheres, cerebellum located under the cerebrum, and brainstem that connect the cerebrum and cerebellum to the spinal cord [2].

The second layers that followed the brain are the meninges. The meninges are made up of three membranes surrounding the brain and spinal cord including dura mater, arachnoid mater, and pia mater. Dura mater is outer most membrane of the meninges and it represent a thick, tough, and inextensible membrane. Arachnoid mater is a thin membrane that covers the brain and pia mater, it is made up of elastic connective tissues, in addition to being an avascular and does not receive any innervation. Pia mater is the inner most layer of the meninges that represent a thin envelope that covers the brain and the spinal cord; this layer is made up of fibrous tissue and reached with blood vessels [2].

The layer that comes after the meninges and protect it from the harms is the skull. This layer is the skeletal structure of the head. In addition, the muscle is the layer located under the skull layer in the head and known as axial muscle. The axial muscle is followed by the fat layer. This layer is made up of connective tissue (cells, fibers, and fluid) with adipocytes containing nuclei, receptors, and lipid droplets of fat. The fat layer is covered by a skin layer that tends to be the largest and most important organ. The skin covers the entire body and protect it from the external environment, regulates the body temperature and control the amount of water excreted from the epidermis. This layer is composed of 3 main layers' epidermis, dermis, and hypodermis [2].

2 Literature Review

Similar research has been done previously to investigate the effect of wireless network with specific frequency on the human health [3]. For instance, A. M. Siti Rokiah *et al.* studied the specific absorption rate in human head due to electromagnetic exposure to 4G signals at frequencies of 1800 MHz and 2600 MHz. The simulations, conducted on CST software, utilized SAM model and antenna rotated by various angles and located near the model's ear at different locations as shown in Fig. 2 [4].

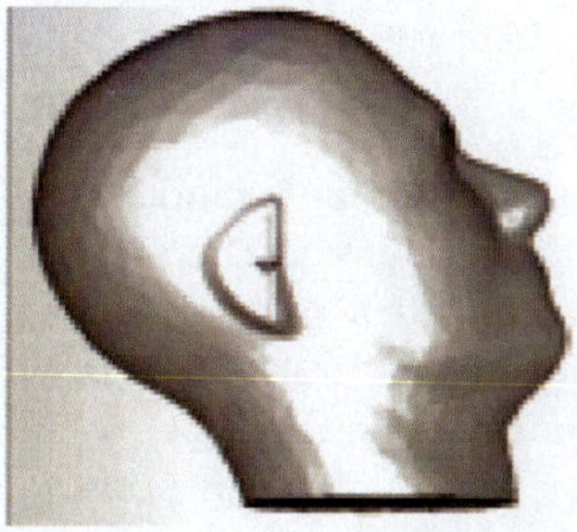

Fig. 2. SAM Phantom model with dipole antenna [4].

J. S. Margish *et al.* reported a study for the 5G users targeting temperature elevation in the human head model resulted due to the radio frequency at 28 GHz emitted by the

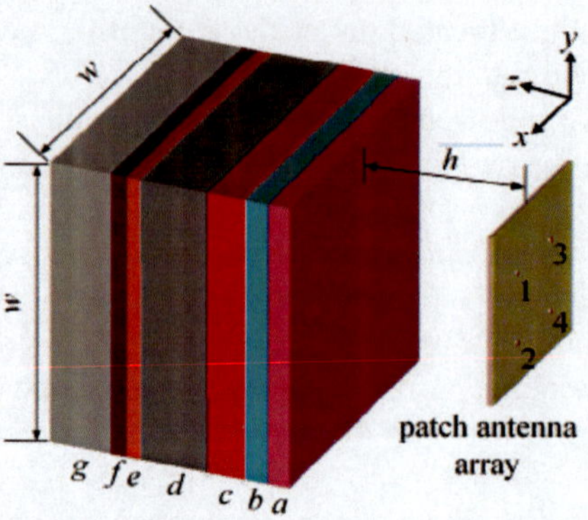

Fig. 3. Multi-layered head model [4].

equipment applications. The head model is made up of skin, fat, muscle, skull, dura, cerebrospinal fluid and brain as demonstrated in Fig. 3 [5].

A. Sahar Aqeel *et al.* reported an SAR simulation, conducted on CST, using SAM model, with antenna located near the model at different distances with different frequencies equal to 900 MHz and 1800 MHz to study the exposure of radiofrequency signals on the human head and come up with adequate safety precautions as demonstrated in Fig. 4 [6].

Fig. 4. Homogenous human head model [5].

W. He *et al.* performed a simulation on a modeled human head consisting of skin, fat, skull, dura mater, cerebrospinal fluid and brain, with antenna facing the model at different distances, to evaluate the SAR resulted from the radiations emitted by the antenna at different frequencies ranged from 100 kHz to 10 GHz. This is to show the radiations on the CSF as demonstrated in Fig. 5 [1].

In this paper we are interested in simulating the effect of 5th generation of the human head on CST studio suite software using a modeled head to analyze the effect of the radiations emitted by the phone within the human head and analyze the SAR "Specific Absorption Rate" results. In addition, the thermal rise of the layers of the head due to these emitted radiation beside the heat source resulted by the processing of the electronic component of the phone itself is also reported.

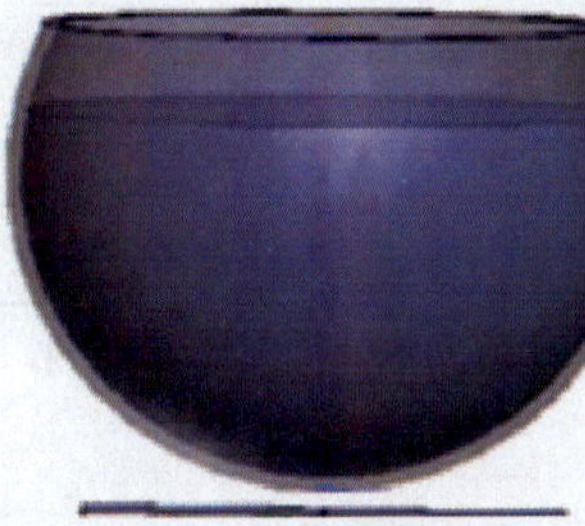

Fig. 5. Human head model with antenna [1].

3 Methodology

The head model used in the study is made up of multilayers including the brain, arachnoid, dura mater, skull, muscle, fat, skin, cartilage, and skin of the pinna, and it is designed on CST studio suite software. The process of building up such model is divided into several steps. First, build up the layers by aligning a squared layer on the surface of the phone followed by setting up the characteristics, dimensions, and thickness of the skin of pinna in the setting of the created layer, then align another square layer on the surface of the skin of pinna layer, and setting up the characteristics and dimensions of the cartilage in the setting of this created layer. Same steps are applied for the rest of the layers in an order manner to build up the rest of the model. Note that the dimensions of the layers are same as listed in Table 1.

Second, the mesh cells of the model are adjusted knowing that as the number of mesh cells increases the smaller structural features will be and more accuracy of computation are obtained. This will be at the cost of the computation time to complete the simulation.

In this step, the mesh cells are going to be adjusted by performing several simulations. The results of each simulation are observed to check the difference between the results and choose the optimal parameters for the meshing cells. In this research, due to the limitation in the server, the minimum accepted mesh cells that give an accurate result and can rely on was chosen.

Third, modify the designed head by changing the boundaries of the model and the thickness of the brain layer depending on the observation of the radiation by how it travels through the model in depth and how it spreads in it. The thickness of the brain layer is decreased to 4 mm, and the boundaries are widening in a way that the spread of the radiation can be seen clearly enough as demonstrated in Fig. 6.

Fourth, a heat source to the simulation to check the liability of the method is added. This step is done by running two simulations with and without a heat source at a frequency equal to 1.5 GHz for 10 s to consume time and observe the change at the same time.

Finally, we run a simulation on the modified model with heat source equals to 1.64 W which represents the heat resulted from a normal power consumption of AP including the heat of the battery, at frequency equals to 28 GHz for 60 s which is the minimum call duration. The duration of the simulation can be increased to monitor the effects for longer time.

Table 1: Essential parameters for each layer in the model.

Tissue	Permittivity	Conductivity (s/m)	Thermal Conductivity (w/km)	Blood Flow (W/Km^3)	Density (kg/m^3)	Heat Capacity (j/kg c°)	Thickness (mm)
Brain	18.59	25.86	0.54	40000	1041	3630	140
Arachnoid Matter	28.19	43.80	0.60	0	1007	4096	0.04
Dura Matter	19.50	24.74	0.50	1125	1130	3364	0.27
Skull	7.51	8.88	0.40	1850	1190	1313	7.1
Muscle	24.44	33.61	0.49	1141	1090	3421	7
Fat	3.70	1.70	0.25	1700	916	2348	2
Skin	18.71	26.19	0.42	9100	1109	3391	1.6
Cartilage	13.2	20	0.49	9000	1100	3568	0.8
Skin of Pinna	18.71	26.19	0.42	9100	1109	3391	0.2

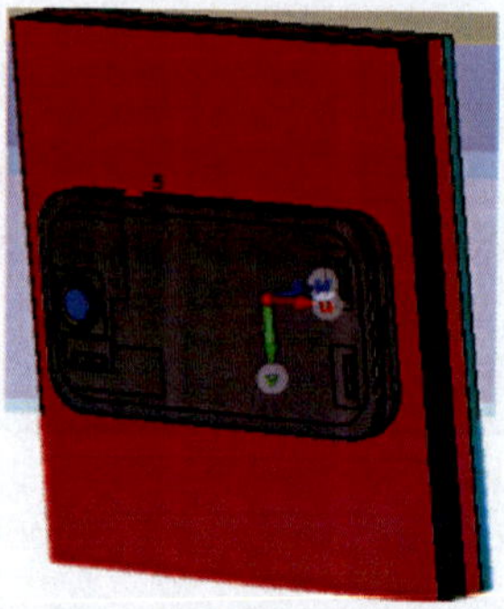

Fig. 6. Modified model.

4 Results and Discussion

After conducting a simulation on the modified head model at frequency equal to 28 GHz with heat source, the duration of the simulation is set to 60 s, the SAR is calculated at 1 g and 10 g in addition to the thermal temperature.

The temperature at 60 s was 36.868 °C, while the $SAR1g$ was 1.20554 × 10^{-17} W/Kg, as $SAR10g$ was 2.9656 × 10^{-17} W/Kg as seen in Figs. 7, 8, and 9.

After going through process of modifying the geometry of the model, checking at what value the mesh cells going to be set and verifying the heat source method, two simulations were conducted on the modified model at frequency equal to 28 GHz. The First simulation was with battery at power 1.64 W and another simulation was conducted without a heat source. The results show that the thermal temperature is elevated by

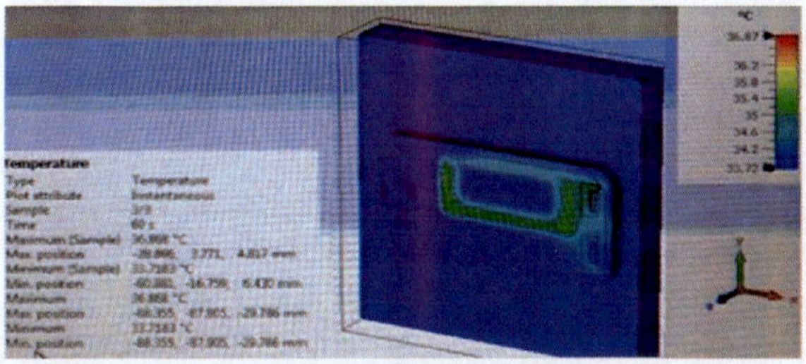

Fig. 7. Temperature at time 60 s with heat source.

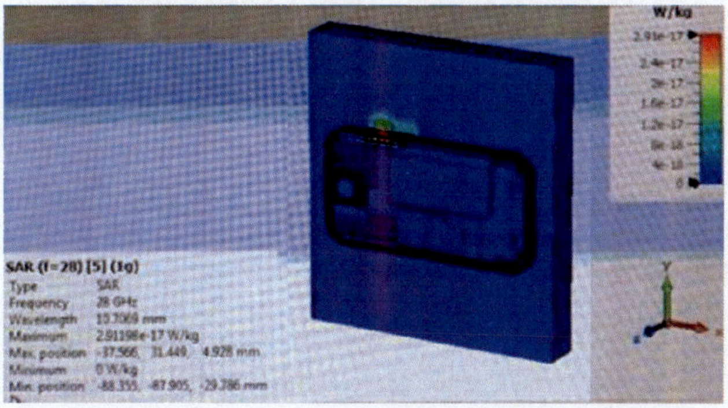

Fig. 8. *SAR*1*g* Value.

Fig. 9. *SAR*10*g* Value.

1.1148 °C due to radiation exposure and additional rise by 1.7532 °C due to the heat source, as in total the thermal temperature elevated by 2.868 °C. Consequently, the temperature elevation is depending on the SAR resulted by emission of the radiations from the port of the phone besides the heat generated by the battery. The results for each of both simulations show maximum SAR and minimum SAR, the maximum SAR represents the maximum measured SAR at the skin of pinna directly under the port location, while the minimum *SAR* represents the measured SAR at the brain layer. Both two measured SARs have very low values in comparison to the permitted SAR which is equal to 1.6 W/Kg set by the FCC standard "federal communication commission".

The obtained results show that the 5G at 28 GHz is safe to be used since the thermal rise is under the maximum accepted thermal rise 2 to 2.5 °C on the outer surface. Not only but also, the SAR is under the maximum value a person can absorb. In addition, the heat generated by the battery will affect the thermal temperature, by causing a rise by 1.7532 °C.

Note that the SAR decreases as the measurement point moves away from the location of the port. Besides, as going deeper in the layers the SAR decreases drastically to reach a null value at the brain layer.

5 Conclusion

In this paper, we have reported a simulation conducted on a simple designed 3D head model using CST suite studio software to study the effect of the heat exerted by the phone on the head t in a parallel manner to the effect of the of 5G radiation under EMF exposure from the antenna of the phone at 28 GHz. The outcomes of the simulation are to investigate these effectors on the SAR and thermal temperature of the head and how it affects the human health in one way or another.

The results show, the SAR and the thermal elevation resulted due to the 5G radiation exposure is below the maximum accepted values set for each person, thus the 5th generation is safe to be used. On the other hand, the heat exerted by the phone due to the processing of the electronics part of the phone cause a rise in the thermal temperature.

References

1. He, W., Xu, B., Gustafsson, M., Ying, Z., He, S.: RF compliance study of temperature elevation in human head model around 28 GHz for 5G user equipment application: simulation analysis. IEEE Access **6**, 830–838 (2018). https://doi.org/10.1109/ACCESS.2017.2776145
2. Scanlon, V.C., Sanders, T.: Essentials of anatomy and physiology. FA Davis (2018)
3. Awada, B., et al.: Simulation of the effect of 5G cell phone radiation on human brain. In: International Multidisciplinary Conference on Engineering Technology (IMCET), Beirut (2018)
4. Rokiah, A.M.S., Hafizuddin, M.M., Muzammil, J., H. A. W. N.: A study of specific absorption rate in human head due to. Indon. J. Electr. Eng. Comput. Sci. **13**(3), 1161–1166 (2019)
5. Margish, J.S., Gaurav, J.R.: Analysis of SAR induced in human head due to the exposure of non-ionizing radiation. Ijert **5**(2), 4 (2016)
6. Abdulrazzaq, S.A., Jabir, S.A.: SAR simulation in human head exposed to RF signals and safety precautions. Int. J. Comput. Sci. Eng. Technol **3**(9), 334–340 (2013)

A Short-Range Free-Space Optical Communication System for Space-Assembled Microsatellites

Demetrio Iero[1](✉), R. Carotenuto[1], M. Merenda[1,2], and F. G. Della Corte[3]

[1] Department of Information, Infrastructure and Sustainable Energy Engineering, University "Mediterranea" of Reggio Calabria, 89124 Reggio Calabria, Italy
demetrio.iero@unirc.it

[2] Center for Digital Safety and Security, Austrian Institute of Technology Gmbh, Vienna, Austria

[3] Department of Electrical Engineering and Information Technologies, University of Napoli Federico II, 80125 Napoli, Italy

Abstract. A free-space optical transmission system for applications that require a proper data communication channel between docked microsatellite modules is presented. The structure and the design of the optical communication interfaces are described. The system acts as a bridge between the CAN (Controlled Area Network) busses on the microsatellites, allowing communication between circuits in different separated modules. This avoids a physical connection between adjacent modules and electromagnetic disturbances that radio-frequency communication systems can generate. The system uses low-cost Commercial-Off-The-Shelf (COTS) components that can reduce significantly the operative costs. A prototype has been built in a format compatible with CubeSat satellites and successfully characterized and tested.

1 Introduction

The use of micro- and nano-satellites has become important in recent years due to the possibility to access affordably at auxiliary and optimized launch opportunities [1]. In particular, the CubeSat initiative [2] has opened up new opportunities providing a standard for satellite architectures reducing cost and development time, and increasing the accessibility to space [3].

Assembly of multiple CubeSats in space into large structures can be valuable for scientific communities. The docking of satellite components in orbit allows structures of considerable size to be easily assembled over time. Docking operations, however, have the necessity of a proper data communication channel between modules.

This paper reports the design of a communication system based on a free-space optical channel. There are various reasons for this choice, but they mainly concern the objective of avoiding a physical electrical connection between the two objects and reducing as much as possible the need for radio-frequency systems [4, 5], which could induce electromagnetic disturbances. The aim is therefore to study the possibility of creating a free-space optical (FSO) transmission system [6–9], that works at low distances

R. Berta and A. De Gloria (Eds.): ApplePies 2022, LNEE 1036, pp. 179–185, 2023.
https://doi.org/10.1007/978-3-031-30333-3_23

(few centimeters) and low data rates (few Megabits) based on Commercial-Off-The-Shelf (COTS) components and to develop a prototype photonic interface between, for example, two CubeSats. The use of COTS components in the context of mini- and micro-satellites can significantly reduce mission costs.

The requirements of the optical communication system are related both to the operating conditions that will be on board the satellite in terms of operating temperature and distance between the units that need to communicate, the voltage and power levels, and the type of communication subsystem on board the satellite.

The realized system makes use of specific microcontrollers and microchips for communication on a CAN (Controller Area Network) bus [10], a low-speed serial standard widely used in the automotive and satellite communication field. The CAN protocol is in use on spacecraft since the 1990s and different missions have demonstrated that this bus can be successfully used in the space environment as a reliable and low-cost communication system [11, 12]. This FSO system allows connecting the CAN busses of two separate subsystems, putting them in communication and allowing the exchange of state information and diagnostic data at a low data rate, up to 1 Mbps, between nodes on different satellite modules. Transmission at higher data rates can be supported but is out of the scope of this work.

The realized prototype will be integrated into a complete satellite platform as part of the Italian MIUR project PON PM3 "Piattaforma Modulare Multi-Missione" which has the overall objective to define a 50 kg class platform with the possibility of housing multiple interoperable payloads for miniaturized satellites. The platform also includes a high data rate interface and a wireless power transfer system to power the docked modules [13, 14].

The paper is organized as follows. Section 2 describes the structure of an optical communication system in free space and the components that have been used for the design of the optical communication interfaces and prototypes. Section 3 reports the characterization of the electronic boards and transceivers. Finally, Sect. 4 presents the conclusions.

2 System Description

2.1 System Architecture and Components

The structure of the FSO comprises a transmitter and a receiver. The transmitter consists of two parts: an interface circuit and a driver circuit. They convert the incoming electrical signal into an optical signal suitable for transmission in free space. The modulated optical signal then propagates in free space until it reaches the receiver, where a photodetector converts the optical signal back into an electrical signal. The receiver in turn consists of the optical detector and of a signal conditioning circuit that makes the output of the detector match the input present at the transmitter.

The commercial optical transceiver chosen for the application is the AFBR-59F1Z manufactured by Avago Technologies [15]. The component is designed for use on a plastic optical fiber (POF) and incorporates an optical transmitter and receiver within a metal housing for shielding against electromagnetic interference (EMI). The transceiver allows a baud rate up to 125 MBd/s when used with optical fiber. The transmitter is a fully integrated device that consists of an LED with an emission spectrum centered at 650 nm, controlled by a driver capable of receiving a Low Voltage Positive Emitter Coupled Logic (LVPECL) or Low Voltage Differential Signalling (LVDS) differential signal. The optical receiver, amplify, and converts the optical signal received from a PIN diode into a differential signal. The transceiver was used in a different operating condition from the one for which it was designed by the manufacturer, i.e. free-space optical communication instead of fiber optics. Its outer metal casing was eliminated, giving access to two separate devices comprising the optics and the transmitter or receiver circuitry.

As the device works on differential voltage levels, to interface it with a microcontroller, it was necessary to use LVDS drivers and receivers between the microcontroller's port and the transceiver. The DS90LV011AQ driver by Texas Instruments was chosen, which is suitable for low power, low noise, and high transmission rates. The matching receiver chosen is the DS90LT012AQ.

The realized photonic interface is shown in Fig. 1. Each of the two boards, one placed on the main CubeSat, and the other on the docked module, will both transmit and receive optical signals. Each board includes a microcontroller that encodes/decodes and transmits/receives the optical signal through the transceiver and interface circuits. The microcontroller also manages the interfacing to the CAN-based bus communication interface of the satellite, relaying the CAN packet between the two satellite parts at a data rate of 1 Mbps. The CAN packets are repacked, encoded with Manchester encoding, and transmitted serially to the transceiver.

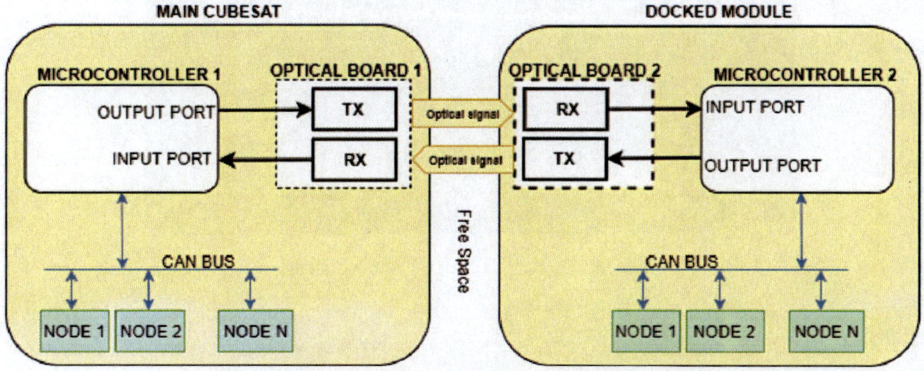

Fig. 1. The architecture of the optical interface boards.

The Manchester encoding [16] has been chosen to allow the exchange of self-clocking signals that can be easily coded/decoded and also prevents the transceiver from going into shutdown mode when long sequences of the same bit are transmitted without signal transitions.

The hardware used consists of Microchip's PIC18LF26K83 microcontrollers [17] with an integrated CAN module operating at 1Mbit/s and Texas Instrument's SN65HVD233D CAN transceiver that implements the physical layer of the CAN bus.

2.2 Prototypes

The prototypes, developed at the DIIES Electronics Laboratory of the Mediterranea University of Reggio Calabria, consist of a main board in PC104 format for CubeSat compatibility [18] and an auxiliary board containing the small optical transceiver that can be mounted on the satellite outer panel. The main board is shown in Fig. 2.

The heart of the low data rate interface is the PIC18LF26K83 microcontroller, which includes a CAN module operating and a Manchester-encoded serial interface used to transmit and receive data to the optical transceiver.

The optical transceivers are located on a separate board (Fig. 3) connected to the main board via a dedicated connector on which the LVDS differential bus lines run. On the main board are LVDS-TTL converters to adapt the differential signal used by the optical transceiver to the microcontroller, which uses TTL-type electrical signals.

Fig. 2. Picture of PC104 main board.

Fig. 3. Optical Boards during the test phase.

The microcontroller receives the signal from the CAN bus, encodes it in Manchester format, and sends it to the optical board. The system is full-duplex and also performs the opposite conversion, receiving the signal from the optical card and forwarding it on the CAN bus. This configuration gives maximum flexibility to the system, allowing it to send and receive data on both sides at maximum capacity.

The main board also includes optimal additional circuits with FPGA (Field Programmable Gate Array) integrated circuits that can enable operation tests at higher data rates, up to 100 Mbit/s.

The total cost of this FSO platform is extremely low, with the total cost for the components and the four boards that constitute the system below 300 euros.

3 Experimental Analysis

The setup used for testing is shown in the diagram in Fig. 4. It involves the use of an additional CAN node that generates a data packet and forwards it on the CAN bus configured to operate at 1 Mbps; the packet is intercepted by the first PC104 board. The PC104 board receives the signal on the CAN bus, encodes, and transmits it to the optical board which ensures the transmission in the free space. The optical signal is received by a second optical board which is connected to a second PC104 board that receives the data, decodes, and forwards it to the CAN bus on the second side of the satellite system. A CAN node is connected to the second CAN bus, which receives the packet and displays the content on a PC terminal via a serial connection.

The main board card receives the data on the CAN bus and recodes it into a format suitable for transmission via the optical transceiver, as described in Sect. 2. The data packet, encoded according to the Manchester standard, contains:

- 2 synchronization bytes identifying the start of the packet
- 1 byte identifying the type of CAN packet (e.g. standard, extended)
- 4 bytes identifying the data (CAN packet ID)
- 1 byte identifying the number of data bytes of the CAN packet
- 8 bytes of data
- 1 control byte (checksum)
- 1 byte identifying the end of the packet.

The packet received on the second optical transceiver is then decoded and re-packed in accordance with the format prescribed by the CAN protocol and forwarded to the CAN bus on the second side of the system.

The received and decoded signal is then displayed on a PC to evaluate the correct reception of the packets. Transmitted and received signals have also been acquired and monitored by means of a logic state analyzer to verify the proper functioning of the system. Data bytes transmitted in each test packet comprise a fixed part and a variable part representing a unique progressive packet number.

The tests, performed under natural light and artificial neon light conditions, have shown that the system successfully forwards all the packets sent from the first CAN node to the nodes on the second CAN bus within a distance of 13 cm between the two aligned transceivers. At longer distances, the transceiver pair does not allow to reconstruct correctly the signal resulting in bit loss or checksum errors that lead to the discarding of received packets.

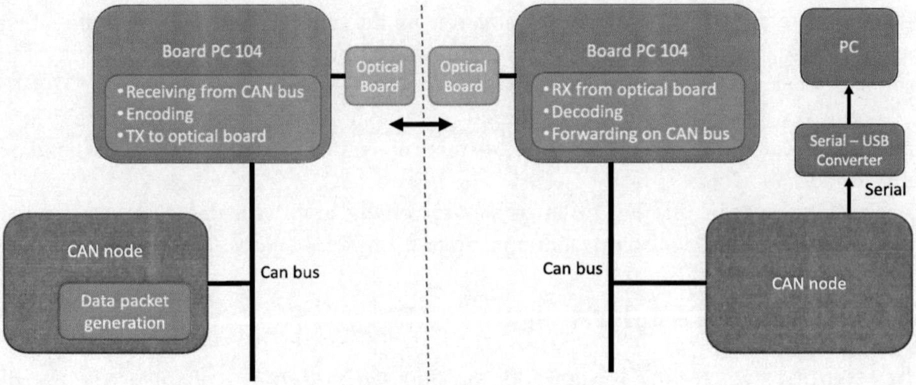

Fig. 4. Test setup block diagram.

The realized interface allows communication even at distances between the two optical transceivers of 13 cm, and thus enables communication to be established before the actual docking of the two satellite modules, allowing useful data to be exchanged also during the approach phase in order to complete the docking correctly.

The system has been tested at different temperatures in the range from −40 °C to 110 °C inside a climatic chamber and exhibited a low Bit Error Ratio (BER) up to 80 °C, clearly limited by the temperature ranges of the commercial components used in the project, in particular the AFBR-59F1Z transceiver. In a space environment, the boards must be adequately protected and isolated from temperatures out of operating range.

This low data rate communication prototype will be integrated into a complete micro-satellite platform and additional tests are expected to be performed to evaluate the performance, reliability, and correct integration of the different sub-systems.

4 Conclusions

A short-range free-space optical transmission system for microsatellite applications has been designed and successfully realized and tested. The system allows to virtually connect the CAN busses on different modules docked to the main satellite, without a physical connection and avoiding the use of radio-frequency systems that could generate electromagnetic disturbances. The prototype, based on off-the-shelf components to minimize production costs, has been successfully tested, showing a reliable communication interface up a distance of 13 cm and a data rate of 1 Mbps for the CAN bus.

Acknowledgments. This research was supported by the Italian MIUR Project PON PM3 ARS01_01181.

References

1. Sweeting, M.N.: Modern small satellites-changing the economics of space. Proc. IEEE **106**, 343–361 (2018)

2. Villela, T., Costa, C.A., Brandão, A.M., Bueno, F.T., Leonardi, R.: Towards the thousandth CubeSat: a statistical overview. Int. J. Aerosp. Eng. **2019**, 1–13 (2019)
3. Crusan, J., Galica, C.: NASA's CubeSat launch initiative: enabling broad access to space. Acta Astronaut. **157**, 51–60 (2019)
4. Merenda, M., Iero, D., Della Corte, F.G.: CMOS RF transmitters with on-chip antenna for passive RFID and IoT Nodes. Electronics **8**, 1448 (2019)
5. Iero, D., Felini, C., Merenda, M., Della Corte, F.G.: RF-powered HF-RFID analog sensors platform. In: De Gloria, A. (ed.) ApplePies 2016. LNEE, vol. 409, pp. 85–91. Springer, Cham (2017). https://doi.org/10.1007/978-3-319-47913-2_11
6. Kaushal, H., Jain, V.K., Kar, S.: Free Space Optical Communication. Springer India, New Delhi (2017)
7. Della Corte, F., et al.: Temperature sensing characteristics and long term stability of power LEDs used for voltage vs. junction temperature measurements and related procedure. IEEE Access **8**, 43057–43066 (2020)
8. Chaudhary, S., Amphawan, A.: The role and challenges of free-space optical systems. J. Opt. Commun. **35** (2014)
9. Iero, D., Merenda, M., Polimeni, S., Carotenuto, R., Della Corte, F.G.: A technique for the direct measurement of the junction temperature in power light emitting diodes. IEEE Sens. J. **21**, 6293–6299 (2021)
10. Robert Bosch GmbH: CAN Specification 2.0. http://esd.cs.ucr.edu/webres/can20.pdf
11. Scholz, A., Hsiao, T.-H., Juang, J.-N., Cherciu, C.: Open source implementation of ECSS CAN bus protocol for CubeSats. Adv. Space Res. **62**, 3438–3448 (2018)
12. Plummer, C., Roos, P., Stagnaro, L.: CAN bus as a spacecraft onboard bus. In: European Space Agency, (Special Publication) ESA SP, pp. 473–483 (2003)
13. Iero, D., Carotenuto, R., Merenda, M., Pezzimenti, F., Della Corte, F.G.: Performance evaluation of Silicon and GaN switches for a small wireless power transfer system. Energies **15**, 3029 (2022)
14. Merenda, M., et al.: Open-source hardware platforms for smart converters with cloud connectivity. Electronics **8**, 367 (2019)
15. Avago Technologies: AFBR-59F1Z Datasheet. https://docs.broadcom.com/doc/AV02-4107EN
16. Forster, R.: Manchester encoding: opposing definitions resolved. Eng. Sci. Educ. J. **9**, 278–280 (2000)
17. Microchip Technology Inc.: PIC18LF26K83 Datasheet. https://ww1.microchip.com/downloads/aemDocuments/documents/MCU08/ProductDocuments/DataSheets/PIC18%28L%29F2526K83-Data-Sheet-DS40001943C.pdf
18. PC/104 Embedded Consortium: PC/104 Specification Version 2.6. https://pc104.org/wp-content/uploads/2015/02/PC104_Spec_v2_6.pdf

A Low-Area, Low-Power, Wide Tuning Range Digitally Controlled Oscillator for Power Management Systems in 28 nm CMOS Technology

M. Mestice(✉), G. Biondi, G. Ciarpi, D. Rossi, and S. Saponara

Department of Information Engineering, University of Pisa, Pisa, Italy
marco.mestice@phd.unipi.it

Abstract. Nowadays, in the world of high-performance computing, saving energy when great computing power is not needed is a must-to-have feature. This usually involves the implementation of Power Management Systems (PMS) to apply power saving polices such as frequency scaling. In particular, for this feature, the actuators of PMS are usually implemented with Phase- or Frequency-Locked Loops, which should occupy a small area and exhibit a low-power consumption. Additionally, they should be able to generate a wide range of frequencies in the order of a few GHz with a fine granularity of a few hundreds of MHz. Since the core of such loops is a tunable oscillator, in this work we present a pseudo-differential Ring Digitally Controlled Oscillator (DCO) implemented with a standard 28 nm CMOS technology to be used in PMS. The proposed DCO features a well-balanced behavior between the noise performance and a wide tuning range, a low-area, and a low-power consumption.

1 Introduction

A requirement in today's world of consumer electronics is to generate high-performance high-frequency clock signals. Beside the actual clock generation, many modern applications need tunable features over a wide range of frequencies. Frequency hopping in wireless communication and power management policies in processors and digital systems are excellent examples of these kinds of applications [1, 2]. The tunability and controllability of the clock usually involves the use of a Controlled Oscillator, e.g., Voltage Controlled Oscillators (VCO) or Digitally Controlled Oscillators (DCO), inserted in feedback loops, such as Phase-Locked Loops (PLL) [3, 4] or Frequency-Locked Loops (FLL) [5, 6]. In this work, we present a pseudo-differential Ring DCO implemented with a standard 28 nm CMOS technology to be used in clock generators for the system level power management in computing applications. The target architecture is a Ring-DCO since, given the application, the area of the clock generator should be limited. Indeed, beside the restrained area that characterizes Ring-Oscillators (RO) in general, a digital control signal lets the feedback loop to be digital as well, leading to a further reduced area of the whole system. Moreover, together with the low area, a wide frequency range

R. Berta and A. De Gloria (Eds.): ApplePies 2022, LNEE 1036, pp. 186–195, 2023.
https://doi.org/10.1007/978-3-031-30333-3_24

from hundreds of MHz to 3 GHz, and frequency resolution below 100 MHz were targeted to be suitable for the chosen applications [6, 7]. Furthermore, the design effort was also focused on reducing the power consumption for the given frequency and frequency resolution constraints. The paper is organized as follows: in Sect. 2 the proposed DCO is described, the design choices are reported from the architectural ones to the schematic ones, and the results from PVT simulations are presented; in Sect. 3 the layout design is summarized, and the post-layout simulations are presented. Finally, in Sect. 4 a comparison with the state-of-the-art DCOs is reported, and in Sect. 5 the conclusions are written.

2 Proposed DCO

ROs consist of delay lines in which the last output is connected to the first input. The operating frequency of an N-stages RO is [8]:

$$f = \frac{1}{2Nt_D} \tag{1}$$

where t_D is the delay of the single delay element and can be written as:

$$t_D = \frac{C_L \Delta V}{I_D} \tag{2}$$

where C_L is the load capacitance, ΔV is the output voltage swing, and I_D is the driving current to the load. Looking at Eqs. 1 and 2, different approaches to control the output frequency can be identified: i) varying the number of delay cells, i.e., changing the number of stages N thanks to multiplexers or three-state buffers. Even though this solution is an easy-to-implement solution, generally it is characterized by a large frequency drift and large area occupation; ii) Varying the load of the single delay element, i.e., changing C_L in Eq. 2 by changing the nodes' capacitance [9]. The capacitor bank may be implemented with Metal-Insulator-Metal (MIM) capacitor or MOS capacitor, which can be easily digitally controlled; iii) varying the output current of the delay elements, i.e., changing I_D in Eq. 2 [10]. Indeed, by changing the current which charges and discharges the nodes' capacitance, the oscillation frequency can be controlled. Compared to thc capacitor bank solution, this one requires less area. However, the design of a controllable current source with a monotonic behavior for a wide tuning range is challenging [11].

Apart from the way the frequency can be controlled, ROs can also differ in the implementation of the delay elements. Indeed, an RO can be single-ended, differential or pseudo-differential. Single-ended Ring Oscillators offer many advantages, such as design simplicity, rail-to-rail output swing and low area occupation, but show a great sensitivity to PVT variations, common mode noise, and ripple of the supply voltage. On the other hand, differential [12] or pseudo-differential [13] ROs exhibit higher noise immunity, even though they need more area and power consumption. In this work, a pseudo-differential RO was selected as it represents a good trade-off between the single-ended and the fully differential solutions. Indeed, the pseudo-differential solution shares the advantages in term of noise and disturbances rejection with the fully differential one.

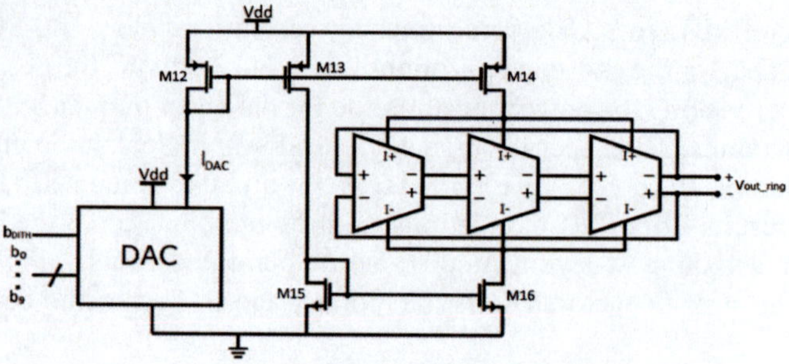

Fig. 1. Proposed DCO architecture.

However, it features lower power consumption and area occupation, similar to a single-ended solution. Figure 1 illustrates the proposed DCO. A Digital to Analog Converter (DAC) generates a current that is mirrored in the delay elements of the RO. As anticipated in the introduction, the DCO was designed to reach three primary goals: i) low power ii) low area, and iii) a frequency resolution below 100 MHz over a frequency range between 500 MHz and 3 GHz. The main goals in designing the DAC, instead, were to obtain a monotonic characteristic, to reduce the power consumption, and to contain the PVT variations. In the next subsections the detailed description of the DAC and of the Current Controlled Oscillator that compose the DCO is reported.

2.1 Pseudo Differential Ring Oscillator

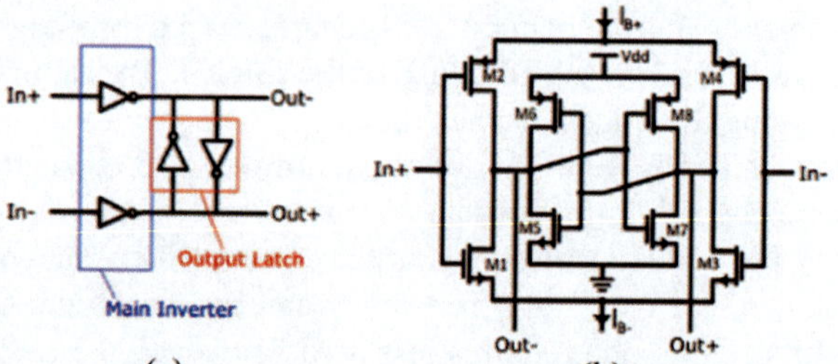

(a) (b)

Fig. 2. Delay element of the RO: (a) digital logic view; (b) schematic level view.

Since the technology targeted for the implementation of the DCO (i.e., the 28 nm TSMC technology) can work at frequencies well above the frequency range needed for this work, the design of the RO is focused on the power and area reduction, achieving good noise performance and, most of all, a wide tuning range. Therefore, a three stages solution was selected. As shown in Fig. 2a, every basic element of the RO is composed by two inverters and one latch that implement a pseudo-differential architecture. Apart from implementing the delay element, the positive feedback formed by the latch improves the rising and falling edges of the output signal, leading to a better noise performance. The schematic of the basic elements of the DCO is shown in Fig. 2b. The input signals drive the main inverters (M1, M2 and M3, M4), which in turn drive the positive feedback latch composed by M5, M6, M7, and M8. The output frequency of the RO realized with these delay elements is determined by the strength ratio between the input inverters and the latch. The strength of the input inverters is controlled by the current generated by the DAC, while the strength of the latch is determined by the ratio W/L of M5, M6, M7, and M8. A high ratio between the latch strength and the inverter strength would lead to an improved noise performance thanks to the increased strength of the positive feedback. On the other hand, it would also lead to a degradation of the tuning range, which would be narrower. Therefore, the DAC's current and the sizes of the main inverters and of the latch were chosen to obtain a wide tuning range, without degrading the noise performance, and minimizing the power consumption. In particular, concerning the power consumption, the current value was chosen to achieve the highest target frequency (above 3 GHz) in all the technology corners and to compensate for the parasitic effects occurring after the layout phase [14, 15]. The DAC's current (I_{DAC} in Fig. 1) chosen with these considerations was, therefore, about 10 mA, shared between all the stages.

2.2 DAC Architecture

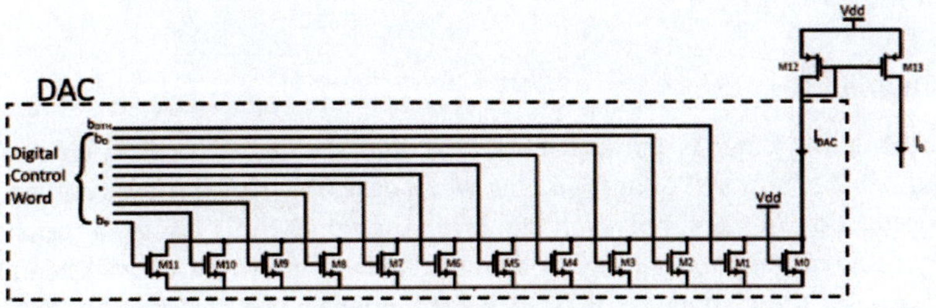

Fig. 3. DAC schematic.

Since the oscillator core is controlled by a current, a current steering DAC was designed, as shown in Fig. 3. It consists of 10 weighted current sources (M2–M11), each one controlled by one bit of a 10-bit digital word. An additional current source (M1) was added to implement a dithering feature since this technique [6] is often exploited in all-digital loops to increase further the DCO's resolution. The output current is given by Eq. 3. Ideally, each current I_n is given by Eq. 4, where w_n equals 2^n and I_0 is a unit current. Therefore, to enhance the matching between the current sources, the weight of every transistor was implemented with w_n equal transistors connected in parallel, obtaining the output current of Eq. 5, where I_0 is the unit current produced by the unit nMOS (i.e., M1 and M2). This current is then collected by a pMOS transistor (M12), which belongs to the mirror that, in turn, supplies the current to the elements of the RO.

$$I_{DAC} = \sum b_n I_n \tag{3}$$

$$I_n = w_n I_0 \tag{4}$$

$$I_{DAC} = I_0 + b_{DITH} I_0 + b_0 I_0 + b_1 2^1 I_0 + \ldots + b_9 2^9 I_0 \tag{5}$$

The output of the DAC is not connected directly to the ring-oscillator to increase the monotonicity of the characteristic. Indeed, if the DAC's output had been connected directly to the RO, the effective capacitance seen from the transistors of the RO would have depended on which transistor of the DAC was on [11]. Therefore, also the delay of the elements of the RO would have been influenced by the capacitance of the DAC's transistors switched on, leading to a possible non-monotonic behavior. Transistor M0, which is always ON, determines the smallest current the DAC can supply, and therefore, it establishes the lower limit of the output frequency range, while the maximum output frequency is determined by all the other transistors together. Therefore, M0 was sized to obtain the current needed by the RO for the lower limit of the frequency range. All the other transistors, instead, were sized to obtain the maximum current chosen during the design of the RO as described in the previous subsection (i.e., 10 mA). In particular, they were divided in binary weighted current sources composed by identical replicas of the unit-current transistor, whose channel is 200 nm wide and 1 μm long with two fingers. With this sizing the unit-current transistor's contribution is 10 μA, and the DAC current vs input digital word characteristic shows a minimum slope of 2 μA/code, an average slope of 5 μA/code, and maximum slope of 10 μA/code depending on the Digital Control Word.

2.3 Simulation Results

In Fig. 4, the tuning range is shown for both the typical case, i.e., typical process corner, 27 °C, and 0.9 V of supply voltage, and the worst case, i.e., slow process corner, 125 °C, and 10% reduction in the supply voltage. Even though there is a loss in the maximum frequency, the required tuning range is achieved also in the worst-case. As can be seen, a compression of the characteristic is seen for higher values of the input digital word. This is mostly due to the DAC and, in particular, to the limited resistance of M12, as well as of M0–M11 [16]. Indeed, varying the Digital Control Word causes a variation in the output node resistance that goes from $r_P || r_0$ for the LSB to $r_P || (r_0/N)$ for N transistors activated, where r_P is the resistance seen towards the drain of M12, i.e., 1/gm, and r_0 is the resistance seen towards the output node of the DAC. For values of N above 400, r_0/N is comparable with 1/gm and, therefore, the effect described is seen.

In Fig. 5a, the power consumption in the tuning range (i.e., as function of the digital input word) is shown. As expected, the power consumption reaches its maximum of about 7 mW at the highest frequency. In Fig. 5b, the frequency resolution as function of the input digital word is drawn.

The resolution is better than the 100 MHz-constraint in all the tuning range, and it can be further reduced at system level by exploiting the dithering feature. As the tuning range characteristic is compressed for high values of the Digital Control Word, for the same reason the derivative of the frequency resolution tends to zero at higher frequencies. Finally, the phase noise at 1.5 GHz of output frequency resulted to be −83.4 dBc/Hz @ 1 MHz.

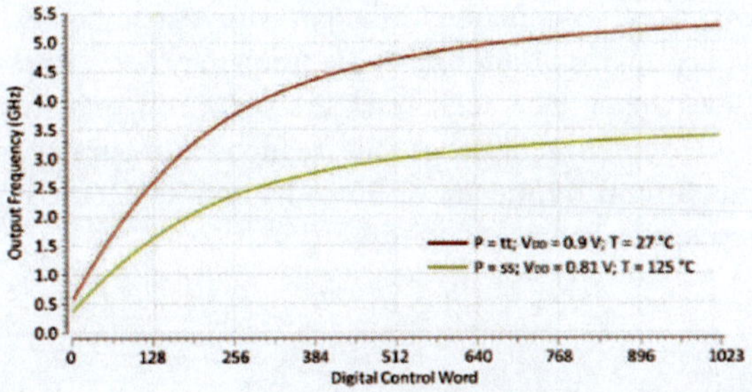

Fig. 4. DCO's tuning range in the typical and worst corner.

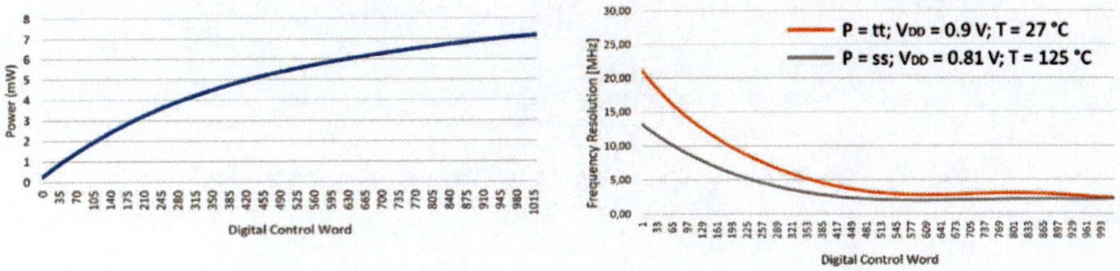

(a) (b)

Fig. 5. (a) Power consumption vs Digital Control Word; (b) Frequency resolution vs Digital control Word in the typical and worst case.

3 DAC and RO Layout

3.1 Layout Design

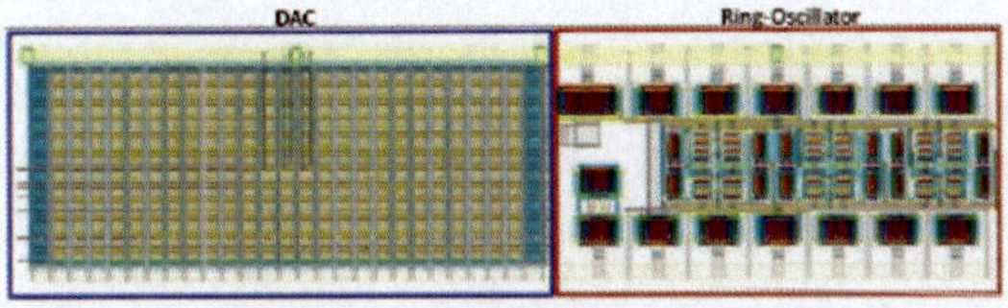

Fig. 6. Layout of the DCO.

The layout of the DCO is depicted in Fig. 6. The layout of the DAC was designed following a common centroid approach. While for the DAC the classic transistors were exploited for the design, the rf-ones, which include a deep N-well and guard rings to reduce the substrate disturbances, were instead used for the RO, given their superior performance and their more detailed model at high-operating frequencies. The total DCO's area is 5083.7 μm^2.

3.2 Post Layout Simulation Results

Figure 7 shows the tuning range of the DCO for pre-layout simulation and post layout simulation for the entire DCO. There is a noticeable difference in terms of frequency loss between the two cases. This is due to the parasitic capacitances and resistances of the metal paths that play an important role in the frequency reduction. Indeed, according to Eqs. 1 and 2 if the capacitance seen from the output of one single cell of the RO increase, then there is a reduction of the output frequency. In our design, the DCO shows a reduction of the maximum frequency of 36%, from 5.28 GHz to 3.35 GHz, yet still reaching the required tuning range.

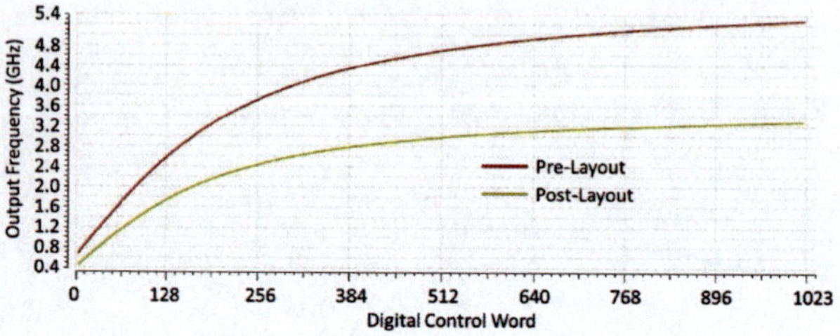

Fig. 7. Pre- and Post-layout simulated tuning range of the DCO.

4 State-of-the-Art Comparison

Table 1 reports a summary of the proposed DCO's characteristic compared to the current state-of-the-art DCOs in similar frequencies ranges and realized in the 28 and 65 nm CMOS technologies. The DCO proposed in this work shows a good trade-off between area, power, and noise performance, and it has a really wide tuning range of 2.8 GHz and 10 bits of resolution. Compared to the DCO presented in [17] the noise performance and power consumption are worse, but in [17] a LC-tank topology has been exploited, leading to an area 8 times greater. Moreover, only 730 MHz of tuning range has been achieved, which is almost 4 times smaller than the tuning range obtained by the DCO presented in this work. In [18], instead, a Ring-DCO in 28 nm is presented with a slightly better power consumption and an area comparable to our DCO. However, even though it has a good 9-bit resolution, the tuning range is limited to few hundred of MHz, and it shows worse noise and jitter performance. In [13] a DCO in 65 nm is reported instead. It achieves better power consumption with a 1.9 GHz tuning range, a 9-bit resolution, and with good noise performance. However, there, no layout is proposed, and no post-layout simulations have been performed. Finally, the DCO in [19] has 16 bits of resolution and occupies an astonishing small area compared to the DCO proposed in this work, but its tuning range is limited to only 400 MHz, and it shows worse noise performance. In conclusion, this work shows the widest tuning range among the state-of-the-art DCOs. Even though such tuning range has been obtained at the expense of the area occupation, the power consumption, and the Phase Noise, this solution is competitive with the sate-of-the-art DCO also from these points of view.

Table 1. State-of-the-art Comparison

	This work[a]	[17][b]	[18][a]	[13][a]	[19][a]
Technology	28 nm	28 nm	28 nm	65 nm	65 nm
Supply Voltage	0.9 V	0.8 V	1 V	1.2 V	1.2 V
Control bits	10 + 1 dithering	12[c]	9[d]	9	16
Frequency range (GHz)	0.5–3.3	3.95–4.68	1.13–1.54	2.7–4.6	1.8–2.2
Area (μm^2)	5083.7	42000	5616	/	370
Simulated/Measured Frequency	1.5 GHz	4.6 GHz	1.3 GHz	4 GHz	2.2 GHz
Power (mW)	1.476	0.35	0.84	1	1.6
Phase noise @ 1 MHz (dBc/Hz)	−83.4	−109.5	−74	/	−78,9
Cycle-to-cycle jitter (rms)	0.46 ps	/	2.3	0.53 ps	/

[a] Simulated - [b] Measured – [c] 4 bits for coarse tuning in binary code and 255 bits for fine tuning in thermometric code – [d] 7 bits for coarse tuning in thermometric code and 6 bits for fine tuning in binary code.

5 Conclusions and Future Work

In this work, we have presented a DCO designed for the power management policies in computing applications. For these kinds of applications, a wide frequency range is needed, and, indeed, the target frequency range for this work was from hundreds of MHz to 3 GHz. As shown by the post-simulation results, this constraint has been comfortably achieved with a frequency resolution well below the required 100 MHz, which can be further reduced by exploiting the dithering feature implemented in the DAC. Also, the layout of the DAC and of the RO have been presented. The layout was a critical design step because of the parasitic capacitances that reduce the oscillation frequency. The design choices at architectural and circuit levels have been demonstrated to be a good trade-off in terms of noise performance, area, and power consumption with respect to the current state-of-the-art DCOs. Indeed, the use of an RO has led to a low area and low power solution, and the pseudo-differential implementation has brought to good noise performance, leading therefore to a well-balanced solution between a single-ended and a fully differential implementation.

Given the push towards ever more advanced technology nodes in digital applications and in high-performance processors, where PMS are usually employed, future developments will be the implementation of the proposed solution in a 12 nm FinFet technology. This upgrade may lead not only to a reduction of the area occupation, but also to an improvement of the performance. Moreover, next to the technology scaling, further future work could involve the substitution of the mirror between the DAC and the RO (transistors M12, M13, M14, M15, and M16 in Fig. 1) with a current magnifier. Even though a well-matched current magnifier would need more design effort with respect to a simple current mirror, this solution could lower the DAC's current leading to a reduction of the total power consumption as well.

Acknowledgments. Work partially supported by European Union's Horizon 2020 research and innovation program under grant agreement No 101036168 (*European Process Initiative SGA2*), and by MIUR with the *Dipartimenti di Eccellenza 2018-2022 Crosslab* project.

References

1. Lojko, B.: A contribution to the design of a frequency synthesizer for fast frequency-hopped spread-spectrum systems. In: 17th International Conference Radioelektronika (2007)
2. Benini, L., et al.: A survey of design techniques for system-level dynamic power management. IEEE Trans. Very Large Scale Integr. Syst. **8**(3), 299–316 (2000)
3. Gardner, F.M.: Phaselock Techniques. Wiley, New York (2005)
4. Mestice, M., et al.: Analysis and design of integrated blocks for a 6.25 GHz Spacefibre PLL. Sensors **20**, 4013 (2020)
5. Golestan, S., et al.: Single-phase frequency-locked loops: a comprehensive review. IEEE Trans. Power Electron. **34**(12), 11791–11812 (2019)
6. Bellasi, D.E., et al.: Smart energy-efficient clock synthesizer for duty-cycled sensor SoCs in 65 nm/28nm CMOS. IEEE Trans. Circuits Syst. I Regul. Pap. **64**(9), 2322–2333 (2017). https://doi.org/10.1109/TCSI.2017.2694322
7. Zhang, X., et al.: An evaluation of {Per-Chip} nonuniform frequency scaling on multicores. In: 2010 USENIX Annual Technical Conference (USENIX ATC 2010) (2010)
8. Rabaey, J.M.: Digital Integrated Circuits: A Design Perspective. Prentice-Hall, Inc., Upper Saddle River (1996)
9. Andreani, P., et al.: A digitally controlled shunt capacitor CMOS delay line. Analog Integr. Circ. Sig. Process **18**, 89–96 (1999)
10. Suman, S., Sharma, K.G., Ghosh, P.K.: Analysis and design of current starved ring VCO. In: 2016 International Conference on Electrical, Electronics, and Optimization Techniques (ICEEOT) (2016)
11. Maymandi-Nejad, M., et al.: A digitally programmable delay element: design and analysis. IEEE Trans. Very Large Scale Integr. (VLSI) Syst. **11**(5), 871–878 (2003)
12. Jalil, J., et al.: CMOS differential ring oscillators: review of the performance of CMOS ROs in communication systems. IEEE Microwave Mag. **14**(5), 97–109 (2013)
13. Gorji, J., et al.: A 2.7 to 4.6 GHz multi-phase high resolution and wide tuning range digitally-controlled oscillator in CMOS 65 nm. In: 2016 24th Iranian Conference on Electrical Engineering (ICEE), 2016, pp. 1694–1699 (2016)
14. Ciarpi, G., et al.: Design and characterization of 10 Gb/s and 1 Grad TID-tolerant optical modulator driver. IEEE Trans. Circ. Syst. I Regul. Pap. **69**(8), 3177–3189 (2022)
15. Monda, D., et al.: Design and verification of a 6.25 GHz LC-Tank VCO integrated in 65 nm CMOS technology operating up to 1 Grad TID. IEEE Trans. Nucl. Sci. **68**(10), 2524–2532 (2021)
16. Razavi, B.: The current-steering DAC [A Circuit for All Seasons]. IEEE Solid-State Circ. Mag. **10**(1), 11–15 (2018). https://doi.org/10.1109/MSSC.2017.2771102
17. Levinger, R., et al.: A 3.9-4.7 GHz 0.35 mW DCO with −187.4 dBc FoM in 28nm CMOS. In: 2018 13th European Microwave Integrated Circuits Conference (EuMIC), pp. 194–197 (2018)

18. Bisiaux, P., et al.: Design of a 1.5 GHz low jitter DCO ring in 28 nm CMOS process. In: 2020 European Conference on Circuit Theory and Design (ECCTD), pp. 1–5 (2020)
19. Selvaraj, S., et al.: Low phase noise, high resolution digitally-controlled ring oscillator operating at 2.2 GHz. In: 2020 9th International Conference on Modern Circuits and Systems Technologies (MOCAST), pp. 1–4 (2020)

Sorting of Live/dead *Escherichia Coli* by Means of Dielectrophoresis for Rapid Antimicrobial Susceptibility Testing

A. di Toma, G. Brunetti, N. Sasanelli, M. N. Armenise, and C. Ciminelli(✉)

Optoelectronics Laboratory, Politecnico di Bari, Via Orabona 4, 70125 Bari, Italy
caterina.ciminelli@poliba.it

Abstract. According to the World Health Organization (WHO) forecasts, AntiMicrobial Resistance (AMR) will represent the leading cause of death worldwide in the next decades. To prevent this phenomenon, rapid antimicrobial susceptibility testing is needed to guide the choice of the proper antibiotic. In this context, we propose a chip-scale system, mainly based on a microfluidic channel combined with a pattern of engineered electrodes, to efficiently test an antibiotic on a bacteria sample. The use of dielectrophoretic (DEP) forces enable the sorting of live/dead bacteria, such as *Escherichia Coli*, with an efficiency larger than 99% for rapid monitoring of the antimicrobial susceptibility at the single-bacterium level.

1 Introduction

AntiMicrobial Resistance (AMR) represents the ability of bacteria to withstand antibiotic treatments [1–3]. This worldwide phenomenon is strictly correlated to the overuse and misuse of antibiotics. At the same time, bacteria and other pathogens are always evolving so as to increase their resistance to antibiotic treatments. The rising use of antibiotics, together with the reduction in the number of new antibiotic molecules, leads to an increase in the mortality rate due to several bacterial infections, such as pneumonia, tuberculosis, and gonorrhea [4–6]. A rapid and efficient antimicrobial susceptibility testing method is needed to properly direct the healthcare staff toward the most powerful antibiotic.

In this context, the plate-count method is well-established in detecting bacteria and studying their evolution under antibiotic treatments [7, 8]. Since its operation is based on the growth of bacteria on an agar plate and on their detection by fluorescence microscopy, this approach results time-consuming (24–72 h), and also it needs expert staff to interpret the results. These performance clashes with the need for fast antibiotic prediction to counteract the spread of hazardous bacterial infections, such as sepsis. Other techniques are based on the monitoring of bacteria metabolic activity, such as Electrochemical Impedance Spectroscopy [9] and impedance flow cytometry [10]. However, these techniques require a very high concentration of cells within the bacteria sample to ensure a notable impedance change. In order to overcome these limitations, several photonic devices and systems have been proposed [1, 11, 12], aiming at performing rapid

R. Berta and A. De Gloria (Eds.): ApplePies 2022, LNEE 1036, pp. 196–202, 2023.
https://doi.org/10.1007/978-3-031-30333-3_25

antimicrobial susceptibility testing (<1 h), also by using a small amount of bacteria concentration.

In order to fulfill the aforementioned requirements, the use of dielectrophoresis (DEP) has been investigated, aiming also to develop a chip-scaled device. The DEP is a phenomenon in which a force is exerted on a dielectric particle when it is hit by a non-uniform electric field, that polarizes particles with respect to the surrounding medium. If the polarizability of the particle is greater than the fluid one, the particle experiences positive DEP (p-DEP), flowing towards the area with the largest electric field. On the other hand, if the polarizability of the particles is smaller than the surrounding medium, the particle is affected by negative DEP (n-DEP) and it goes where the electric field is weaker [13–15]. Therefore, the DEP results are suitable for particle manipulation, by properly using the physical and electrical properties of cells and medium, in terms of conductivity and permittivity.

To rate the efficiency of the antibiotic under test, a DEP-based microfluidic channel has been designed by exploiting the different properties of live/dead bacteria, e.g. *Escherichia Coli*, under antibiotic treatment. In fact, since the cell membrane, which is normally impermeable and highly insulating, typically becomes permeable after cell death and attracts ions from the cytoplasm, with a resulting increase in the dead cell membrane conductivity [16, 17], the p-DEP and n-DEP have been used to sort live/dead bacteria with very high efficiency, making the proposed device as a promising tool for rapid antimicrobial susceptibility testing.

In literature, other DEP-based devices have been proposed to perform rapid antibiotic susceptibility testing [18, 19]. In particular, a quadrupole configuration has been proposed to measure antibiotic efficiency through the frequency response of p-DEP, with an overall operating time of 2 h and an applied voltage of 10 V_{pp} [18]. In addition, interdigitated DEP electrodes combined with an impedance analyzer have been proposed to measure the bacteria concentration [19]. As the trapped bacteria concentration increases, the impedance decreases, with a median efficiency of 95.9% and applied voltage of 15 V_{pp}. Finally, the proposed device results faster than competing technologies, with low power consumption and high efficiency, allowing a rapid and reliable analysis.

2 DEP-Based Microfluidic Channel

The proposed DEP-based device consists of a sidewall PDMS microfluidic channel, with a 1D pattern of metal electrodes on both channel edges, forming the so-called interdigitated castellated configuration (see Fig. 1). In order to preserve all physical and electrical properties of the target bacteria (*Escherichia Coli*), the *Luria Bertani broth* (LB broth) has been chosen as the medium. The LB broth proves to be isotonic and isosmotic with the bacteria under test. The non-uniform electric field is generated by driving the top/bottom electrodes with a voltage $+V_p/-V_p$.

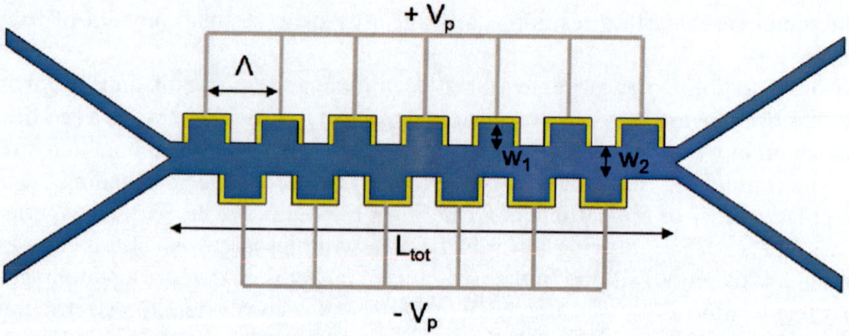

Fig. 1. Castellated DEP-based microfluidic channel (yellow regions: electrodes).

The whole device has been designed, in terms of applied voltage V_p, length L_{tot}, width of inner channel w_2, width/height of electrodes w_1, and their period Λ, aiming at maximizing the sorting efficiency and, at the same time, reducing the power consumption. In order to guarantee the vital lead time between antibiotic and bacteria samples (in the order of dozens of minutes), a PDMS coil is assumed at the device input port.

Since the proposed device aims to sort live/dead bacteria by using their different physical and electrical properties, the metabolic activity of *Escherichia Coli* under antibiotic treatment has been deepened. The live bacterium has been analyzed by using a three-shell spherical model, with a radius of 500 nm (Fig. 2(a)) [16], while, since its external membrane is destroyed by antibiotic, the dead bacterium has been modeled as a one-shell cell [17] (Fig. 2(b)). The properties of the live/dead bacteria and medium are summarized in Table 1. According to the Clausius-Mosotti theory [16, 17], the operating frequency has been set to 2 MHz, to ensure that live and dead bacteria are subjected to p-DEP and n-DEP forces, respectively.

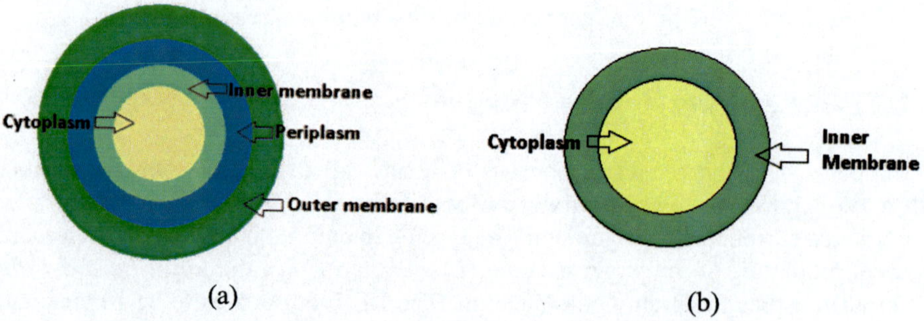

Fig. 2. Topological model of live (a) and dead (b) *E. Coli* (a) [16, 17].

Table 1. Live and dead E. coli properties at 2 MHz [1, 16, 21].

Property	Value @ 2 MHz
Density live (kg/m^3)	1116.6 [21]
Density dead (kg/m^3)	1184.3 [21]
Outer membrane dielectric constant	10 [16]
Inner membrane dielectric constant	5.5 [16]
Periplasm dielectric constant	60 [16]
Cytoplasm dielectric constant	108 [16]
Outer membrane conductivity (S/m)	10^{-6} [16]
Inner membrane conductivity (S/m)	10^{-12} [16]
Periplasm conductivity (S/m)	3.2 [16]
Cytoplasm conductivity (S/m)	0.22 [16]
Fluid dielectric constant	80 [16]
Fluid conductivity (S/m)	0.35 [16]
Cytoplasm dielectric constant dead EC	60 [1]
Cytoplasm conductivity dead EC (S/m)	0.09 [1]

3 Performance Evaluation

The device has been designed with transient analysis by using 3D *Finite Element Method* (FEM) approach, aiming at finding the best compromise between power consumption, large efficiency, and operating time. To reach these goals, $w_1 = 40$ µm, $w_2 = 5$ µm, $L_{tot} = 540$ µm, and $\Lambda = 80$ µm have been designed.

Figure 3 reports the trend of the efficiency η by varying the applied voltage $|V_p|$. The efficiency η is useful to rate the proposed device, expressed as the concentration ratio between the desired particles and the undesired ones at a single outlet. In particular, the device has been designed so that live (dead) bacteria, that are subjected to p-DEP (n-DEP), flow towards the upper (bottom) outlet.

The values in Fig. 3(a) have been calculated as the average of several simulation results, in order to take into account stochastic simulation effects. A voltage $|V_p|$ equal to 3.2 V ensures an average efficiency of 99.8%, with an operating time of about 4 s, making the proposed device suitable for rapid antimicrobial susceptibility testing.

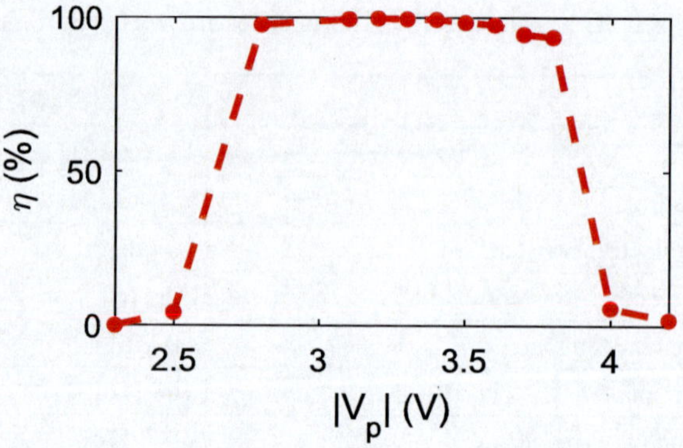

(a)

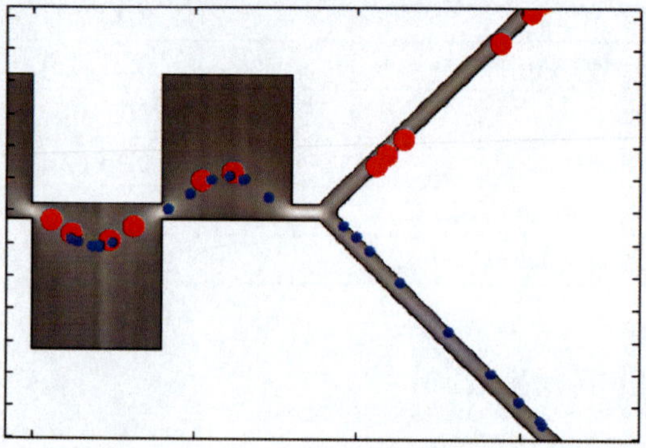

(b)

Fig. 3. (a) Efficiency η vs. applied voltage $|V_p|$ @ 2 MHz; (b) zoom at the microfluidic channel output ports for $|V_p| = 3.2$ V (red and blue dots represent live and dead bacteria, respectively).

4 Conclusions

Here, we have demonstrated the feasibility of dielectrophoresis to separate live/dead bacteria, and then, contrast the AMR phenomenon. A DEP-based interdigitated castellated microfluidic channel has been proposed to achieve a highly efficient live/dead bacteria sorting (>99%) with low power consumption (=6.4 V) and small operating time (≈4 s), making the proposed device suitable for rapid antimicrobial susceptibility testing, useful to support the healthcare staff tackling the AMR worldwide phenomenon. In order to make the device functional, it would be useful to add a feedback loop to reinject the particles in the inlet channel, with a resulting increase in the efficiency of the sorting logic.

References

1. Brunetti, G., Conteduca, D., Armenise, M.N., Ciminelli, C.: Novel micro-nano optoelectronic biosensor for label-free real-time biofilm monitoring. Biosensors **11**(10), 361 (2021)
2. Stewart, P.S., Costerton, J.W.: Antibiotic resistance of bacteria in biofilm. Lancet **358**, 135–138 (2001)
3. Andersson, D.I., Hughes, D.: Antibiotic resistance and its cost: is it possible to reverse resistance? Nat. Rev. Microbiol. **8**, 260–271 (2010)
4. Chang, H.H., Cohe, T., Grad, Y.H., Hanage, W.P., O'Brien, T.F., Lipsitch, M.: Origin and proligeration of multiple-drug resistance in bacterial pathogens. Microbiol. Mol. Biol. Rev. **79**, 101–116 (2015)
5. "World Health Statistics 2020, World Health Organization: Geneva, Switzerland, (2020)
6. "Antimicrobial Resistance: Tackling a Crisis for the Health and Wealth of Nations, London, UK (2014)
7. Arabski, M., et al.: The use of lysozyme modified with fluorescein for the detection of Gram-positive bacteria. Microbiol. Res. **170**, 242–247 (2015)
8. Dickson, R.P., et al.: Enrichment of the lung microbiome with gut bacteria in sepsis and the acute respiratory distress syndrome. Nat. Microbiol. **1**, 16113 (2016)
9. Delcour, A.H.: Outer membrane permeability and antibiotic resistance. Biochim. Biophys. Acta **1794**, 808–816 (2009)
10. David, F., Hebeisen, M., Schade, G., Franco-Lara, E., Di Berardino, M.: Viability and membrane potential analysis of bacillus megaterium cells by impedance flow cytometry. Biotechnol. Bioeng. **109**, 483–492 (2012)
11. Conteduca, D., Brunetti, G., Dell'Olio, F., Armenise, M.N., Krauss, T. F., Ciminelli, C.: Monitoring of individual bacteria using electro-photonic traps. Biomed. Optics Express **10**, 3463–3471 (2019)
12. Petrovszki, D., et al.: An integrated electro-optical biosensor system for rapid, low-cost detection of bacteria. Microelectron. Eng. **239–240**, 111523 (2021)
13. Kim, D., Sonker, M., Ros, A.: Dielectrophoresis: from molecular to micrometer-scale analytes. Anal. Chem. **91**, 277−295 (2019)
14. Rahman, N.A., Ibrahim, F., Yafouz, B.: Dielectrophoresis for biomedical sciences applications: a review. Sensors **17**, 449 (2017)
15. Pethig, R.: Review article—dielectrophoresis: status of the theory, technology, and applications. Biomicrofluidics **4**, 022811 (2010)
16. Bai, W., Zhao, K.S., Asami, K.: Dielectric properties of E. coli cell as simulated by the three-shell spheroidal model. Biophys. Chem. **122**, 136–142 (2006)
17. Subramanian, S., Tolstaya, E.I., Winkler, T., Bentley, W.E., Ghodssi, R.: An integrated microsystem for real-time detection and threshold-activated treatment of bacterial biofilms. ACS Appl. Mater. Interfaces **9**, 31362–31371 (2017)
18. Chung, C.-C., Cheng, F., Chen, H.-M., Kan, H.-C., Yang, W.-H., Chang, H.-C.: Screening of antibiotic susceptibility to β-lactam-induced elongation of gram-negative bacteria based on dielectrophoresis. Anal. Chem. **84**, 3347–3354 (2012)
19. Del Moral-Zamora, B., et al.: Combined dielectrophoretic and impedance system for on-chip controlled bacteria concentration: application to Escherichia coli. Electrophoresis **36**, 1405–1413 (2015)
20. Garcia, P.A., et al.: Intracranial nonthermal irreversible electroporation: in vivo analysis. J. Membr. Biol. **236**, 127–136 (2010)

21. Jen, C.P., Chen, T.W.: Selective trapping of live and dead mammalian cells using insulator-based dielectrophoresis within open-top microstructures. Biomed. Microdevices **11**, 597–607 (2009)
22. Castellarnau, M., Errachid, A., Madrid, C., Juarez, A., Samitier, J.: Dielectrophoresis as a tool to characterize and differentiate isogenic mutants of Escherichia coli. Biophys. J. **91**, 3937–3945 (2006)
23. Lewis, C.L., Craig, C.C., Senecal, A.G.: Mass and density measurements of live and dead gram-negative and gram-positive bacterial populations. Appl. Environ. Microbiol. **80**, 3622–3631 (2014)

Short Contributions

Soluble Mandrel Technology to Produce Parts in Composite Material for Formula 1

Jacopo Agnelli[1(✉)], David Benedetti[1], and Nicholas Fantuzzi[2]

[1] Carbon Dream S.p.A., Via Fausto Melotti, 16, 50028 Barberino Tavarnelle, FI, Italy
jacopo.agnelli@tiscali.it

[2] DICAM Department, University of Bologna, Viale del Risorgimento, 2, 40136 Bologna, Italy

Abstract. The composite production of parts where it is necessary to create hollow structures, such as -for example- air ducts and passages for wiring, does not pass through the traditional model-mold-piece, or direct mold-piece approach, mainly due to the complexity of the shapes to be made and therefore of the equipment to be engineered and produced. On the contrary, thanks to the modern 3D printing technology, or fast prototyping, it is possible to quickly generate the core of the object to be made, on which to proceed with the lamination.

1 Introduction

Since the volume of composite parts is constantly increasing in various sectors and in particular in the Formula 1, racing and aerospace sectors, the efforts to define and use more convenient and suitable processes for this purpose have increased significantly in the last years [1–9]. Recent trends aimed at reducing weight to replace traditional fillers (such as honeycomb and foam) in structural applications (sandwich), as well as the greater demand for parts with complex geometries and suitable for functional integration, have led to the definition of new technologies for the construction of hollow structures and related production processes, replacing complex molds broken down into many tooling blocks, which are difficult to make, preserve and maintain.

Soluble materials have been used for some years for the construction of particular cables, such as conduits, but also in structural elements such as crossbeams and beams, fittings and pressure vessels [10, 11]. Particularly complicated geometries with undercuts, variations in section and diameter or multiple branches require the use of soluble cores for their production. This type of innovative production process is suitable for use also for higher production volumes, including where a high level of automation is required (such as filament winding, automated fiber positioning, resin transfer molding (RTM) and processes pressing in a hot platen press).

New materials for the construction of the mandrels, resistant to pressure within certain limits, and with an adequate sealing system, allow to create soluble cores with adequate rigidity and surface quality, with reasonable washout (melting) performance, even on long lengths and sections narrow transverse. The innovative design methodologies allow a rapid definition of the geometries of the mandrels to be made, which can also be generated in high volumes, by means of semi-automatic and economical

R. Berta and A. De Gloria (Eds.): ApplePies 2022, LNEE 1036, pp. 205–209, 2023.
https://doi.org/10.1007/978-3-031-30333-3_26

production processes. Versatile shapes in terms of handling, such as workable blocks and 3D printed mandrels, make them suitable for use especially in the production of prototypes and small series [12–17].

1.1 Soluble Mandrels

Ancient lost wax casting has been used successfully for thousands of years and is still the choice for metal casting but also for some fiber-reinforced polymeric parts, i.e. composite materials, which is what will be covered in this paper. Soluble materials have been used for some years for the construction of particular cables, such as conduits, but also in structural elements such as crossbeams and beams, fittings and pressure vessels. Particularly complicated geometries with undercuts, variations in section and diameter or multiple branches require the use of soluble cores for their production.

This type of innovative production process is suitable for use also for higher production volumes, including where a high level of automation is required (such as filament winding, automated fiber positioning, resin transfer molding (RTM) and processes pressing in a hot platen press). New materials for the construction of the mandrels, resistant to pressure within certain limits, and with an adequate sealing system, allow to create soluble cores with adequate rigidity and surface quality, with reasonable washout (melting) performance, even on long lengths and sections narrow transverse.

The innovative design methodologies allow a rapid definition of the geometries of the mandrels to be made, which can also be generated in high volumes, by means of semi-automatic and economical production processes. Versatile shapes in terms of handling, such as workable blocks and 3D printed mandrels, make them suitable for use especially in the production of prototypes and small series.

1.2 Features

To fabricate hollow, jointless, one-part composite structures, you need mandrels with key features:

- suitable for the production processes of traditional composites, i.e. stable at the polymerization conditions in the autoclave (temperature and pressure)
- robust and light
- capable of having undercuts or changes in cross section
- easy to wash off with water-based solutions
- no requirement for temperature or pressurization of the washing medium
- no special disposal procedures
- no damage to the composite laminate
- less expensive, in series production, than solutions that involve the use of molds

The soluble mandrels save time, and therefore manpower, allowing the entire production surface to be used for the manufacture of composite parts, rather than for the manufacture of the mandrels themselves. This basic solution is suitable to produce mandrels in almost any conceivable shape. The limitations for manufacturing and use are mainly a combination of very thin (<8 mm), very long (>5 m) and very complicated shape. De-mouldability and sealing application remain key factors, regardless of whether manufacturing a mandrel is possible or not.

1.3 Applications

Typical applications are:

- hollow stiffening structures/sections for aircraft
- air/cable ducts for Formula 1 and aerospace
- high pressure tanks/vessels.
- manifolds
- air intakes
- other automotive (Fig. 1).

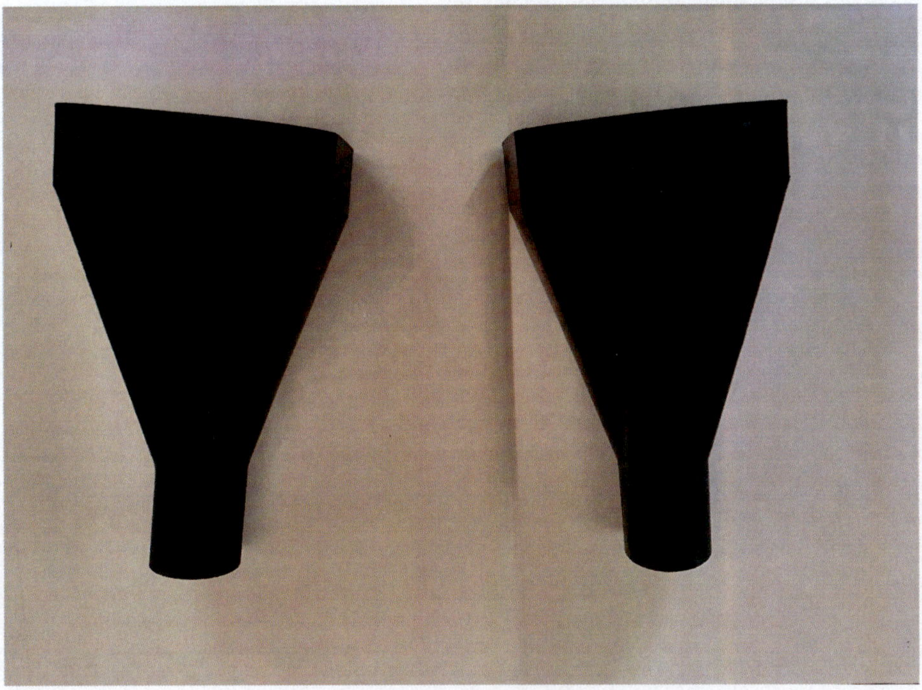

Fig. 1. One example of product of Carbon Dream spa

1.4 Prototypes and Small Series

In addition to the production processes of standard composites, in large volumes and in series, the soluble mandrels allow a great versatility in the realization of prototypes and small series. In the following paragraphs, applications not adopted by Carbon Dream, such as processes for high production volumes and other particular applications, will be omitted.

Main advantages:

- ease of access to the production technology of mandrels (3D printers)
- outsourcing of mandrel production
- versatility and speed in the implementation and updating of changes, such as the optimizations that are needed in prototyping
- rapid availability of the mandrels, compared to the production of traditional molds

Requirements:

- easy to use
- fast and flexible design modification
- realization without or at low cost of tools
- availability of CAD geometric models only
- confidentiality constraints to produce parts

Technologies available for making mandrels:

- pressing of versatile materials into boards
- CNC machined materials
- combination of machined blocks
- rapid prototyping

1.5 Rapid Prototyping

The most popular method for obtaining single or small series mandrels is via additive manufacturing. The finished mandrel is obtained directly from the CAD file through 3D printing with a wide range of performing materials, without tooling costs. However, special surface preparation is required to ensure a smooth finish by harmonizing the discrete intervals of material layer deposition in 3D printing. Tolerances in the order of ± 0.5 mm can be achieved. The tolerance and the achievable dimensions depend on the evolution and the processing volume of the printer used.

2 Conclusions

The use of soluble mandrels for hollow structures allows for the flexible and cost-effective production of customer-specific composite parts with complicated geometry. Suitable for prototypes, small series and high-volume production, different classes of soluble core materials and the evolution of geometries help to find the optimal and cost-effective part manufacturing solution. The requirements of modern, automated manufacturing methods and thermoplastic matrices can be met with further development of existing materials and techniques for lost core.

Acknowledgments. The R&D project S.MO.P.RA.S.CA.F1 "Developing master models rapid prototyping for super car and Formula 1" was funded by the Italian Ministry of Industry MISE with a public incentive.

References

1. Ding, Y., Lan, H., Hong, J., Wu, D.: An integrated manufacturing system for rapid tooling based on rapid prototyping Robot. Comput. Integr. Manuf. **20**(4), 281–288 (2004)
2. Chung, K.-C., Shu, M.-H., Wang, Y.-C., Huang, J.-C., Lau, E.M.: 3D printing technologies applied to the manufacturing of aircraft components. Mod. Phys. Lett. B **34**, 2040018 (2020)
3. Forcellese, A., Simoncini, M., Vita, A., Di Pompeo, V.: 3D printing and testing of composite isogrid structures. Int. J. Adv. Manuf. Technol. **109**(7–8), 1881–1893 (2020). https://doi.org/10.1007/s00170-020-05770-4
4. Türk, D.A., Triebe, L., Meboldt, M.: Combining additive manufacturing with advanced composites for highly integrated robotic structures. Procedia CIRP **50**, 402–407 (2016)
5. Tosto, C., et al.: Additive manufacturing of plastics: an efficient approach for composite tooling. Macromol. Symp. **389**, 1900069 (2020). https://doi.org/10.1002/masy.201900069
6. Wang, Y.-C., Chen, T., Yeh, Y.-L.: Advanced 3D printing technologies for the aircraft industry: a fuzzy systematic approach for assessing the critical factors. Int. J. Adv. Manuf. Technol. **105**(10), 4059–4069 (2018). https://doi.org/10.1007/s00170-018-1927-8
7. Parandoush, P., Lin, D.: A review on additive manufacturing of polymer-fiber composites. Compos. Struct. **182**, 36–53 (2017). ISSN 0263-8223
8. Agnelli, J., Benedetti, D., Fantuzzi, N., Saponara, S.: The exploitation of sustainable composite materials for the manufacturing of high-efficient electric cars. In: Saponara, S., De Gloria, A. (eds.) ApplePies 2021. LNEE, vol. 866, pp. 300–309. Springer, Cham (2022). https://doi.org/10.1007/978-3-030-95498-7_42
9. Fantuzzi, N., Bacciocchi, M., Benedetti, D., Agnelli, J.: The use of sustainable composites for the manufacturing of electric cars. Compos. Part C Open Access **4**, 100096 (2021)
10. Jing, X., Chen, S., An, J., Zhang, C., Xie, F.: Thermoplastic Mandrel for manufacturing composite components with complex structure. Aerospace (2021). mdpi.com
11. Lombardi, J.L., Vaidyanathan, K., Artz, G., Gillespie, J., Yarlagadda, S.: A water soluble mandrel material for fabricating complex polymer composite components. In: Proceedings of the SAMPE 2001, Long Beach, CA, USA, 6–10 May 2001, vol. 46, pp. 1316–1319 (2001)
12. Ngo, T.D., Kashani, A., Imbalzano, G., Nguyen, K.T.Q., Hui, D.: Additive manufacturing (3D printing): a review of materials, methods, applications and challenges. Compos. Part B Eng. **143**, 172–196 (2018). ISSN 1359-8368
13. Wang, Y., Zhou, Y., Lin, L., Corker, J., Fan, M.: Overview of 3D additive manufacturing (AM) and corresponding AM composites. Compos. Part A Appl. Sci. Manuf. **139**, 106114 (2020). ISSN 1359-835X
14. Love, L.J., et al.: The importance of carbon fiber to polymer additive manufacturing. J. Mater. Res. **29**(17), 1893–1898 (2014). https://doi.org/10.1557/jmr.2014.212
15. Black, S.: A growing trend: 3D printing of aerospace tooling. Composites World (2015)
16. Schniepp, T.: Design guide development for additive manufacturing (FDM) of composite tooling. In: SAMPE Conference Proceedings, Long Beach, CA, 23–26 May 2016, pp. 2259–2269 (2016)
17. Türk, D.A., et al.: Additive manufacturing with composites for integrated aircraft structures. In: SAMPE Conference Proceedings, Long Beach, CA, 23–26 May 2016, pp. 1404–1418 (2016)

A Reconfigurable 2D-Convolution Accelerator for DNNs Quantized with Mixed-Precision

Luca Urbinati(✉) and Mario R. Casu(✉)

Department of Electronics and Telecommunications,
Politecnico di Torino, Turin, Italy
luca.urbinati@polito.it, mario.casu@polito.it

Abstract. Mixed-precision uses in each layer of a Deep Neural Network the minimum bit-width that preserves accuracy. In this context, our new Reconfigurable 2D-Convolution Module (RCM) computes $N=1$, 2 or 4 Multiply-and-Accumulate operations in parallel with configurable precision from 1 to $16/N$ bits. Our design-space exploration via high-level synthesis obtains the best points in the latency vs area space, varying the size of the tensor tile handled by our RCM and its parallelism. A comparison with a non-configurable module on a 28-nm technology shows many reconfigurable Pareto points for low bit-width configurations, making our RCM a promising mixed-precision accelerator for inference.

Keywords: 2D-Convolution · Reconfigurable Accelerator · Mixed-Precision

1 Introduction

Low bit-width quantization is used in Deep Neural Networks (DNNs) to satisfy memory and latency constraints of embedded edge devices. Mixed-precision aims to quantize each DNN layer with the minimum bit-width [1] that preserves accuracy [2]. DNN accelerators started to support multiple precisions. For example, UNPU [3] uses serial multipliers and supports from 1 to 16 bits; DNPU [4] uses look-up table-based reconfigurable multipliers that support 4-/8-/16-bit multiplication; Bit Fusion [5] supports multiple input/weight pairs bit precisions (8/2, 4/4, 2/8 and 8/8) by composing and decomposing 2-bit multipliers.

In this context, our new Reconfigurable 2D-Convolution Module (RCM) uses Multiply-and-Accumulate (MAC) units with Sum Together (ST) multipliers [6] to process in one shot N (activations, weights) pairs with up to $16/N$ bits, where N is 1, 2 or 4. Thus, N–1 MAC operations are saved compared to a non-configurable 16-bit multiplier. We define the supported configurations as 16x, 8x and 4x. The same precisions are used by both Envision [7] and our previous Reconfigurable Depth-wise Convolution accelerator [8]. However, Envision's Sum Separate (SS) multiplier requires an external partial product addition. Moreover, our RCM supports a variable kernel size to satisfy many different convolution layers, from the Point-wise layers of MobileNets to the first layer of ResNetV1.

R. Berta and A. De Gloria (Eds.): ApplePies 2022, LNEE 1036, pp. 210–215, 2023.
https://doi.org/10.1007/978-3-031-30333-3_27

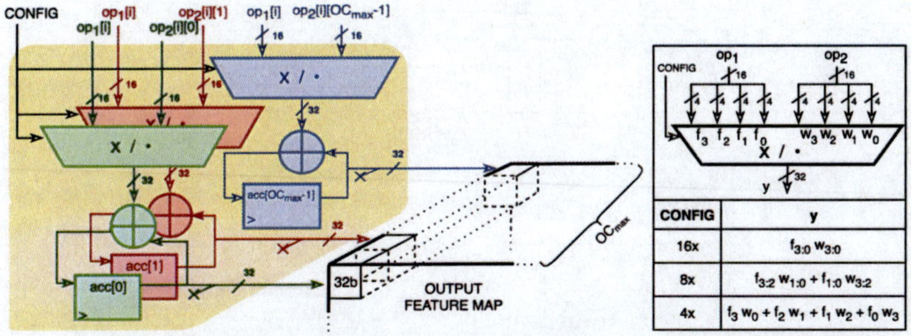

Fig. 1. RCM MAC Unit array with reconfigurable multipliers.

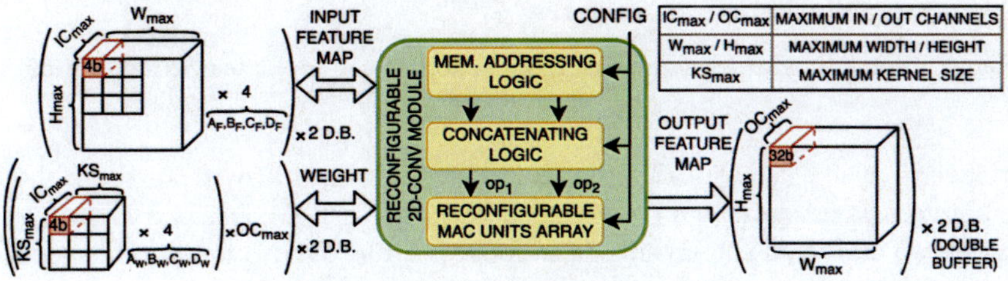

Fig. 2. Overview of the RCM.

We report the results of a Design-Space Exploration (DSE) of Latency vs Area. We quickly explored sixty-four RCM variants with High-Level Synthesis (HLS), varying the maximum number of supported input and output channels, and compared them with the corresponding non-configurable Standard 2D-Convolution Module (SCM) designs in a CMOS FDSOI 28-nm technology.

2 Hardware Architecture

The RCM's MAC Unit array, shown in Fig. 1, has as many ST reconfigurable multipliers as the maximum number of output channels that the RCM can process in parallel, $\mathrm{OC_{max}}$. Each MAC receives op_1 and op_2 16-bit operands from the input and weight buffers, respectively, unpacks them in four 4-bit values to match the internal reconfigurable multiplier arrangement, and accumulates partial results in a register. The table in Fig. 1 shows the three operations done by one reconfigurable multiplier according to the CONFIG signal. When a convolution between a kernel and an input receptive field is completed, the accumulated result is cast to 32-bit and stored in an output buffer for successive computations.

Figure 2 is an overview of the RCM, which includes a memory with double buffers made of four 4-bit SRAMs for input features and weights, each with size $(\mathrm{W_{max}} \times \mathrm{H_{max}} \times \mathrm{IC_{max}})$ and $(\mathrm{KS_{max}}^2 \times \mathrm{IC_{max}} \times \mathrm{OC_{max}})$, respectively. We refer to these as A_F, B_F, C_F and D_F for features, and A_W, B_W, C_W and D_W for weights. The output memory is a single 32-bit SRAM of size $(\mathrm{W_{max}} \times \mathrm{H_{max}} \times \mathrm{OC_{max}})$.

The size of the feature-map and weight tensors of a layer can exceed the size of the memory buffers of our RCM, which requires to iterate over multiple tiles.

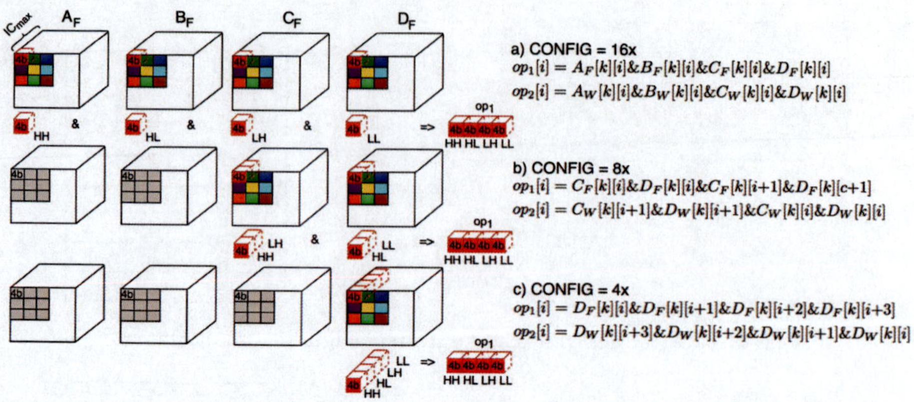

Fig. 3. Memory addressing and concatenation of the input feature-map buffer.

Therefore, a wise selection of the buffers size is essential to strike the right balance between latency, which decreases with larger buffer size, and area. Toward this goal, we analyzed the layers of the most popular DNNs for classification and object detection supported by multiple commercial edge devices and their development platforms. These include ResNetV1/V2, MobileNetV1/V2 (and the SSD and SSD-Lite versions), YOLO-V2/V3/V4/Tiny, and EfficientNet-B0. Based on our survey, we set $\mathrm{KS}_{\mathrm{max}} = 7$, because some networks require a 7×7 kernel (e.g., ResNetV1), and $\mathrm{W}_{\mathrm{max}} = 18$ ($= \mathrm{H}_{\mathrm{max}}$), to limit the RCM iterations when computing large layers and to limit the buffers area. For the parameters $\mathrm{IC}_{\mathrm{max}}$ and $\mathrm{OC}_{\mathrm{max}}$, we performed the DSE outlined in Sect. 3.

The RCM memories will be filled by an embedded processor or DMA engine as follows. In the 16x case, the input feature and weight tiles are split into 4-bit chunks and stored from the most to the least significant into A_F-D_F and A_W-D_W, respectively. In the 8x case, the *two* 4-bit chunks and stored in C_F-D_F and C_W-D_W. Finally, in the 4x case each element is stored in D_F and D_W.

Our RCM requires a particular memory addressing and concatenating logic to feed all the MAC units in parallel. Let us refer to the toy example in Fig. 3. Here, filters with shape $3 \times 3 \times \mathrm{IC}_{\mathrm{max}}$ create a 3×3 receptive field on a feature-map tile. Figure 3 shows how a tile is read to obtain op_1. Similarly, a weight tile is read to get op_2. The example refers to only one of the filters and one MAC unit out of $\mathrm{OC}_{\mathrm{max}}$, but it applies of course to all filters and MAC units with different weights. Index $i \in \{0, \ldots, \mathrm{IC}_{\mathrm{max}} - 1\}$ denotes the input channel, and index $k \in \{0, \ldots, \mathrm{KS} \times \mathrm{KS} - 1\}$ the receptive field. The three configurations use a different addressing scheme as indicated in the same figure.

In general, the number of MAC cycles to get $\mathrm{OC}_{\mathrm{max}}$ output *pixels* is $\mathrm{IC}_{\mathrm{max}}/N$. The theoretical speedup achievable by the RCM would be N, but due to the control logic overhead the actual speedup $s(N)$ is lower than N, as shown in Fig. 4. OW ($\leqslant \mathrm{W}_{\mathrm{max}}$), IC ($\leqslant \mathrm{IC}_{\mathrm{max}}$) and KS ($\leqslant \mathrm{KS}_{\mathrm{max}}$) are three run-time, tile-dependent configuration parameters that correspond to the output width, the number of input channels and the kernel size of the tiles processed by the RCM, respectively; $\mathrm{IC}/N \times \mathrm{KS}^2$ are the useful clock cycles, while $o_1 = 2$ and $o_2 = 5$

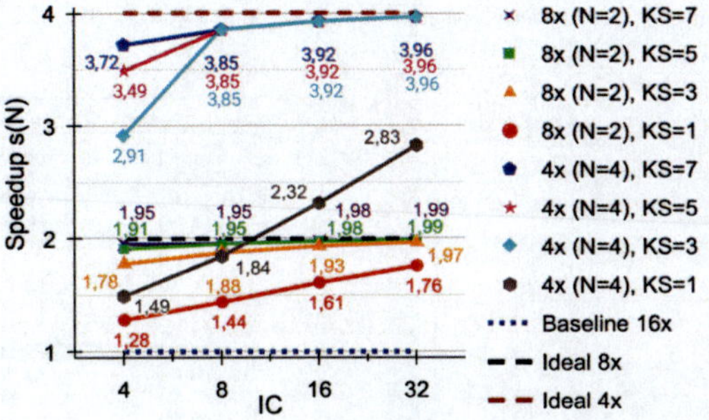

$$s(N) \simeq \frac{o_1 + \mathrm{OW} * (o_2 + \mathrm{IC} * \mathrm{KS}^2)}{o_1 + \mathrm{OW} * (o_2 + \mathrm{IC}/N * \mathrm{KS}^2)}$$

IC	INPUT CHANNELS
KS	KERNEL SIZE
N	PARALLEL DOT-PRODUCTS
OW	OUTPUT WIDTH
o_1, o_2	CONTROL LOGIC OVERHEADS

Fig. 4. RCM speedup for OW = 18.

are those responsible for the control logic overhead. The curves in Fig. 4 show that the speedup tends to saturate as IC approaches 32. KS = 1 is the worst case because of the dominant contribution of the overheads.

3 Experimental Results

More than 90% of the analyzed 2D-convolution layers have input and output channels multiple of 4, 8, 16 and 32. Therefore, we performed a DSE using Catapult HLS by sweeping $\mathrm{IC}_{\mathrm{max}}$ and $\mathrm{OC}_{\mathrm{max}}$ in $\{4, 8, 16, 32\}$ and the operating clock frequency f_{clk} from 400 to 1000 MHz (200-MHz steps). To reduce latency, at the expense of area, we used the HLS unrolling directive applied to the output channels loop, and the memory interleave directive applied to the weight memories. We synthesized the RTL netlists generated by Catapult HLS with Synopsys Design Compiler. We compared our RCM with an SCM based on standard 16-bit multipliers, which extend the operands sign for low precision configurations.

We analyzed the performance of RCM and SCM over two different 2D-convolution layers. The first is the most frequent layer among the selected DNNs: $(16 \times 16 \times 256)$ as feature-map tensor (padding included) and $(3 \times 3 \times 256 \times 256)$ as weight tensor. The second is the last point-wise layer of MobileNetV1: $(7 \times 7 \times 1024)$ for inputs and $(1 \times 1 \times 1024 \times 1024)$ for weights. Since the results for both layers are similar, due to space limitations we report the results of the DSE of Latency vs Area for the first case only, in Fig. 5. The latency is the total number of clock cycles multiplied by the clock period. The table adjacent to each plot contains the sorted Pareto-points in ascending order of area and reports input and output channels for each point. From Fig. 5 we observe:

- In the 16x case (Fig. 5a), as expected, all Pareto point are of SCM type, because the RCM points suffer from the area overhead of a more complex memory addressing logic and of the ST multipliers. Thus, in the following we only consider the 8x (Fig. 5b) and 4x (Fig. 5c) cases.
- $\mathrm{OC}_{\mathrm{max}}$ significantly affects area and latency: as $\mathrm{OC}_{\mathrm{max}}$ increases, the size of weight and output memories increase, but the number of MAC units grow and more output channels can be computed in parallel. Indeed, all the points with larger area and lower latency have a high value of $\mathrm{OC}_{\mathrm{max}}$.

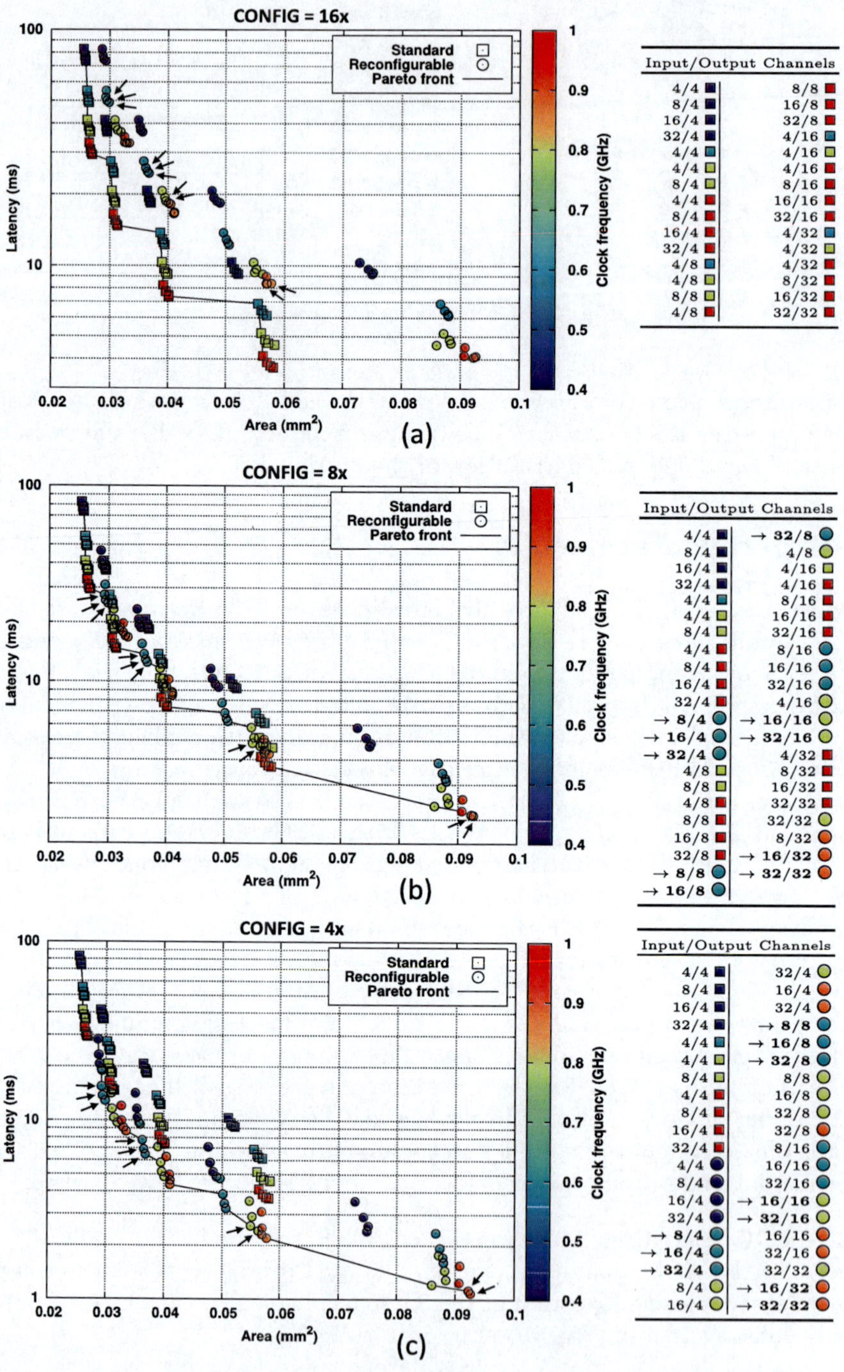

Fig. 5. Latency vs Area DSEs for CONFIG 16x (a), 8x (b) and 4x (c).

- For area $< 0.06\,\text{mm}^2$ increasing IC_{max} reduces latency more than increasing OC_{max} for a given area increase. However, to reduce latency below 2–3 ms (e.g., area $\geq 0.06\,\text{mm}^2$) OC_{max} must increase up to 32 and IC_{max} must saturate.
- The low frequency RCM solutions ($f_{clk} = 400\,\text{MHz}$) in the Pareto curve occur only for area $< 0.03\,\text{mm}^2$ in the 4x case. However, they are not worth when latency is the goal, as a higher clock frequency ($\geqslant 600\,\text{MHz}$) leads to a significant better latency for a marginal area increase.
- For 8x and 4x, 39% and 73% of Pareto points are reconfigurable, respectively.

Since 8-bit precision is enough for many DNNs and the trend is to go below 8-bit [9], a designer willing to use our RCM can choose among many Pareto points, as shown in Fig. 5. For example, those marked with an arrow ($\rightarrow$) are optimal in the 8x and 4x case, and close enough to optimal in the 16x case, making them suitable to DNNs requiring variable precision in their layers.

4 Conclusion

We presented a Reconfigurable 2D-Convolution Module for Heterogeneously Quantized DNNs synthesized in a CMOS FDSOI 28-nm technology from a high-level description. The results of the design-space exploration show many Pareto points, especially for low-precision configurations, which dominate the non-reconfigurable counterparts. In the future, we plan to combine this accelerator and our previous Reconfigurable Depth-wise Module [8] into an SoC, hence providing a complete solution to accelerate mixed-precision DNNs in hardware.

References

1. Anwar, S., et al.: Fixed point optimization of deep convolutional neural networks for object recognition. In: IEEE ICASSP (2015)
2. Vasquez, K., et al.: Activation density based mixed-precision quantization for energy efficient neural networks (2021). https://doi.org/10.23919/DATE51398.2021.9474031
3. Lee, J., et al.: UNPU: an energy-efficient deep neural network accelerator with fully variable weight bit precision. IEEE J. Solid-State Circ. **54**(1), 173–185 (2019)
4. Shin, D., et al.: 14.2 DNPU: an 8.1TOPS/W reconfigurable CNN-RNN processor for general-purpose deep neural networks. In: IEEE ISSCC (2017)
5. Sharma, H., et al.: Bit fusion: bit-level dynamically composable architecture for accelerating deep neural network. In: IEEE ISCA (2018)
6. Mei, L., et al.: Sub-word parallel precision-scalable MAC engines for efficient embedded DNN inference. In: IEEE AICAS (2019)
7. Moons, B., et al.: 14.5 envision: a 0.26-to-10TOPS/W subword-parallel dynamic-voltage-accuracy-frequency-scalable convolutional neural network processor in 28 nm FDSOI. In: IEEE ISSCC (2017)
8. Urbinati, L., Casu, M.R.: A Reconfigurable depth-wise convolution module for heterogeneously quantized DNNs. In: IEEE ISCAS (2022)
9. Choi, J., et al.: Accurate and efficient 2-bit quantized neural networks. In: MLSys (2019)

Check for updates

Diagnostic Analytics for Pixelated Particle Detectors: A Case Study

Werner Florian Samayoa[1,2,3](✉), Bruno Valinoti[1,2,3], Romina Molina[1,2,3], Luis G. García[1,2,3], Maria Liz Crespo[1,3], Sergio Carrato[2], Andres Cicuttin[1,3], and Stefano Levorato[3]

[1] MLAB, The Abdus Salam International Centre for Theoretical Physics, Strada Costiera, 11, 34151 Trieste, TS, Italy
{wflorian,bvalinot,rmolina,lgarcia1,mcrespo,cicuttin}@ictp.it

[2] Dipartimento di Ingegneria e Architettura, Università degli Studi di Trieste, v. Valerio 6/1, 34127 Trieste, TS, Italy
carrato@units.it

[3] Sezione di Trieste, Istituto Nazionale di Fisica Nucleare, Trieste, Italy
stefano.levorato@ts.infn.it

Abstract. We present a method for diagnostics analysis for pixelated particle detectors. The method is based on extracting information from the detector in the form of model parameters by using a representative mathematical model. To illustrate the procedure we analyzed real experimental data obtained with the electromagnetic calorimeter ECAL2 of the COMPASS experiment at CERN. Having observed the data, the typical pulses were fitted with a mathematical model. Heat maps were drawn to visualize the distribution of the mean values of each of the fitted parameters. This data visualization technique is useful for highlighting areas with similar behavior and detecting abnormal responses in single cells.

Keywords: Diagnostic Analytics · Pixelated Detector · Modeling · Instrumentation · Data Acquisition Systems

1 Introduction

Most particle physics experiments are based on a combination of two main types of particle detector: trackers and calorimeters. Each provides information on the position, time, or energy of the particles [1]. Because the quality of this information is crucial for correctly reconstructing the event that produced the detected particles, there is a need to design and deploy detectors that are more precise and accurate. For pixelated detectors, this implies an increase in the number of individual channels and front-end electronics with higher operating frequencies. In the case of calorimeters, high-energy resolution is also desired. This can only be achieved by using low-noise electronics with large dynamic ranges. Given that it is not feasible to store all the data generated by the detectors, a rigorous criterion must be applied to select the data worth storing. The solution to this problem is based on the concept of trigger and event [2].

R. Berta and A. De Gloria (Eds.): ApplePies 2022, LNEE 1036, pp. 216–221, 2023.
https://doi.org/10.1007/978-3-031-30333-3_28

All the captured data are useful if the detectors undergo proper calibration and maintenance. During calibration, configurable parameters are set to accurately represent the measured physical phenomena [3]. It is in the best interest of users that the calibration remains valid for as long as possible, to avoid incorrect measurements or detector malfunctions. Various monitoring and diagnosis approaches can be implemented depending on the available data and nature of the detector. The drifts of the detector from the calibration can be tracked using a mathematical model that describes the output pulse of the detector. The parameters of such model offer indirect measurements of different aspects of the detector. By studying how each parameter changes, it is possible to uncover degradation in different parts of the detector and inhomogeneities. Experts can then study this in depth to adequately address these issues.

In this article, we present a case study based on the pixelated electromagnetic calorimeter ECAL-2 of the COMPASS experiment [4] to illustrate the diagnostic analytics method derived from mathematical modeling and data analysis.

2 ECAL-2 Calorimeter

The detector consists of a matrix of 64×48 independent cells covering an area of $2.44 \times 1.83\,\mathrm{m}^2$ [5]. Each cell consists of scintillating material coupled with a photomultiplier tube (PMT). Three detector cells with different radiation hardness and energy resolution were used to construct the calorimeter. The outer part is made up of 1,332 TF1 lead glass scintillators called GAMS [6]. In the middle, 848 radiation-hardened GAMS are implemented (GAMS-R) [7]. Finally, the core is composed of 888 Shashlik modules. The light guides inside each of the cells direct the photons to FEU-84-3 photomultipliers. The distribution of the three types of modules depends on the amount of radiation that each module can hold. Taking into account the intensity of the hadron beam and the duty cycle, Shashlik modules are expected to stand for nearly 20 years [5], while GAMS and GAMS-R stand for several years without significant degradation. To accommodate the passing of non-interactive beam particles there is a hole of 2×2 in the central part of the detector.

Photons from a detected particle are collected by a PMT that generates a small current pulse proportional to the energy of the particle and a small constant dark current [8]. The pulse is integrated, filtered, and amplified by a custom analog shaper. The output of the shaper is then digitized on the MSADC (Mezzanine Sampling ADC [9]) by two interleaved 12-bit 40 Msps ADCs resulting in 80 Msps effectively.

A trigger control system composed of several detectors performs the filtering and selection of the events based on the trajectory, momentum, and polarization of the particles. Each time the detected event meets the criteria, a trigger signal is generated and transmitted to the calorimeter front-end electronics. For every trigger, a 32-sample segment is acquired and stored for offline analysis. Each segment contains a pulse that corresponds to the detection of a particle.

2.1 Calibration, Alignment, and Monitoring

To calibrate the detector, a 40 GeV electron beam is used to directly stimulate every channel. When the amplitude of the detected photons is taken into account, the PMT high voltage is adjusted. The calibration is monitored between spills using an LED system that simultaneously stimulates the channels through optical fibers. This process is useful to correct for drifts in the energy of individual channels and is essential for the operation of the detector. However, we propose a different method that allows one to extract additional information.

3 Diagnostic Analytics

We preset a method that focuses on the extraction of physically relevant parameters from the shape of the pulses. Using a mathematical model and curve fitting functions it is possible to obtain the values of the parameters that describe the correctly calibrated pulse. These parameters are related to the physical aspects of the detector. By studying the parameters and their variations, it is possible to detect drifts in the behavior of the channels beyond the energy. To illustrate this procedure, a raw data set from the COMPASS 2009 hadron run [4] was used for the analysis and characterization of the detector.

3.1 Pulse Modeling

A data set containing around 1.2 million segments corresponding to triggered events throughout the calorimeter was selected.

Once the typical pulse is identified, a suitable model and the correct value of the parameters need to be determined. In this case, the model corresponds to a second order semi-Gaussian analog filter [10]. Considering that the experimental signals are the superposition of an ideal signal with a secondary smaller pulse plus noise, we can assume that the residuals of individual fittings represent the actual noise in the trace. Taking all this into account, the model is defined as follows.

$$F(t, t_0, \tau, \beta, a, k, t_1) = \begin{cases} \beta, & t < t_0 \\ \beta + a\left(\frac{e(t-t_0)}{2\tau}\right)^2 e^{-\frac{(t-t_0)}{\tau}}, & t_0 \leq t < t_0 + t_1 \\ \beta + a\left(\frac{e(t-t_0)}{2\tau}\right)^2 e^{-\frac{(t-t_0)}{\tau}} + \\ ka\left(\frac{e(t-t_0-t_1)}{2\tau}\right)^2 e^{-\frac{(t-t_0-t_1)}{\tau}}, & t_0 + t_1 \leq t \end{cases} \tag{1}$$

where β accounts for the constant offset due to the dark current of the PMT and other factors. The amplitude is denoted by the parameter a, the arrival time by t_0, and the exponential time by τ. In addition, the parameter k defines the relative amplitude of the secondary pulse with respect to the main pulse. Lastly, t_1 defines the secondary pulse arrival time as a delay relative to t_0.

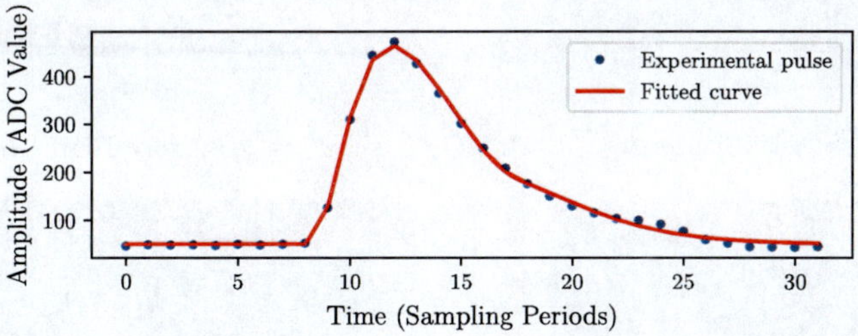

Fig. 1. Typical experimental pulse fitted with the exponential model

The result of the fit of the mathematical model is shown in Fig. 1. From this plot, it can be seen that the model is good enough to describe the pulse, which is essential; otherwise, the parameter values would not accurately describe the physical aspects of the detector.

4 Results and Discussion

Using the defined model, all pulses from the experimental data set are fitted. Heat maps are produced by plotting the average value of each parameter for each channel to study the inhomogeneities of the instrument. Color gradients and discontinuities can indicate defects or anomalies related to a specific aspect of the instrument. Figures 2 show several heat maps corresponding to parameters a, t_0, tau, β, k, and t_1.

As previously mentioned, the detector has a monitoring system to correct drifts from the energy calibration. This method only considers the energy of the pulses, leaving out important aspects that can affect the operation of the detector. Some of these aspects are brought to light through the model parameters.

From the heat maps, it is clear that the calorimeter cannot be treated as a homogeneous detector. Furthermore, all model parameters are associated with a specific physical aspect of the calorimeter and its front-end electronics. Consequently, anomalies or discontinuities in the heat maps provide valuable information not only for diagnostic purposes but also for possible optimizations or refinements of the instrument.

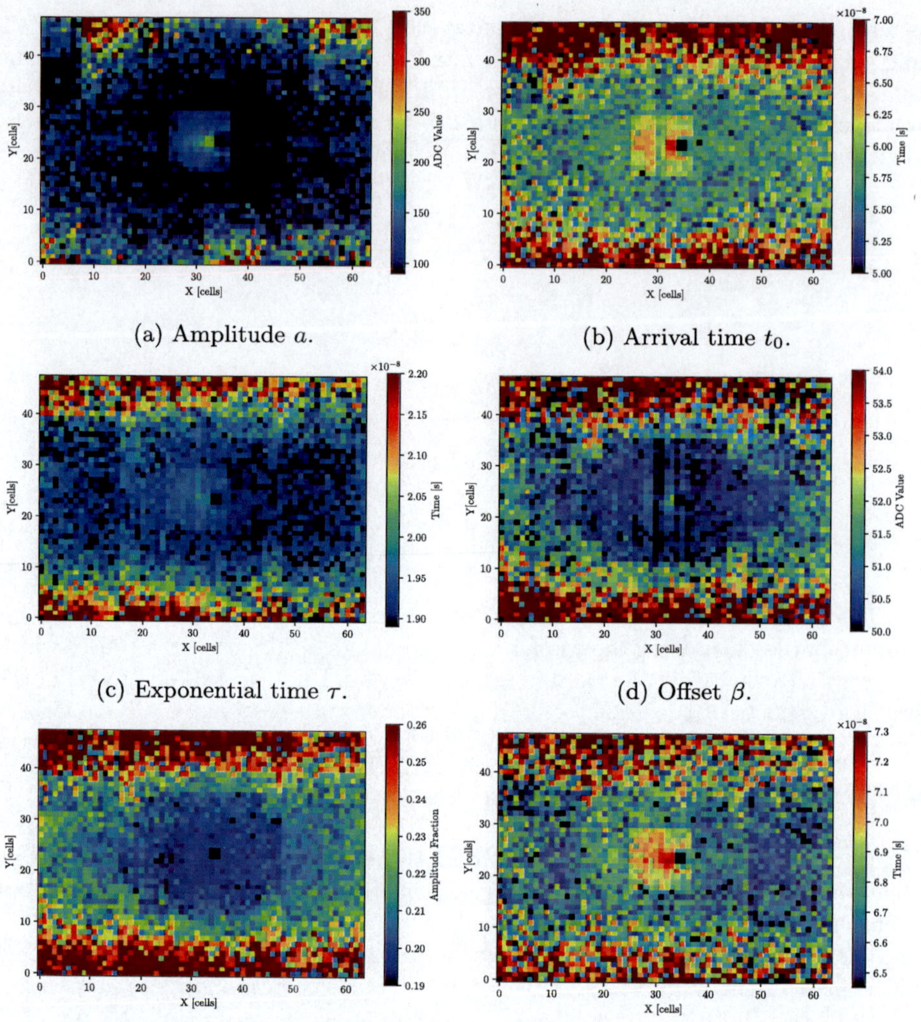

(a) Amplitude a.

(b) Arrival time t_0.

(c) Exponential time τ.

(d) Offset β.

(e) Secondary pulse amplitude fraction k.

(f) Secondary pulse delay time t_1.

Fig. 2. Heat maps of the calorimeter displaying the mean value of each parameter.

5 Conclusions

Modern experiments have called for bigger and faster detectors. In the case of pixelated detectors, an increase in size means more independent channels and faster front-end electronics. Having thousands of independent detectors poses a great challenge for calibration, monitoring, and maintenance.

With a reliable monitoring and maintenance method, the health and calibration of the detector can be preserved for longer periods. The ECAL-2 calorimeter in COMPASS already counts with a monitoring system that tracks the energy of each channel. We present a method of diagnostic analysis that allows for extracting more information by means of a mathematical model of the expected pulses. Given that the model parameters are related to the physical aspects of the detector, studying how each parameter changes helps to understand which components are deteriorating. By drawing heat maps, it is possible to evidence inhomogeneities in the detector and explore spatially how each parameter varies.

Since the calorimeter cannot be considered a complete uniform detector, a localized analysis will give more significant values for determined regions. These regions can be drawn by following the sharp edges present on each of the parameters, allowing for an intermediate level of abstraction. Taking into account these regions and the significance of each parameter, experts could make more informed decisions when diagnosing a malfunction in specific channels or regions.

References

1. Abbon, P., Albrecht, E., et al.: The compass experiment at CERN. Nucl. Instrum. Methods Phys. Res. Sect. A **577**(3), 455–518 (2007). https://www.sciencedirect.com/science/article/pii/S0168900207005001
2. Huber, S., Friedrich, J., Ketzer, B., et al.: A digital trigger for the electromagnetic calorimeter at the COMPASS experiment. IEEE Trans. Nucl. Sci. **58**, 1719–1722 (2011). https://cds.cern.ch/record/1605088
3. Abbon, P., Adolph, C., et al.: The compass setup for physics with hadron beams. Nucl. Instrum. Methods Phys. Res. Sect. A **779**, 69–115 (2015). https://www.sciencedirect.com/science/article/pii/S0168900215000662
4. Austregesilo, A.: The compass hadron spectroscopy programme (2012)
5. Polyakov, V.: Radiation hard shashlik calorimeter for compass experiment, October 2010. http://goo.gl/4lJ8uG
6. Binon, F., Buyanov, V., et al.: Hodoscope multiphoton spectrometer GAMS-2000. Nucl. Instrum. Methods Phys. Res. Sect. A **248**(1), 86–102 (1986). https://www.sciencedirect.com/science/article/pii/0168900286905012
7. Kobayashi, M., Prokoshkin, Y., et al.: Radiation hardness of lead glasses TF1 and TF101. Nucl. Instrum. Methods Phys. Res. Sect. A **345**(1), 210–212 (1994). https://www.sciencedirect.com/science/article/pii/0168900294909903
8. Akopyan, M.V., Medved', S.A., Pishchal'nikov, Y., Skripachev, O.V.: Stand for testing photomultipliers FEU 84-3 (1988). http://inis.iaea.org/search/search.aspx?orig_q=RN:21028877
9. Mann, A., Konorov, I., Paul, S.: A versatile sampling ADC system for on-detector applications and the advanced TCA crate standard. In: 2007 15th IEEE-NPSS Real-Time Conference, pp. 1–5 (2007)
10. Fairstein, E.: Linear unipolar pulse-shaping networks: current technology. IEEE Trans. Nucl. Sci. **37**(2), 382–397 (1990)

Developing a Toolchain for Synthetic Driving Scenario Datasets

Marianna Cossu(✉), Riccardo Berta, Alessio Capello, Alessandro De Gloria, Luca Lazzaroni, and Francesco Bellotti

Department of Electrical, Electronic and Telecommunication Engineering (DITEN), University of Genoa, Via Opera Pia 11a, 16145 Genova, Italy
marianna.cossu@edu.unige.it, {berta,adg,franz}@elios.unige.it

Abstract. Formalization of driving scenarios is key to define the operational design domain (ODD) of Automated Driving Functions (ADF). Training machine learning (ML) requires huge datasets, that are costly to produce. We propose a toolchain to generate driving scenario video-clip datasets based on the state-of-the-art CarLA driving simulator engine. Scenarios are randomically generated based on a set of parametric features, that are specified by the user. The variability includes both environmental and scenario-specific aspects. As an initial experiment, we have generated a dataset with 200 samples for each one of the 6 implemented classes. The tool is able to achieve a generation rate of about 130 scenarios (7 s. long each) per hour. The tool includes a verification module, which checks the successful completion of each sample.

Keywords: Driving scenarios · driving scenarios generator · synthetic datasets · automated driving · CARLA driving simulator

1 Introduction

Introducing on-road automatic driving functions (ADF) requires that all possible situations are correctly managed in the targeted operational design domain (ODD). This is ever more frequently done by verifying the proper functioning in a set of logical driving scenarios, that abstract the behavior of vehicles and other road actors in various driving situations [1]. A driving scenario describes, in a defined time horizon, the maneuvers of multiple entities such as vehicles, pedestrians and other traffic players, and is typically formalized through formats being defined by industrial initiatives such as OpenSCENARIO [2].

Machine learning (ML) is playing a key role in the ADF's implementation for automotive control (e.g., [3–5]). In order to learn and understand high-level meaningful information from data, supervised ML systems need labeled datasets for the target application. High quality real-world sensor datasets are fundamentals but very expensive and need long time for the realization. At present, we are aware of one publicly available dataset, the Prevention Dataset, that is labeled with lane change scenarios [6]. This problem can be partially addressed using virtual simulations. Simulations can be particularly useful

R. Berta and A. De Gloria (Eds.): ApplePies 2022, LNEE 1036, pp. 222–228, 2023.
https://doi.org/10.1007/978-3-031-30333-3_29

for different research cases, they allow the users to have a great generation controllability and variability (e.g., for weather conditions, etc.), they can obviate the problem of recording dangerous scenarios like the ones involving accidents or corner cases (which are very expensive and dangerous to record in the real world). Synthetic datasets could be a good complement for training real-world automotive systems, by increasing the number and variety of samples of real-world datasets, particularly contributing to corner cases.

This paper presents the development of a toolchain for efficiently generating synthetic driving scenario datasets. The main requirements for this system, coming from our experience in industrial research projects (e.g., [7, 8]) include: ability to create various instances of a significant set of scenario types, ability to include an environment variability (e.g., different weather conditions, types of cars and towns) and ease of use.

As simulation environment, we have chosen Car Learning to Act (CarLA) an open-source simulator for autonomous driving research. The simulation platform also include useful digital assets (urban layouts, buildings, vehicles) and supports flexible specification of sensor suites and environmental conditions [9–11].

CarLA's scenario runner supports OpenSCENARIO v1, onto which the previous version of our toolchain was based [12]. In order to increase flexibility and efficiency in preparing high-variability scenarios, which is key for datasets, we decided to resort to more generic.xml files describing variable scenarios.

The remainder of the paper is organized as follows: Sect. 2 explains the user parameters specification and the architecture of the toolchain; Sect. 3 illustrates the obtained results, the difficulties encountered during the tool development and the limitations of the simulator used; Sect. 4 presents some key concluding remarks.

2 System Architecture

This section describes the system architecture to implement a workflow for generating synthetic datasets fulfilling requirements defined by users through a JSON configuration file. The tool consists of a series of two main blocks (Fig. 1). The first one is the Scenario Generator, which receives the scenario configuration parameters by the users and creates the corresponding descriptor instances. The descriptor instances are used by Scenario Runner's block, which runs the appropriate simulation using the CarLA's engine. Successful simulations are finally saved as.mp4 clips.

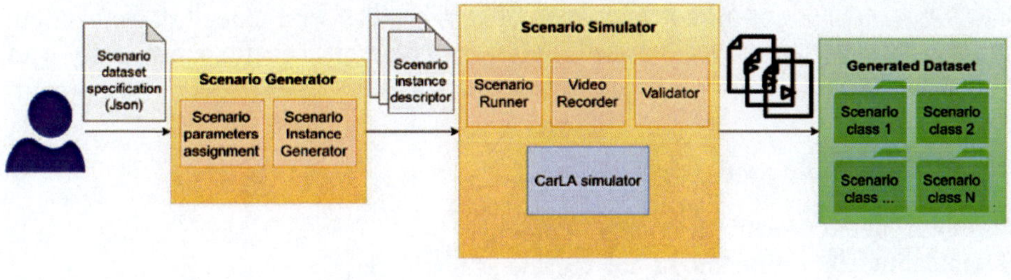

Fig. 1. Overall system architecture

2.1 User Parameter Description

A user can specify in a.JSON configuration file the desired parameters for the scenario instances to be generated. Parameters are single values or statistic parameters (mean and variance) out of which instances of scenarios will be generated. Table 1 reports the common parameters, such as the cities (CarLA provides 10 towns), type of vehicle considered (e.g., car, truck, etc. - 30 different vehicle models can be chosen), time of day (night, day, etc.) and weather (clear, foggy, etc.). CarLA provides 14 combinations of time and weather conditions (Fig. 2). For each parameter, the user can specify a range or one or more values and the system takes care of generating all the combinations. Table 2 reports the scenarios-specific parameters.

Table 1. Common scenario parameters

Parameter		Values	Description
Traffic	Vehicle types	e.g., truck, motor bike, etc.	Type of vehicles used during simulation
	Number of vehicles	Mean (μ), variance (σ^2), upper bound (UB) and lower bound (LB)	Number of traffic vehicles surrounding the main action
Environment	Weather	clear, cloudy, mid rainy fog, hard rain fog, etc.	8 different weather conditions
	Time of day	night, day... Sunrise, sunset	7 different time of day conditions
	Town	01, 02,... 10	CarLA map asset includes 10 different towns
Ego	Speed	μ, σ^2, UB and LB	Ego vehicle's speed

2.2 Scenario Generator

Scenario generator processes the user-defined configuration file for the given scenario type and builds the scenario instance descriptor accordingly. One.xml descriptor is automatically generated for each combination of city, weather, and time of day condition. The same.xml descriptor may be used for different types of scenario, consistently with the configuration files of these scenario types.

2.3 Scenario Simulator

Starting from the Scenario Runner tool provided by CarLA, we realized a tool capable of simulating eight different ad-hoc created scenario classes (cut-in front and behind, cut-out front and behind, lane change, following a lead vehicle, break, free ride) which can generate up to 13 different scenarios. These classes describe the behavior of each

Table 2. Scenario-specific parameters

Parameter		Values	Description
Cut	Time headway	μ, σ^2, UB, LB	Trigger value for the action
	Duration	μ, σ^2, UB, LB	Time to complete the cut
	Action	In, out, both	Execute cuts-in, cuts-out or both (random choice)
	Longitudinal	Behind, front, both	The cut is executed in front of the vehicle, behind, or both
	Lateral	Left, right, both	The cut is executed from left, cut from right or both
Lane change	Duration	μ, σ^2, UB, LB	Time to complete the cut
	Direction	Left, right, both	Left-to-right, right-to-left or both
Brake	Time reaction	μ, σ^2, UB, LB	Ego brake time reaction
	Start time headway	μ, σ^2, UB, LB	Trigger value for the action
	Brake intensity	range [0.2, 1]	Intensity of the leading vehicle brake
Follow	Headway	μ, σ^2, UB, LB	Distance between ego and leading vehicle
	Duration	μ, σ^2, UB, LB	Duration of the scenario
Free ride	Duration	μ, σ^2, UB, LB	Duration of the scenario

Fig. 2. Example of different weather conditions implemented in CarLA

actor in the scene and define the validation conditions to classify a scenario run as a success or a failure. Failure conditions may be a crash or an off-road. Only successful scenario instances are stored as clips in the dataset. The first step of the behaviour of each scenario instance is the spawning of the vehicles in random waypoints compatible with the current scenario class, followed by reaching the desired speed of the vehicles. For instance, lane change samples can be created only in three lane road areas.

3 Results and Discussion

The developed toolchain is being used in the generation of sample instance for each scenario types, using different configurations. Specular types (e.g., changing left to right) can be done very easily. Early results are promising and seem to confirm the validity of the approach. All the scenario types are generated. Particularly we have developed a small dataset (see Table 3) with 6 scenario types: brake front, cut-in front and behind and cut-out front and behind, lane change (Fig. 3). Some of the instances were automatically discarded because they completed incorrectly. We argue that this problem is essentially due to the fact that, in each simulation, the various vehicles, start from zero speed and have different pick-ups, which impedes an immediate correct control. To limit this problem, we add a "avoid collision" behavior to the vehicles to force them to brake when they are too close to the vehicle ahead. The really problematic scenario is the lane change, because the CarLA's dynamic path planning creates an unrealistic zigzag in curves, and we currently can detect this behavior only when vehicles go into a wrong lane. There is the need to update the Scenario Validator to detect all those movements and discard those clips. Significant issues have been overcome during the development. These include, for instance, the implementation of a realistic braking behaviour; the algorithm to find proper positions for spawning vehicles; the algorithm to manage the traffic behaviour which must be in the context (the user can statistically specify the number and type of vehicles in the surrounding) but should not disturb the execution of a specific scenario (each scenario class has its own traffic management module).

Table 3. Generator performances for 6 different driving scenario classes.

Driving scenario type	Successful samples	Deleted videos	Mean sample action duration	Generation time
Brake+reaction	200	12	3 s	1 h 30 min
Ego lane change	200	98	4 s	1 h 35 min
Cut out behind	200	48	3 s	1 h 20 min
Cut in front	200	63	3 s	1 h 50 min
Cut out front	200	75	3 s	1 h 30 min
Cut in behind	200	56	3 s	1 h 30 min

Fig. 3. Snapshots from the currently implemented scenarios

4 Conclusions and Future Work

We have presented a system to generate driving scenario video-clips datasets. Scenarios are generated parametrically as specified by the user. The variability includes both environmental and scenario-specific aspects. We are planning to develop a function that automatically checks the expected feasibility and warns the user about possible corrections of unfeasible scenario parameters. As an initial experiment, we have generated a 1200 sample (6 class) dataset. The tool is able to achieve a generation rate of about 130 scenarios (7 s. long each) per hour. The rate depends on several parameters, particularly the vehicle speeds, since the simulation starts from zero speed and it takes the vehicles some time to reach their target values.

In the future, we committed to generate scenario datasets based on statistically-defined parameter values. We expect that these synthetic samples will be useful to complement real-world datasets, especially by covering corner cases, that would be difficult to have otherwise. On the sensor side, we intend to exploit the CarLA's capabilities, particularly by integrating more cameras (of different types and placed in different positions), lidars, proximity sensors and radars. We also hope that CarLA will support OpenSCENARIO V2 which would allow better effectiveness and efficiency.

References

1. Weber, H., et al.: A framework for definition of logical scenarios for safety assurance of automated driving. Traffic Inj. Prev. **20**, S65–S70 (2019). https://doi.org/10.1080/15389588.2019.1630827
2. ASAM OpenSCENARIO. https://www.asam.net/standards/detail/openscenario/. Accessed 21 July 2022
3. Elallid, B.B., Benamar, N., Hafid, A.S., Rachidi, T., Mrani, N.: A comprehensive survey on the application of deep and reinforcement learning approaches in autonomous driving. J. King Saud Univ. – Comput. Inf. Sci. (2022). https://doi.org/10.1016/j.jksuci.2022.03.013
4. Korbmacher, R., Tordeux, A.: Review of pedestrian trajectory prediction methods: comparing deep learning and knowledge-based approaches (2021). http://arxiv.org/abs/2111.06740, https://doi.org/10.48550/arXiv.2111.06740
5. Kiran, B.R., Sobh, I., Talpaert, V., Mannion, P., Sallab, A.A.A., Yogamani, S., Pérez, P.: Deep reinforcement learning for autonomous driving: a survey (2021). http://arxiv.org/abs/2002.00444
6. Izquierdo, R., Quintanar, A., Parra, I., Fernández-Llorca, D., Sotelo, M.A.: The PREVENTION dataset: a novel benchmark for PREdiction of VEhicles iNTentIONs. In: 2019 IEEE Intelligent Transportation Systems Conference (ITSC), pp. 3114–3121 (2019). https://doi.org/10.1109/ITSC.2019.8917433
7. Bellotti, F., et al.: Managing big data for addressing research questions in a collaborative project on automated driving impact assessment. Sensors. **20**, 6773 (2020). https://doi.org/10.3390/s20236773
8. Cirimele, V., et al.: The fabric ICT platform for managing wireless dynamic charging road lanes. IEEE Trans. Veh. Technol. **69**, 2501–2512 (2020). https://doi.org/10.1109/TVT.2020.2968211
9. Team, C.: CARLA. http://CarLA.org/. Accessed 21 July 2022

10. Dosovitskiy, A., Ros, G., Codevilla, F., Lopez, A., Koltun, V.: CARLA: an open urban driving simulator (2017). http://arxiv.org/abs/1711.03938, https://doi.org/10.48550/arXiv.1711.03938
11. CARLA: Car Learning to Act — An Inside Out – ScienceDirect. https://www.sciencedirect.com/science/article/pii/S1877050921025552. Accessed 21 July 2022
12. Motta, J., et al.: Developing a synthetic dataset for driving scenarios. In: Saponara, S., De Gloria, A. (eds.) ApplePies 2021. LNEE, vol. 866, pp. 310–316. Springer, Cham (2022). https://doi.org/10.1007/978-3-030-95498-7_43

Ticketing Systems for Smart Public Transportation: Tools at the User Side

Antoni Martínez-Ballesté(✉), Nicolás Villalobos, Edgar Batista, Pablo López-Aguilar, and Agusti Solanas

School of Engineering, Universitat Rovira i Virgili, 43007 Tarragona, Spain
antoni.martinez@urv.cat

Abstract. Smart transportation systems are an integral part of the smart cities of tomorrow. With the proliferation of miniaturised sensors, IoT devices and 5G communication technologies, plenty of opportunities are yet to be developed to make transport systems more convenient, from the user side, and more cost-efficient and sustainable from the service providers side. Among the many actors involved in this domain, ticketing systems are paramount to access public transportation, such as trains, metros or buses. However, these systems must cope with a number of strong security and privacy requirements. This article overviews the current landscape of tools for a secure deployment of the user side of ticketing systems in public transportation.

Keywords: Smart transportation · Ticketing systems · Security · Lightweight cryptography · Smart card

1 Introduction

In today's society, the transportation of people and the problems it entails are gaining importance. Despite the effect of the COVID-19 pandemic on the popularisation of teleworking, commuting and its effects on large urban areas are still an issue. Apart from collapses and delays that workers may suffer due to the traffic or accidents, the mass movement of vehicles is a problem for the environment. Large cities and metropolitan areas are betting on public transportation (or mass transit) to the detriment of private automobiles. Moreover, metropolitan and regional governments are putting efforts into integrating transportation means into a multimodal transport system.

Ticketing systems are a linchpin in the success of mass transit. These consist of two parts: (i) the *user* side (*e.g.* the tickets for using the transportation system) and (ii) the *service* side (*i.e.* ticket machines, turnstiles...). At the user side, transportation tickets have evolved from printed cards (controlled by ticket inspectors) and magnetic stripe cards (typically read and written at turnstiles), to smart cards and smartphone apps (*i.e.* mobile ticketing), including the electronic purchase of tickets: users can currently buy tickets in advance through

R. Berta and A. De Gloria (Eds.): ApplePies 2022, LNEE 1036, pp. 229–234, 2023.
https://doi.org/10.1007/978-3-031-30333-3_30

websites or apps, import them into digital wallets or even print them. In state-of-art *smart ticketing* systems, user tickets are stored in the chip of a smart card, or in a smartphone app. Such systems clearly benefit the environment and do not suffer from read/write errors, as in the case of magnetic stripe cards. In addition, they naturally fit into multimodal transportation since, in general, allow passengers to seamlessly hop on and off buses, trains, bicycles, and the like. On the whole, smart ticketing aims at encouraging people to use public transportation because of its convenience: *i.e.* does away with the need for cash, can decrease the time it takes to board transport, and, all in all, play a key role in emerging models, *e.g.* Mobility as a Service (MaaS) [1].

1.1 Security, Privacy and Ticketing Systems

Fare collection is a crucial aspect of ticketing systems, which must fulfil some security requirements [4], namely: *integrity* (*i.e.* tickets shall not be manipulated and its verification should be possible by all parties), *unforgeability* (*i.e.* tickets can only be issued by authorised authorities), and *non-overspending.* In addition, ticketing systems must also cope with *fairness* (*i.e.* if a user presents a valid ticket, the service provider shall provide the service linked to that specific ticket), *portability*, *flexibility* (*i.e.* allow the use of tickets in different transport means within the same city) and *availability.*

Certainly, modern ticketing systems must be designed, developed and deployed considering security of systems, networks and the information. Moreover, the privacy of users (*i.e.*, location tracking, transportation habits...) must be tackled to mitigate the Big Brother effect. In order to achieve such deployments and the aforementioned properties, cryptographic tools are needed. Nevertheless, the constrained resources in some of the actors involved in the ticketing system must be considered.

1.2 Contribution and Plan of the Paper

In this paper we overview the techniques and tools currently available in the user side of smart ticketing systems. The rest of the paper is organised as follows. Section 2 introduces the actors involved: smart cards and cryptographic protocols, Sect. 3 reviews outstanding current proposals and, finally, Sect. 4 concludes the paper. Addressing a general approach to security (DDoS attacks, physical layer attacks, countermeasures, etc.) and privacy aspects is out of the scope of the paper.

2 Tools at the User Side

In this section we address the elements that play a key role in the user side of a smart ticketing system, *i.e.* fare collection and ticketing validation.

2.1 Smart Cards

Plain old bank cards had to be swiped through a POS (Point of Sale) so the data embedded in the magnetic stripe could be read. Later, chips were introduced and, currently, thanks to NFC (*Near Field Communication*, which evolved from RFID, *i.e.* radio-frequency identification), the physical contact between card and reader is not required. The use of smart card technology has expanded to a variety of fields beyond payments, *e.g.* transportation ticketing systems. *Contactless smart card* communications are ruled by the ISO/IEC 14443 standard which is divided into four main sections: physical characteristics; initialisation and anti-collision; transmission protocol; radio frequency power and signal interface. Regarding the latter, cards must operate at 13.56 MHz frequency and support communication range up to 10 cm. Two deployments under the aforementioned standard are recalled next:

- **MIFARE** is a proprietary technology owned by NXP Semiconductors. Their smart cards are based on ISO/IEC 14443. MIFARE Classic (launched in 1994) introduced a proprietary encryption algorithm and authentication protocol called Crypto1. MIFARE Ultralight and DESFire were developed to protect data at low cost. The latter was designed for multi-application smart card solutions in access, loyalty program, payments, as well as public transportation. After some security flaws were found [5], it was superseded by MIFARE DESFire EV1.
- **FeliCa** is a proprietary technology created by Sony, which has become the standard smart ticketing system in Japan. It also conforms to ISO/IEC 14443.

Smart card ecosystems' security is assessed following the ISO/IEC 15408 *Common Criteria for Information Technology Security Evaluation*, which assigns the smart card platform a *Evaluation Assurance Level* (EAL). MIFARE DESFire last version achieved EAL5+ level, whereas FeliCa achieved EAL6+, which means high levels of formal verification and testing. Manufacturers' websites publish documentation that focuses on the security of the communication between the reader and the card.

2.2 Smart Card Emulation

Since NFC technology is steadily being embedded into smartphones, these devices can communicate through this technology with other NFC devices, *e.g.* another smartphone to share files, a POS to make payments or ticket validators to collect fares. Initially, the secure data related to payments (*e.g.* the credit card information) had to be stored in the so-called *Secure Element* (SE) in the smartphone: a tamper-resistant microprocessor-based element, typically the SIM card or a secure chip.

As an alternative, *Host Card Emulation* (HCE) emulates the SE, allowing the smartphone to act as a smart card from the contactless reader perspective, without the presence of an actual smart card or SE. Telecom providers and smartphone manufacturers control the access to the SE chip, thus limiting what systems could make use of it. With HCE there is a change in the paradigm: it is open to integration with other applications such as storing transport passes and holding multiple cards and wallets. Both technologies allowed major companies in the smartphone market to enable payments using not only smartphones but also other devices like smart watches (*e.g.* Apple Pay, Samsung Pay, Google Pay). However, using HCE entails some security concerns [8] and, hence, the use of cryptography and secure communication protocols and techniques like *tokenisation* (in a nutshell, no real sensitive payment data but a surrogate value is stored on the smartphone) must be considered [2]. Nevertheless, in the public transportation arena, smartphone apps entirely replacing smart cards is quite unrealistic: not everyone has a smartphone and occasional users like tourists may be refrained from installing apps due to unexpected roaming costs.

2.3 Lightweight Cryptography

Traditional cryptographic algorithms were designed for desktop and server environments where processing power and energy consumption were not a concern. With the rise of the IoT and embedded systems, the necessity arises to explore lightweight cryptography (LWC) algorithms that consider a number of aspects, namely power consumption, latency (how long it takes to perform a task), throughput (the rate the plaintext is processed by the algorithm) and resources (in terms of *Gate Equivalences*, GE).

NIST, the USA's National Institute of Standards and Technology, considers an 80-bit key length to be the minimum for lightweight cryptography. For enhanced security, 112-bit and longer are recommended. Also, according to ISO/IEC standardization, a lightweight cipher should have a GE value between 1000 and 2000. Some of the most outstanding LWC are PRESENT, which is designed from the well-known Advanced Encryption Standard (AES), CLEFIA, Enocoro and Trivium, which are included in the ISO/IEC 29192 standard. Refer to [7] for a comprehensive description of LWC protocols.

However, all in all, the AES using 128 bit keys has been shown to be suitable to resource constrained devices [3]. Due to its importance as standard, this flavour of AES is the *de facto* protocol used in real frameworks and settings: the aforementioned MIFARE DESFire, MIFARE Plus, MIFARE Ultralight and FeliCa rely on this protocol instead of implementing other LWC proposals. Some of these proposals also consider DES and 3DES; however, since these protocols have been found to be weak, NIST has deprecated their use.

3 Current Smart Ticketing Proposals

In this section, we address some example proposals for smart ticketing in public transportation which are currently in use. Some of their promoting organisations founded the *Smart Ticketing Alliance* (STA), who considers "essential that public authorities and users can be confident in the quality of contactless communication between contactless readers and fare media" [6].

- **ITSO**[1] (*Integrated Transport Smart card Organisation*) is a non-profit organization in the UK, whose specification became the national standard to regulate the interoperability of tickets across public transport operators in that country. Its ecosystem includes smart cards, the POS and the back-office processing system, which provides key management facilities in a secure datacenter, manages lifecycle cryptographic keys, etc. It makes use of AES encryption and MIFARE smart cards. Founding member of STA.
- **Calypso**[2] is an open global security standard proposed by transport operators which operates through contactless smart cards or contactless compatible devices and has been successfully implemented in 25 countries, for instance, it deploys the Paris Navigo public transportation system and others in Portugal, Italy, Mexico, Belgium, Morocco, and Israel. The specification relies on a central system, which tracks transactions, a reloading system that allows to top up cards and adds tickets to them, a validating system that grants access to transport services, and optional devices for controlling purposes such as an inspector checking a passenger has a valid ticket. Regarding cryptography, it also uses AES. Founding member of STA.
- **CiPurse**[3] is another open security standard for public transportation proposed by OSPT (*Open Standard for Public Transportation* alliance), which proposes vendor neutrality and interoperability across vendor systems. It also makes use of AES, and supports payment media such as contactless cards, wearables, whether using SE or HCE. It is used in several countries like Ecuador, Brazil or South Korea.
- **CEPAS** (*Contactless E-Payment Application Standard*) is a Singaporean standard for electronic money stored in a smart card, which proposed the use of 3DES (after an amendment, AES was adopted) and is also intended for public transportation ticketing. It is deployed in the Singapore EZ-Link transportation system.

Other deployments in real settings are not based on the aforementioned frameworks, but make use of standard smart card technologies, for instance the Hong Kong's Octopus and Tokyo's Suica use the FeliCa smart card platform. The New York metro card and Madrid's transportation system make use of MIFARE smart cards. Moreover, some public transportation systems have introduced the use of smartphone apps. Notable examples include: Hong Kong's Octopus, London's Oyster, Japan's Suica, and Singapore's EZ-Link. Finally, note that this list is not exhaustive, since there is an increasing number of smart ticketing systems being enabled [9].

4 Conclusions

The future of smart ticketing has already started and will provide several advantages not only to consumers, but also to transport operators, and all citizens liv-

[1] https://www.itso.org.uk.
[2] https://calypsostandard.net.
[3] https://www.osptalliance.org/cipurse-specifications.

ing in large cities. To this end, in this article we have described the advantages brought by the use of smart ticketing for both passengers and the environment. However, to develop this technology under the concept of "security by design" remains mandatory to guarantee, among other objectives, the integrity, unforgeability, portability and availability of the information. Although dosens of LWC encryption systems are ready to be adopted in ticketing systems, standards make use on the well-known AES-128 encryption. In addition, the gradual implementation of ticketing applications in smartphones would be a positive step towards a more efficient transportation system, although HCE systems still have to be matured from the security perspective. Future work will focus on validating the suitability of LWC in future smart ticketing systems, as well as the study of interoperability between transport providers.

Acknowledgments. This work is supported by the Government of Catalonia under grant 2020PANDE00103 (ACTUA), by URV's Institut de Ciències de l'Educació under grant 07GI2119, by MCIN/AEI/10.13039/501100011033 and "ERDF A way of making Europe" under grant RTI2018-095499-B-C32 (IoTrain), by Agència de Gestió d'Ajuts Universitaris i de Recerca under grants 2017-SGR-896, and by Universitat Rovira i Virgili with project 2021PFR-URV-118.

References

1. Butler, L., Yigitcanlar, T., Paz, A.: Barriers and risks of Mobility-as-a-Service (MaaS) adoption in cities: a systematic review of the literature. Cities **109**, 103036 (2021)
2. Europay Visa Mastercard: Payment Tokenisation. https://www.emvco.com/emv-technologies/payment-tokenisation/. Accessed 12 June 2022
3. Lara-Niño, C.A., Morales-Sandoval, M., Díaz-Pérez, A.: An evaluation of AES and present ciphers for lightweight cryptography on smartphones. In: International Conference on Electronics, Communications and Computers, pp. 87–93 (2016)
4. Mut-Puigserver, M., Payeras-Capellà, M.M., Ferrer-Gomila, J.L., Vives-Guasch, A., Castellà-Roca, J.: A survey of electronic ticketing applied to transport. Comput. Secur. **31**(8), 925–939 (2012)
5. Oswald, D., Paar, C.: Breaking Mifare DESFire MF3ICD40: power analysis and templates in the real world. In: Preneel, B., Takagi, T. (eds.) CHES 2011. LNCS, vol. 6917, pp. 207–222. Springer, Heidelberg (2011). https://doi.org/10.1007/978-3-642-23951-9_14
6. Smart Ticketing Alliance. https://www.smart-ticketing.org. Accessed 11 June 2022
7. Thakor, V.A., Razzaque, M.A., Khandaker, M.R.A.: Lightweight cryptography algorithms for resource-constrained IoT devices: a review, comparison and research opportunities. IEEE Access **9**, 28177–28193 (2021)
8. Umar, A., Mayes, K., Markantonakis, K.: Performance variation in host-based card emulation compared to a hardware security element. In: 1st Conference on Mobile and Secure Services, pp. 1–6 (2015)
9. Union Internationale des Transports Publics (UITP): Demystifying ticketing and payment in public transport. Technical report, UITP, Brussels, Belgium, November 2020

Debris Detection and Tracking Through On-Board LiDAR

Giulio Campiti, Mattia Tagliente, Giuseppe Brunetti, Mario N. Armenise, and Caterina Ciminelli(✉)

Optoelectronics Laboratory, Politecnico di Bari, Bari, Italy
caterina.ciminelli@poliba.it

Abstract. The uncontrolled growth of space debris around the Earth is forcing satellites to increasingly disrupt their operations in order to prevent potentially catastrophic collisions. Currently, decisions on avoidance maneuvers are made using tracking data mainly obtained through ground-based sensors. As the uncertainties in these data largely affect maneuvering rates, solutions are needed to obtain higher quality observations and therefore decrease the rate of unnecessary maneuvers. This paper studies the possibility of enabling satellites to make autonomous observations of space objects at risk of collision by using onboard LiDAR sensors. As space-based observations do not suffer from diffractions and other problems related to the atmosphere, the proposed solution could be an effective means of obtaining more precise risk estimates. An orbital mechanics analysis of typical conjunction dynamics has been performed to derive the required sensor performance.

1 Introduction

As hundreds of thousands of satellites are planned to be launched in the next few years, the risk of in-orbit collisions is going to dramatically increase. Currently, the near-Earth space environment is mainly monitored through ground-based sensors. Approximately 36,000 objects are large enough to be reliably tracked and cataloged by those systems, but they already have a significant impact on space operations. At European Space Agency (ESA), every week hundreds of collision warnings are received for typical satellites in Low-Earth Orbit (LEO) [1], eventually leading to an average of two maneuvers per satellite per year. However, these maneuver and alarm rates are mainly driven by the presence of uncertainty in the position knowledge of the tracked objects, whereas the true risk of collisions in LEO is very low [2]. Therefore, in the context of the rapidly increasing space population, it is fundamental to improve the quality of surveillance data and reduce the frequency of unnecessary maneuvers.

One possibility could be to enable satellites to make autonomous observations of the objects threatening a collision during the orbits prior to the predicted events. The acquired data could be processed directly on board to refine the information received from the ground, e.g., using a Kalman filter where the ground data would serve as an initialized state, as proposed in [3].

R. Berta and A. De Gloria (Eds.): ApplePies 2022, LNEE 1036, pp. 235–241, 2023.
https://doi.org/10.1007/978-3-031-30333-3_31

Space-based observations could be an effective means to improve the accuracy of orbital information because 1) they can be made very close to the predicted Time of Closest Approach (TCA), and 2) they do not suffer from atmospheric issues such as diffractions, aberrations, and turbulences. The former reason is especially important since the closer the orbital data is to TCA, the better the accuracy of the prediction. Indeed, space operators often change maneuver decisions upon receiving new, updated information. Final decisions are currently uplinked during the last available ground-station accesses before TCA, which can be several hours prior to the events. In the following hours, however, satellites would still have various currently unexploited opportunities to observe the other object and refine its positional knowledge.

The use of payloads for surveillance purposes has already been suggested many times in the literature (e.g., [4]), but mainly with the aim of better understanding the debris environment. In this article, using onboard sensors is instead aimed at refining the orbital knowledge of debris objects before potential collisions. In particular, a Laser Imaging Detection and Ranging (LiDAR) system is proposed as a secondary payload for medium to large satellites (> 500 kg), where it would only represent a small percentage of the overall mass. Recent developments have demonstrated the possibility of integrating LiDAR systems into a single Photonic Integrated Circuit (PIC). The advantages of active sensors can thus be combined with a small footprint and mass, making this solution particularly attractive for the envisaged application.

To derive the performance required by the system, the trajectory evolution of thousands of historical conjunction events has been reproduced and analyzed, with a focus to understand the occurrence of observing opportunities for conjuncting objects and characterizing them in terms of duration, relative distances, and other relevant features. These aspects are then discussed and translated into system requirements for the LiDAR.

2 Analysis of Observing Opportunities Before a Potential Collision

2.1 Collection and Propagation of Historical Conjunction Events

Collecting orbital data on past conjunction events was possible through the SOCRATES web service offered by the Center for Space Standards & Innovation (CSSI), which every day publishes a list of forecasted conjunctions for the coming week [5]. Besides reporting general information about each event, such as the predicted minimum miss-distance and TCA, orbital data is also provided for conjuncting objects in the form of Two-Line Elements (TLEs), thus making it possible to propagate the orbits and reproduce the full trajectory evolution.

For a meaningful analysis, only certain events were selected for which the predicted miss-distance was less than 1 km and the TLEs were not generated more than 2 days before TCA, so as to limit propagation errors. Moreover, only conjunctions in LEO were considered, as it is by far the most congested and problematic orbital region. The final database includes data on 12,000 events.

For each event, the objects' trajectory was propagated with the Simplified General Perturbations model 4 (SGP4), using a fine time step and a total time span covering the 2 days leading to each predicted TCA.

2.2 Results: Characteristics of Observing Opportunities

The propagated trajectories were analyzed in terms of relative distance to understand the expected performance of the sensor system. The existence of multiple close passes before the closest approach was observed in almost all events. However, how this would translate into potential observing opportunities for a satellite equipped with a payload depends on the maximum distance at which it is able to distinguish and observe another space object.

Figure 1 shows the probability of having at least one observing window in the orbits prior to TCA depending on the maximum range of the payload. As shown, with visibility around the satellite limited to 50 km, a secondary object could be seen before TCA in just over half cases, while the chances increase to 90% with a maximum range of 250 km.

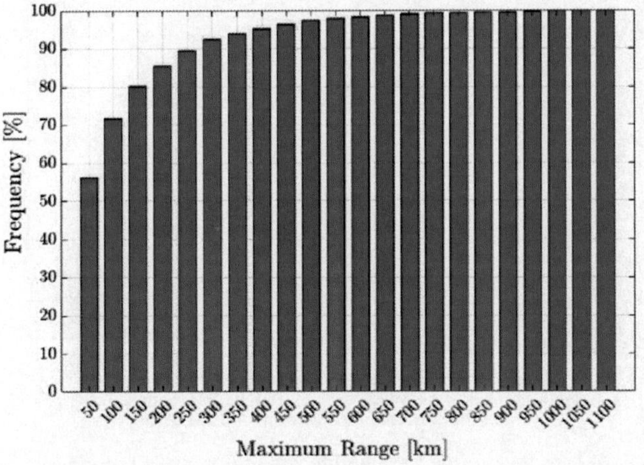

Fig. 1. Probability of having at least one observing opportunity before a potential collision depending on the maximum range of the payload

However, Fig. 1 only shows the probability of being able to see an object before TCA but does not convey information on how long it would remain in the visibility of the payload, which is crucial for obtaining useful observations. It was therefore analyzed for how long a secondary would remain within a relative distance less than the maximum range of the payload, for different range values. For a more realistic analysis, the following practical limitations on object visibility were considered:

1. As the secondary advances towards the main satellite during a close pass (i.e., the relative distance diminishes), the angular rates between the two accelerate significantly, eventually reaching values that make it impossible to continue tracking. Thus, object visibility was considered lost whenever angular rates exceeded 1°/s.
2. Observing opportunities occurring later than one orbit before TCA were not considered, since by that time it is more desirable to have already made a maneuver decision.

3. Secondary objects were considered visible only while advancing relative to the satellite and not as they move away. This is necessary since the angular position between the two objects can change drastically across the point where the minimum distance is reached, and it is therefore impractical that a satellite would reorient its attitude to track the object both while approaching and receding.

The results of the analysis are reported in Fig. 2 for five possible values of the maximum range. With a payload able to see at a distance of 400 km, there is a 75% chance that the satellite would have at least 10 s to observe a secondary object. Instead, for a maximum range of 600 km, this number increases to about 93%.

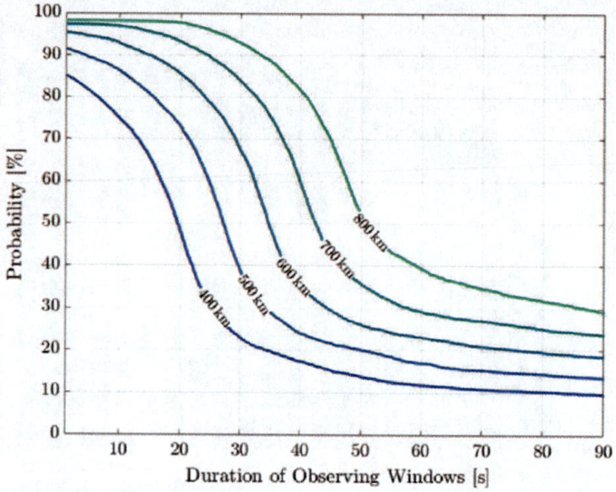

Fig. 2. Probability of how much time the satellite has to observe a secondary object before TCA during a single observing opportunity, for different values of the maximum range of the payload

Finally, it can be relevant to the LiDAR performance to consider the relative speeds involved in the encounters. It was found that the maximum value is almost 16 km/s, and most conjunctions occur with relative velocities between 12 – 15 km/s. More details and additional plots can be found in [6].

3 Spaceborne LiDAR for Debris Detection and Tracking

A spaceborne sensor has to operate in a harsh environment, where the interference caused by external light sources can reach higher power levels than those of the signal backscattered by the targets. Moreover, the detection system must work at an ultra-long distance, where the background represents an additional source of shot noise with a detrimental effect on the Signal-to-Noise Ratio (SNR). A LiDAR system is proposed here for its combination of high immunity to interference and the greatest possible sensitivity, besides requiring less volume and mass compared to other active sensors such as radars. The coherent detection technique makes it conceivable to maintain a minimum SNR threshold of 20 dB, thanks to the inherent rejection of incoherent interferences and the shot-noise limited level of receiver sensitivity. Additionally, it lets the distance and speed of the target be obtained simultaneously.

The first task that any tracking sensor must perform before acquiring data on a target object is naturally to detect it. As previously mentioned, the orbital state of a secondary object is first estimated from the ground – where risky events can be identified up to 7 days in advance – and then uplinked to the at-risk satellite. The object trajectory is automatically propagated on board to find potential observing windows and identify limited regions of space where the secondary can be expected to appear. The total search volume, that should be scanned for one detection, clearly depends on the uncertainty of the orbital information provided from the ground. It can be described by a gaussian probability distribution [7] where the largest value is always along the track direction. During the last 24 h before TCA, the along-track uncertainty in the debris position is typically lower than 1000 m. A circular frame with a radius $r_f = 90$ m has been considered to also ensure technological feasibility. Other important sensor requirements are the range and lateral resolution, set to 10 m to reduce the maneuvers rate. According to [2], if secondary objects' position was always available with an uncertainty of 10 m, a typical Earth-observation satellite such as ESA's Sentinel-2 would have to maneuver just once every 40 years. Therefore, this value would lead to a significant improvement over the current situation (two maneuvers/year). To achieve this resolution for the single cell, a beamwidth of 0.0011° has been obtained following the steps reported in [8]. The combination of maximum payload range and observation time is crucial to detect a large amount of debris. As shown in Fig. 2, the detection probability increases along with the maximum range. However, the feasibility of the system in terms of reasonable output power, source linewidth, and receiver aperture limits the range (R) to 500 km. An observation time of 10 s has been chosen to allow a probability of detection of about 90% (see Fig. 2). Given the frame radius (r_f) and the maximum range (R_{max}), the Field of View (FoV) of the system in the normal direction (V) to that of the apparent motion of the secondary object (FoV_V) can be expressed as:

$$FoV_V = 2 \arcsin\left(\frac{r_f}{R_{max}}\right) \tag{1}$$

The FoV along the direction (H) of the apparent motion of the target (FoV_H) is estimated as a function of the maximum apparent angular rate ($\omega_{r,max} = 1°/\text{s}$, as reported in Sect. 2.2), the time required to scan a single cell (T_{meas}), and the number of individual cells (N_{cell}) needed to cover the whole observation frame:

$$FoV_H = \omega_{\mathrm{r,max}} \cdot N_{cell} \cdot T_{meas}. \tag{2}$$

N_{cell} can be calculated as:

$$N_{cell} = 4\sum_{i=0}^{\lfloor \xi \rfloor} \lceil \sqrt{\xi^2 - i^2} \rceil \tag{3}$$

where ξ is the radius of the observation frame normalized with respect to the side of the base area of the cell, which is in turn defined as the square inscribed in the normal section of the laser beam. Moreover, as T_{meas} is the time it takes to scan a single cell, the product of T_{meas} and N_{cell} should be lower than 10 s. Therefore, the FOV required to fulfill the system requirements is $FOV = FOV_H \times FOV_V = 0.02° \times 10°$. The system level requirements have been summarized in Table 1.

Beam scanning, Frequency Modulated Continuous Wave (FMCW)-based LiDARs exhibit advantages that meet the needs of our application. Since the range resolution is inversely proportional to the modulation bandwidth and the SNR, a chirped waveform generator [8] with a large Time-Bandwidth product guarantees a modulation bandwidth of almost 1 GHz and a range resolution of 10 m [10]. The lateral resolution and FOV requirements could be met by using an NxN Optical Phased Array with an aperture of 6.8 cm and N> 7,700 [11] for transmission. According to an analysis based on the LiDAR link budget equation [12], the optical power of the laser source should be almost 14 W, achievable by using a laser diode array and a receiving OPA with N >53,900. However, it should be noted that an OPA with such performance has not yet been demonstrated.

In summary, a silicon photonics implementation of the described FMCW-based LiDAR system is consistent with current C-band optics knowledge. This could help mitigate the risk of debris by significantly reducing the number of unnecessary maneuvers performed by medium, or larger, satellites.

Table 1. Suggested onboard system for debris detection and tracking functional requirements.

Frame radius [m]	Range [km]	Range resolution [m]	Lateral resolution [m]	FoV [deg]
≥ 90	≥500	≤10	≤10	≥0.12° × 10°

References

1. European Space Agency, "CREAM - ESA's Proposal for Collision Risk Estimation and Automated Mitigation", th. rep. (2019)
2. Setty, S.J., et al.: SLR for space debris monitoring and analysis on requirements and achievable orbital improvements. In: 1st NEO and Debris Detection Conference (2019)
3. Scott, R.L., et al.: On-orbit Observations of conjuncting space objects prior to the time of closest approach. J. Astronautical Sci. **67**(4) (2020)
4. Mamani, J., et al.: LiDAR small satellite for space debris location and attitude determination. In: Proceedings of INCAS (2019)
5. Socrates. CelesTrak. (Retrieved 4 July 2022). https://celestrak.com/SOCRATES/
6. Campiti, G., et al.: Observing opportunities of space conjuncting objects in the orbits Prior to the closest approach. In: Space Debris Risk Assessment and Mitigation Analysis Workshop, ESA/ESOC (2022)
7. Wirth, W.: Radar Techniques Using Array Antennas. IET (2013)
8. Luo, Y., et al.: A review of uncertainty propagation in orbital mechanics. Prog. Aerosp. Sci. **89**, 23–39 (2017)
9. Brunetti, G., et al.: Chip-scaled ka-band photonic linearly chirped microwave waveform generator. Front. Phys. **10**(785650) (2022)

10. Hayes, M.: Statistical Digital Signal Processing and Modeling, John Wiley & Sons (1996)
11. Balanis, C.: Antenna Theory: Analysis and Design. John Wiley & Sons (2016)
12. Tan, K., et al.: Modeling hemispherical refectance for natural surfaces based on terrestrial laser scanning backscattered intensity data. Opt. Express **24**, 22971–22988 (2016)

Check for updates

Automatic IP Core Generator for FPGA-Based Q-Learning Hardware Accelerators

Lorenzo Canese, Gian Carlo Cardarilli, Luca Di Nunzio, Rocco Fazzolari, Marco Re, and Sergio Spanó(✉)

University of Rome "Tor Vergata", Rome, Italy
{canese,cardarilli,di.nunzio,fazzolari,re,spano}@ing.uniroma2.it

Abstract. We introduce a MATLAB-Simulink software able to generate customizable hardware IP cores for the Reinforcement Learning algorithm called Q-Learning. The tool automatically produces the VHDL code and runs both synthesis and implementation for any AMD-Xilinx FPGA using the Vivado software chain. Our automatic generator relies on the "HDL coder" from Mathworks to produce an efficient hardware accelerator based on the state of the art. The model can be customized by the user according to the desired Q-Matrix size and bit-depth for all the algorithm parameters.

1 Introduction

Hardware acceleration of Reinforcement Learning [1] has been a trending topic in the VLSI design research scene [2–4].

One of the most relevant algorithm in the field is Q-Learning [5] which is based on the so-called Q-Matrix. In the last years, several hardware implementations for the aforementioned algorithm have been proposed [6–13]. If all the cited literature is analyzed [14], it can be reached that Spanó et al. work [12] is the state of the art in terms of throughput, hardware resources usage, power consumption, scalability, and flexibility to any use case.

For those reasons, we propose an automatic IP core generator based on the architecture of the cited paper. The software builds a Q-Learning hardware accelerator ready to be implemented on FPGA devices. The application is based on MATLAB-Simulink and relies on the "HDL coder" tool from Mathworks to generate the VHDL code.

Our software is compatible with every AMD-Xilinx FPGA that can be programmed using the Vivado chain on Windows machines.

The remainder of this paper is organized as follows. Section 2 shows the hardware architecture of the Q-Learning accelerator. Section 3 proposed an overview on the MATLAB application. Finally, Sect. 4 draws the conclusions and future developments for our project.

R. Berta and A. De Gloria (Eds.): ApplePies 2022, LNEE 1036, pp. 242–247, 2023.
https://doi.org/10.1007/978-3-031-30333-3_32

2 Hardware Accelerator Architecture

The Q-Learning algorithm [5] update rule is the following:

$$Q_{new}(s_t, a_t) = (1 - \alpha)Q(s_t, a_t) + \alpha[r_t + \gamma_t \max_A Q(s_{t+1}, A)] \quad (1)$$

where Q represents the Q-Matrix value, s and a are the states and actions respectively, r is the reward, α is the learning rate, and γ is the discount factor. t is the iteration index of the algorithm. The capital A stands for the entire row of the Q-Matrix associated to a certain state.

As stated in the previous section, the generated IP core architecture is based on the one from Spanó et al. [12]. The top level view is shown in Fig. 1, please note that vector signals are represented using bold text and thick arrows.

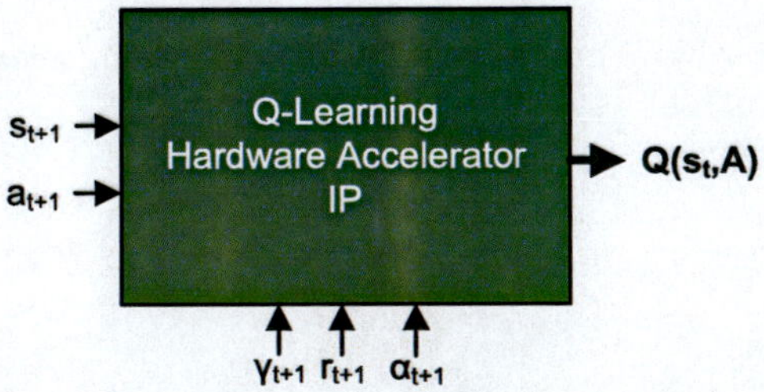

Fig. 1. Top-level view of the Q-Learning accelerator.

The system inputs are the previous mentioned parameters, while the output is the full Q-Matrix row associated to the current iteration t. This feature is meant for the accelerator to match the input of a typical Reinforcement Learning action-policy generator [15].

A more detailed view of the architecture is shown in Fig. 2, a thorough explanation of the system can be found at Spanó et al. [12].

It is worth underlying that the "automatic" implementation results are interchangeable with the "manual" ones obtained in the referred work. For this reason, we suggest to check Spanó et al. [12] for the FPGA implementation data.

3 Application Overview

The developed MATLAB-Simulink application can be downloaded at the following URL: https://www.mathworks.com/matlabcentral/fileexchange/112745-automatic-ip-core-generator-for-q-learning-hardware, its icon is presented in Fig. 3.

When started, the "Automatic IP core generator for Q-Learning Hardware" prompts the user for the architecture settings, as shown in Fig. 4.

The settings are split in three categories:

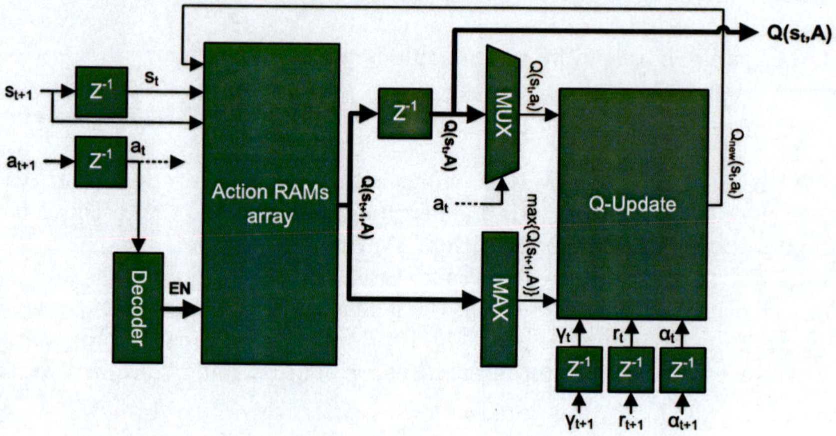

Fig. 2. Hardware architecture of the Q-Learning accelerator.

Fig. 3. Q-Learning IP generator application icon.

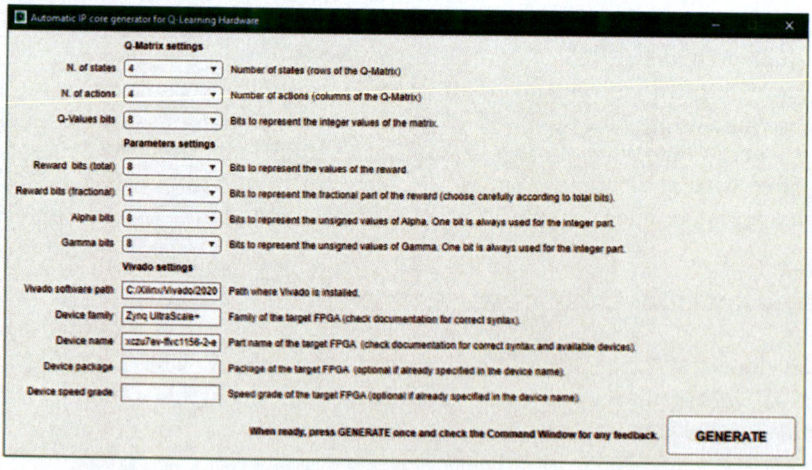

Fig. 4. Q-Learning IP generator application setup window.

- **Q-Matrix settings.** The size of the Q-Matrix can be set as long with the bits used to represent the values of the table. Please note that such values are meant to be integer.
- **Parameters settings.** The reward bit-depth can be set by specifying its full value and the bits reserved to represent the fractional part. Since both α and γ are "up to 1" unsigned numbers, when specifying their depth, one bit will be always reserved to the integer part and the remainder to the fractional part.
- **Vivado settings.** For synthesis purposes, the path of the Vivado installation and the target FPGA details must be specified.

Once the "GENERATE" button has been pressed, the software starts to build the IP core. The window in Fig. 5 appears.

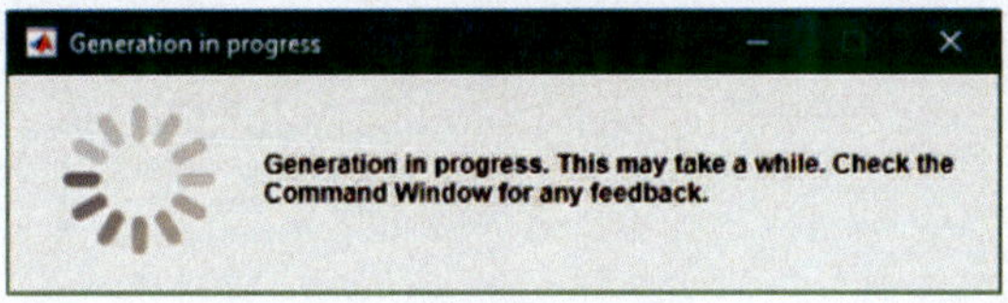

Fig. 5. Q-Learning IP generator application progress window.

At the end of the generation progress, the window in Fig. 6 appears.

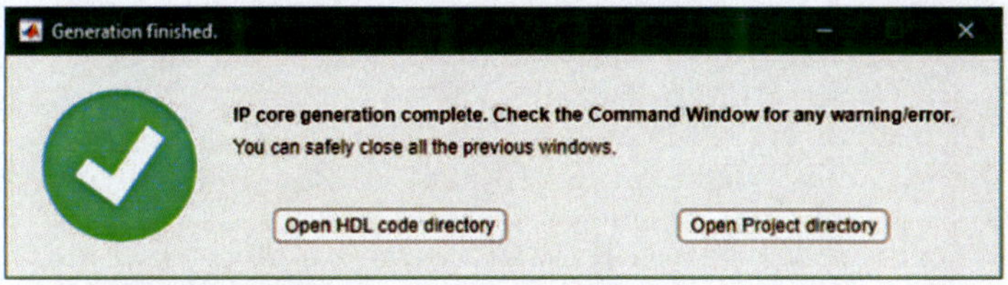

Fig. 6. Q-Learning IP generator application final window.

The user can now chose to view the VHDL code of the accelerator or to open the Vivado project to proceed further (e.g. to generate the FPGA bitstream and to program the device).

4 Conclusion

We introduced a MATLAB-Simulink software able to generate customizable hardware IP cores for the Reinforcement Learning algorithm called Q-Learning. We showed how the tool automatically produces the VHDL code and runs both synthesis and implementation for any AMD-Xilinx FPGA using the Vivado software chain. Our automatic generator relies on the "HDL coder" from Mathworks to produce an efficient hardware accelerator based on the state of the art. The model can be customized by the user according to the desired Q-Matrix size and bit-depth for all the algorithm parameters.

Since this is the first release of the software, we plan to extend the support for Intel FPGAs and ASICs. Moreover, a Verilog version and the inclusion of a Policy Generator are already in progress. We suggest to check for any design update on the official website of the project.

Acknowledgments. The authors would like to thank Advanced Micro Devices, Inc. (AMD) for providing the FPGA software tools with the AMD-Xilinx University Program.

References

1. Sutton, R., Barto, A.: Reinforcement Learning: An Introduction. A Bradford Book, Cambridge (2018)
2. Rothmann, M., Porrmann, M.: A survey of domain-specific architectures for reinforcement learning. IEEE Access **10**, 13753–13767 (2022)
3. Waseem, S.M., Roy, S.K.: Hardware realization of reinforcement learning algorithms for edge devices. In: VLSI and Hardware Implementations Using Modern Machine Learning Methods, pp. 233–254. CRC Press (2021)
4. Su, J.D., Tsai, P.Y.: Processing element architecture design for deep reinforcement learning with flexible block floating point exploiting signal statistics. In: 2020 Asia-Pacific Signal and Information Processing Association Annual Summit and Conference (APSIPA ASC), pp. 82–87. IEEE (2020)
5. Watkins, C.J., Dayan, P.: Q-learning. Mach. Learn. **8**(3), 279–292 (1992)
6. Cardarilli, G.C., et al.: An FPGA-based multi-agent reinforcement learning timing synchronizer. Comput. Electr. Eng. **99**, 107749 (2022)
7. Liu, X., Diao, J., Li, N.: A FPGA-based accelerator implementation for path planning using q_learning algorithm. J. Phys. Conf. Ser. 2245, 012014 (2022). IOP Publishing
8. Cardarilli, G.C., et al.: "MR Q-Learning" algorithm for efficient hardware implementations. In: 2021 55th Asilomar Conference on Signals, Systems, and Computers, pp. 1186–1190. IEEE (2021)
9. Sahoo, S.S., Baranwal, A.R., Ullah, S., Kumar, A.: MemOReL: a memory-oriented optimization approach to reinforcement learning on FPGA-based embedded systems. In: Proceedings of the 2021 on Great Lakes Symposium on VLSI, pp. 339–346 (2021)
10. Baranwal, A.R., Ullah, S., Sahoo, S.S., Kumar, A.: ReLAccS: a multilevel approach to accelerator design for reinforcement learning on FPGA-based systems. IEEE Trans. Comput. Aided Des. Integr. Circuits Syst. **40**(9), 1754–1767 (2020)
11. Meng, Y., Kuppannagari, S., Rajat, R., Srivastava, A., Kannan, R., Prasanna, V.: QTAccel: a generic FPGA based design for Q-table based reinforcement learning accelerators. In: 2020 IEEE International Parallel and Distributed Processing Symposium Workshops (IPDPSW), pp. 107–114. IEEE (2020)
12. Spanó, S., et al.: An efficient hardware implementation of reinforcement learning: the Q-learning algorithm. IEEE Access **7**, 186340–186351 (2019)

13. Da Silva, L.M., Torquato, M.F., Fernandes, M.A.: Parallel implementation of reinforcement learning Q-learning technique for FPGA. IEEE Access **7**, 2782–2798 (2018)
14. Sun, J., Sharma, N., Chakareski, J., Mastronarde, N., Lao, Y.: Hardware acceleration for post-decision state reinforcement learning in IoT systems. IEEE Internet Things J. **9**(12), 9889–9903 (2022)
15. Cardarilli, G.C., et al.: An action-selection policy generator for reinforcement learning hardware accelerators. In: Saponara, S., De Gloria, A. (eds.) ApplePies 2020. LNEE, vol. 738, pp. 267–272. Springer, Cham (2021). https://doi.org/10.1007/978-3-030-66729-0_32

Review of Security Vulnerabilities in LoRaWAN

Junaid Qadir[1](✉), Ismail Butun[2], Paolo Gastaldo[1], and Daniele D. Caviglia[1]

[1] Department of Electrical, Electronic and Telecommunications Engineering and Naval Architecture (DITEN), University of Genoa, 16145 Genoa, Italy
junaid.qadir@edu.unige.it, {paolo.gastaldo,daniele.caviglia}@unige.it
[2] Department of Electrical Engineering and Computer Science, KTH Royal University of Technology, 100 44 Stockholm, Sweden
butun@kth.se

Abstract. The realm of Low Power Wide Area Network (LPWAN) has a paramount influence on the way we work and live. For instance, real-time applications and rapid packet transiting for long-range have now come into practice that was previously considered mysterious. However, euphoria becomes a problem when it comes to security considerations, as low-power devices possess limited processing units that are unable to elucidate robust security algorithms. In this case, the Low Power Wide Area Network (LoRaWAN) stepped into a technological competition that filled the gap by adopting the end-to-end security feature. Though, LoRaWAN protocol entails fundamental security requirements but the implementation matters. This paper presents security analyses in the LoRaWAN networks. In addition, we provide a bibliometric overview of security considerations in LoRaWAN that helps researchers for thorough insights and implementation.

Keywords: IoT · LoRaWAN · Security · Vulnerability · Confidentiality · Integrity · Authenticity · Bibliometric

1 Introduction

The magical appearance of the Internet of Things (IoT) has made communication convenient between physical objects without human intervention. Thus, the word IoT refers to the interconnected devices that detect, collect, and transmit data across the world via existing Internet infrastructure. As per technological prediction by Statista[1], there is expected more than 30.9 billion devices will be connected seamlessly with each other on global Internet.

Several communication protocols have already opened opportunity for IoT devices. For example, ZigBee, Bluetooth, and RFID have revealed their use for IoT resource restricted devices, because of low energy consumption. However,

[1] https://www.statista.com/statistics/1101442/iot-number-of-connected-devices-worldwide.

R. Berta and A. De Gloria (Eds.): ApplePies 2022, LNEE 1036, pp. 248–254, 2023.
https://doi.org/10.1007/978-3-031-30333-3_33

they are unable to continue operation for applications requiring coverage for large distance.

Low Power Wide Area Network (LPWAN) enables communication links over long range with low energy consumption. Popular LPWAN technologies including, Long Range Wide Area Network (LoRaWAN) [1], SigFox [6], Narrowband-IoT (NB-IoT) [7] are the most widely used in different use cases. As these technologies provide communication over several kilometers (km); therefore, security for the transmitted packet is the major concern in LPWAN. SigFox provides communication without having consideration of security features, while the NB-IoT possess basic LTE-encryption. Therefore, LoRaWAN is the most preferred choice as it provides strong end-to-end security. By design, LoRaWAN is highly secured as it considers confidentiality and authenticity using many security keys; nonetheless, the network's implementation matters if these keys are kept insecure or reused the same cryptographic numbers.

LoRa is developed by Semtech Inc., Camarillo, CA, USA, which is a physical layer whereas LoRaWAN is the upper layer of LoRa, which defines the communication protocol and system architecture. Together with LoRa, it enables communication over very long distance on local, national and international (using roaming) level, with extremely low power consumption. LoRaWAN system architecture consists of end-device, gateway, network server, application server, and join server (LoRaWAN v1.1). The end-device uses radio waves to communicate with the gateway and utilizes the chirp spread spectrum (CSS) modulation technique [2], which possess the same characteristic as frequency-shift keying (FSK) modulation used in many legacy wireless communication system. However, it is immune to interference therefore increases the communication range. The end-device can be activated using two different methods such as; Activation by Personalization (ABP) and Over-The-Air Activation (OTAA). The only difference between two activation is as the ABP activation stores the security keys permanently, while the OTAA generates security keys dynamically.

Security in LoRaWAN is evolving as it is a constant target of malicious actors [8]. Several security challenges including replay attacks, bit flipping attacks, key management related attacks that affect confidentiality, integrity, and availability are confronted in the literature. And the LoRa Alliance is constantly enhancing the protocol to ensure it stays ahead of the changing security landscape. This paper discusses security vulnerabilities and privacy issues in LoRaWAN specification. We discuss cybersecurity breaches in LoRaWAN off-the-shelf that exhibits several attacks scenario targeting end-device, gateway, and network server. In the last, a bibliometric overview is given, that provides a thorough insights for researchers and engineers looking to deploy LoRaWAN infrastructure and enhance it's security in the future.

2 Cyber Risks and Threats in LoRaWAN

This section discusses cybersecurity risks and threats analysis in LoRaWAN. Though, LoRaWAN specification has been introduced by employing strong security layers. However, some well-known weaknesses have been pinpointed that come with high risks. Therefore, our aim is to highlight security vulnerabilities and privacy issues in LoRaWAN implementation. Several threats and attacks are follows as below

2.1 Confidentiality

Confidentiality is the practice of maintaining data security using conventional cryptographic encryption techniques. The data is considered to be not confidential if disclosed to the intended audience. The following list includes numerous attacks that compromise LoRaWAN's confidentiality:

- Keys Related Vulnerabilities: LoRaWAN security is heavily dependent on security keys, and the implementation becomes vulnerable if the keys are comprised. There are numerous ways to expose keys that are highlighted in [4] including reverse engineering of device, keys disclosure, device tags, hardcoded keys in open source code, and non random keys etc.
- Plain-text Key Capture: Cerrudo et al. [4] published a white paper and mentioned that the LoRaWAN network can be compromised if the text files containing the keys of the end device are shared on the Internet, or not used hardware security module (HSM).
- Eavesdropping Attack: LoRaWAN employs AES in counter mode to ensure the confidentiality of the packet. However, still the ABP devices are vulnerable to eavesdropping attack as these devices use the same encryption keys for long time. Noura et al. in [10], investigated that if two ciphertexts are encrypted with the same key stream, then the attacker may able to decrypt the message by XORing both ciphertexts and can get the original message.

2.2 Integrity

Integrity is the essential step of cybersecurity as it preserves the data from being added, changed, or deleted during transmission from a source to the destination. Attacks that compromise LoRaWAN integrity are discussed below.

- Bit Flipping Attack: In this attack, an attacker intercepts the cipher message and modify the message by adding, changing, and deleting a single or number of bits. As a result, the application server receives a modified version of the packet. In LoRaWAN, the packet is only encrypted using AES counter mode (CTR-mode) that provides XOR operation instead of shuffling the bits. Therefore, the authors in [11] discuss that LoRAWAN is susceptible to Bit-Flipping attack as the attacker can modify the message between the network and application servers.

- Device Cloning: Due to the low cost, the LoRaWAN end-devices are becoming ubiquitous, and attackers with access to the device physically can clone the firmware. Cloning the firmware can compromise the device and expose it to integrity breaches [9].

2.3 Availability

Availability ensures the presence of the network and system while requested by the user. There are numerous attacks that could jeopardize the availability of LoRaWAN.

- Replay Attack: Replay attack affects denial-of-service (DoS) of the end-device in LoRaWAN, and consists of re sending the capture messages in the edge of the network. Replay attack issue has been resolved in the new version of LoRaWAN, however, the ABP activated devices remain vulnerable to this attack. In LoRaWAN, the end-devices use two counters such as the uplink and the downlink counters. So, these counters increase with every message. And the value resets until it reaches the maximum value [11]. In replay attack, the attacker hands on the message with higher counter value and injects it when the gets start from the 0. In this case, the network server considers the injected message as legitimate and received a false packet from the attacker.
- Wormhole Attack: The authors in [5] discuss to perform the wormhole attack, it is therefore, needed to have a sniffing and a jamming tool to block the packet sent from the end-device. Consequently, the packet gets lost the destination and can exploit it for the whole network in the form of replay for a time being.
- Selective Forwarding Attack: It is a routing related attack which severally affect the network availability. In LoRaWAN implementation, the attacker choose the packet and can selectively forward it in order to block other end-devices in the network [3].

3 Bibliometric Overview

This section presents the bibliometric overview of LoRaWAN. We collect the data from two different major databases i.e., Scopus[2] and WebofKnowledge[3]. Then, we perform the string such as ("LoRa" OR "LoRaWAN") AND ("Security" OR "Cybersecurity"), and collect all values from each database. Finally, the overall value has recorded graphically using Matlab. Figure 1(a) shows publication record started from 2015 until 2022 and the maximum publications have

[2] https://www.scopus.com/.
[3] https://www.webofknowledge.com/.

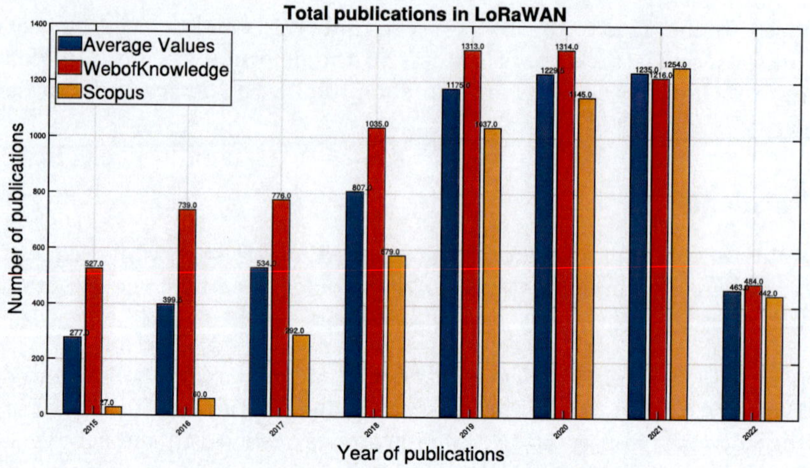

(a) String searched ("LoRa" OR "LoRaWAN")

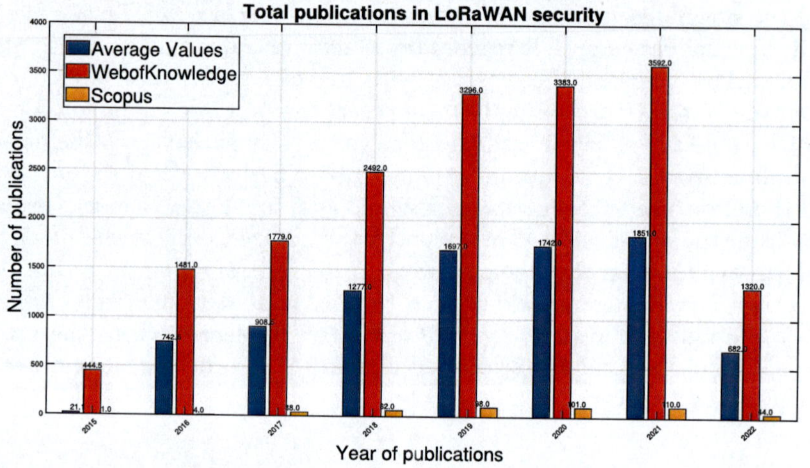

(b) String searched ("LoRa" OR "LoRaWAN")AND("Security" OR "Cybersecurity")

Fig. 1. a) Total number of papers published in LoRaWAN, b) Total papers published in LoRaWAN security.

been recorded in the year of 2020. Furthermore, the number of publications in LoRaWAN security is shown in Fig. 1(b). In addition, Table 1 shows the number of papers that addressed the following attacks.

Since the data were collected in mid 2022, the final numbers on 2022 data are incomplete and should not mislead the reader. The trend within the last decade shows that the 2022 numbers might surpass 2021.

Table 1. Papers dealt with various attacks

String searched	Papers dealt with attacks		Type of documents	
Attacks	Scopus	WebofKnowledge	Article (S+W)	Conference proceedings (S+W)
"LoRAWAN" AND "Key related vulnerabilities"	1	1	(1),(1)	(0),(0)
"LoRAWAN" AND "Plain-text Key Capture"	0	0	(0),(0)	(0),(0)
"LoRAWAN" AND "Eavesdropping Attack"	11	8	(3),(3)	(8),(5)
"LoRAWAN" AND "Bit Flipping Attack"	5	3	(1),(1)	(4),(2)
"LoRAWAN" AND "Device Cloning"	1	1	(0),(0)	(1),(1)
"LoRAWAN" AND "Replay Attack"	35	25	(11),(10)	(24),(15)
"LoRAWAN" AND "Wormhole Attack"	2	1	(0),(0)	(2),(1)
"LoRAWAN" AND "Selective Forwarding Attack"	0	0	(0),(0)	(0),(0)

(S+W) = Scopus +WebofKnowledge

4 Conclusion

LoRaWAN is an emerging protocol that has received widespread acceptance across a variety of useful applications in numerous regions. It advances the packet by keeping in view several security encryption techniques, but there are several flaws that could compromise LoRaWAN's security and privacy. In this paper, we present cybersecurity vulnerabilities of LoRaWAN protocol that previously associated with LoRaWAN implementation. In addition, the bibliometrics overview is presented by providing the number of papers published within the last decade on the cybersecurity of LoRaWAN vs. overall publications in LoRaWAN.

References

1. Alliance, L.: what is lorawan. https://lora-alliance.org/resource_hub/what-is-lorawan/. Accessed March 12 2022
2. Butun, I.: Towards smart sensing systems: a new approach to environmental monitoringsystems by using lorawan. In: 2022 IEEE Zooming Innovation in Consumer Technologies Conference (ZINC). IEEE (2022)
3. Butun, I., Pereira, N., Gidlund, M.: Analysis of lorawan v1. 1 security. In: Proceedings of the 4th ACM MobiHoc Workshop on Experiences with the Design and Implementation of Smart Objects, pp. 1–6 (2018)
4. Cerrudo, M.C., Fayo, E.M., Sequeira, M.: Lorawan networks susceptible to hacking: common cyber security problems, how to detect and prevent them. IOActive, Seattle, WA, USA, White Paper 1 (2020)
5. Chacko, S., Job, M.D.: Security mechanisms and vulnerabilities in lpwan. In: IOP Conference Series: Materials Science and Engineering, vol. 396, p. 012027. IOP Publishing (2018)
6. Lavric, A., Petrariu, A.I., Popa, V.: Long range sigfox communication protocol scalability analysis under large-scale, high-density conditions. IEEE Access **7**, 35816–35825 (2019)
7. Malik, H., Alam, M.M., Le Moullec, Y., Kuusik, A.: Narrowband-iot performance analysis for healthcare applications. Procedia Comput. Sci. **130**, 1077–1083 (2018)
8. Mohamed, A., Wang, F., Butun, I., Qadir, J., Lagerström, R., Gastaldo, P., Caviglia, D.D.: Enhancing cyber security of lorawan gateways under adversarial attacks. Sensors **22**(9), 3498 (2022)

9. Noura, H., Hatoum, T., Salman, O., Yaacoub, J.P., Chehab, A.: Lorawan security survey: issues, threats and possible mitigation techniques. Internet Things **12**, 100303 (2020)
10. Noura, H.N., Salman, O., Hatoum, T., Malli, M., Chehab, A.: Towards securing lorawan abp communication system. In: CLOSER, pp. 440–447 (2020)
11. Yang, X., Karampatzakis, E., Doerr, C., Kuipers, F.: Security vulnerabilities in lorawan. In: 2018 IEEE/ACM Third International Conference on Internet-of-Things Design and Implementation (IoTDI), pp. 129–140. IEEE (2018)

Experiments on Speeding Up the Recursive Fast Fourier Transform by Using AVX-512 SIMD Instructions

Giacomo Sansone(✉) and Marco Cococcioni(✉)

Department of Information Engineering Largo Lucio Lazzarino, University of Pisa, 56122 Pisa, Italy

giacomo.sansone@yahoo.com, marco.cococcioni@unipi.it

Abstract. The *Fast Fourier Transform* is probably one of the most studied algorithms of all time. New techniques regarding hardware and software are often applied and tested on it, but the interest in FFT is still large because of its applications - signal and image processing, numerical computations, etc. In this paper, we start from a trivial recursive version of the algorithm and we speed it up using AVX-512 Single Instruction Multiple Data (SIMD) instructions on an Intel i7 CPU with native support to AVX-512. In particular, we study the impact of two different storage choices of vector of complex numbers: *block interleaving* and *complex interleaving*. Experimental results show that automatic vectorization provides a 10.65% ($\sim 1.12\times$) speedup, while with vectorization done by hand the speedup reaches 33.78% ($\sim 1.51\times$). We have made our code publicly available, which could be helpful for SIMD instructions teaching purposes.

Keywords: Recursive Fast Fourier Transform · SIMD instructions · AVX-512 · complex number arithmetic · complex interleaving/block interleaving · memoization · automatic vectorization

1 Introduction

FFT has been studied far and wide. Every year, new results about its implementation appear, boosting the speed of some of the most famous versions, such as FFTW [1]. Recently, NEC-SX Aurora Vector Engine has been used to test the behaviour of some FFT's implementations on large vector registers (256 double, 16384 bit per register) [2]. That is an important result for our work, since it pushes the usage of SIMD/vectorized architectures. Nevertheless, outside the world of High Performance Computing (HPC), the most available SIMD technology is the AVX-512 extension (see Sect. 3) which is spreading among x86 CPUs, both Intel and AMD. Hence in this work we focus on the latter, with the following contributions:

- we experiment a different and uncommon way to memorize complex numbers;
- we work on a manual vectorization of the FFT, keeping it simple and readable so that it can be used for teaching purposes;
- we measure the performances of different versions, pointing out the advantages of using AVX-512.

R. Berta and A. De Gloria (Eds.): ApplePies 2022, LNEE 1036, pp. 255–263, 2023.
https://doi.org/10.1007/978-3-031-30333-3_34

2 Different Data Layouts: An Overview

We can memorize complex numbers as pairs of floating point numbers, choosing the dimension (single/double precision) best suited for the needs of the computation at hand. Once we have a `complex` structure, by instantiating an array we obtain a sequence of numbers where the real and imaginary parts are staggered. We can call this kind of memorization *complex interleaved* (Fig. 1a).

An alternative is to memorize separately the two components in two arrays of floating point numbers. This is called *block interleaved* memorization (Fig. 1b): that is not common at all since existing software and standards for C/C++ only support the interleaved data format [3], but it could be useful dealing with vector registers and data gathering from memory.

Exploiting a mixture of these memorizations to boost the performance of algorithms on SIMD architectures has already been studied [3], achieving up to 2x performance improvements over state of the art library implementations. Our work will study and compare both types of memorization.

3 The AVX-512 Instruction Set

It has been a while now since computers have had some kind of SIMD extensions. SSE and AVX2 are fundamental in the history of this process, though their usage was limited by the length of their registers, respectively 128 and 256 bits. AVX-512 came out in 2013 as an improvement of the latter, introducing new instructions and providing registers 512 bits long.

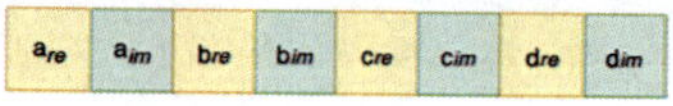

(a) Complex Interleaved

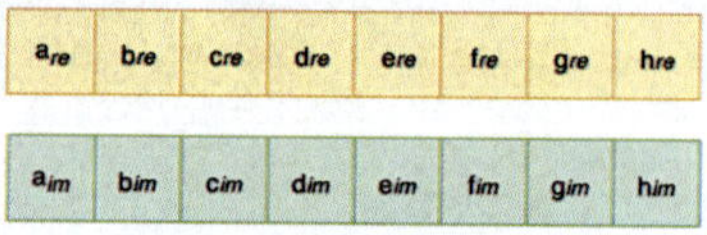

(b) Block Interleaved

Fig. 1. Different memorizations for array of complex numbers

Despite the speedup you can get from these instructions, AVX-512 is not always appropriate: it does not make IO-bound programs faster, as well as programs with complex conditional behaviours, since there are no parallel operations to execute; the tasks which can be boosted because of their parallelism regard AI, cryptography, mathematical computations... Programmers should understand where and when this extension could be useful, in order to gain a speedup which is independent from the algorithm itself.

Browsing the Intel's intrinsics guide[1] is a great starting point: this way one may familiarize with nomenclature and the different available instructions.

[1] https://www.intel.com/content/www/us/en/docs/intrinsics-guide/index.html.

3.1 Loop Unrolling

One of the main usage of SIMD instructions is through *loop unrolling*, which allows to avoid loops or diminish the number of iterations. For instance, suppose one need to compute an *Hadamard porduct* of two arrays of 8 `doubles` each. Exploiting AVX-512, one can proceed in this way:

```
// Declaration of the arrays
double vector1[8];
double vector2[8];
double result [8];
// Load the two arrays
__m512d _vec1 = _mm512_load_pd((void*)vector1);
__m512d _vec2 = _mm512_load_pd((void*)vector2);
// Compute the sum using an AVX-512 intrinsic
__m512d _res  = _mm512_add_pd(_vec1, _vec2);
// Store the result
                _mm512_store_pd((void*)result, _res);
```

In case the length of the arrays was greater than 8 (but, for simplicity, still multiple of 8) one could iterate this snippet $N/8$ times.

3.2 Superword Level Parallelism

SLP is another technique widely adopted by programmers and compilers to perform vectorization. It consists of gathering instructions which are similar but not directly linked, and computing them using SIMD registers. An example of this technique is shown in Sect. 5.2.

4 The Recursive Version of the FFT Algorithm

Radix-2 algorithm is the simplest way to decrease the complexity of the DFT (Discrete Fourier Transform), from $O(N^2)$ to $O(NlogN)$, though nowadays FFT algorithms are thousands of lines of code long (they perform different operations based on different kinds of input and of the available hardware). Furthermore, an iterative version of the algorithm can be way more optimized than a recursive one, which is forced to use the stack an exponential amount of times.

Despite this, our work just aims to experiment with the operation of manual vectorization of the code; for this reason, we looked for an algorithm which is both interesting and useful in real-world applications: the recursive FFT is simple and follows the mathematical expression provided by Cooley and Tukey in 1965 [4]. It was easier to get into, but feasible enough to test AVX-512 capabilities and the two types of memorization as shown above. As most of the real-world use cases of the FFT, we will just consider input whose length is a power of 2.

4.1 Twiddle Factor: How Memoization Can Help

The first thing we observed about the trivial version of FFT we used was the enormous usage of trigonometric functions to calculate the roots of the unity. For each call of the function on an input of length N, one have to use all the roots of order N, expressed as

$$e^{-j\frac{2\pi k}{N}} = \cos(\frac{2\pi k}{N}) - j\sin(\frac{2\pi k}{N}) \quad k \in \{0, 1, ..., N-1\}$$

The trigonometric functions are tremendously time-consuming for CPUs. We made a simple benchmark[2] of the FFT with an input of 2^{13} both using and avoiding the computation of these roots: the latter was two times faster than the former.

An immediate observation is that once one computed the roots of order N, the following calls of the function with an input of the same length can use them again. This technique is well known in literature as *twiddle factor* [5].

The needed roots can be both calculated before executing the algorithm or computed them on the way and saved for further use. That is what we did: we introduced a *look-up table* where we saved the results of the computation for N, so that we could access them later. The idea recalls the *memoization* of dynamic programming.

5 The Vectorized Version of the Recursive FFT

We made two different versions of the AVX-512 FFT, one with a complex-interleaved memorization (called CI_AVX), another with a block-interleaved memorization (called BI_AVX). The C++ source code of both the versions has been made public available and can be downloaded from this link: https://github.com/pcineverdies/FFT-AVX-512.

5.1 Link to Hadamard Product

The main element to vectorize the function is to exploit the radix-2 expression of the FFT: given an input X of size $N = 2^M$, its DFT is equal to

$$DFT(X)_k = DFT(X_e)_k + e^{-j\frac{2\pi k}{N}} \cdot DFT(X_o)_k \quad k \in \{0, 1, ..., N-1\}$$

where $DFT(X_e)$ is the DFT of the even terms of X, $DFT(X_o)$ is the DFT of its odd terms. As we compute these two elements using recursion, the result is made by the element-wise product between the vector of the roots and $DFT(X_o)$, added to $DFT(X_e)$. That is an interesting result, since element-wise product can be easily vectorized with both the memorizations of complex numbers. An intuitive idea of the process is shown in Fig. 2 (that figure is inspired by one found in [3]).

[2] The machine we used has an Intel® Xeon®, 15 GiB of RAM and Ubuntu 4.15.0-171 as OS.

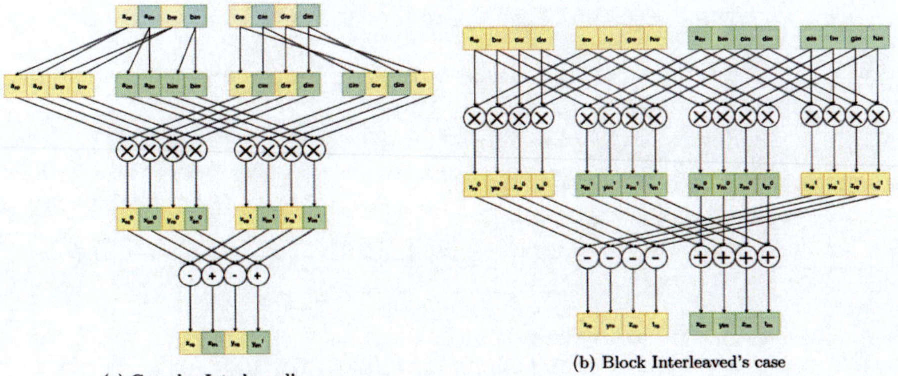

(a) Complex Interleaved's case

(b) Block Interleaved's case

Fig. 2. Element-wise product using two memorizations.

5.2 Base Cases of Recursion

As a consequence of the DFT's expression, when the input is a sequence of 4 or 2 elements everything can be brought back to additions and subtractions between real and imaginary parts of the input. That is why we can avoid a recursion up to a sequence of length 1 (the DFT of a number is the number itself), and stop at a length of 4. We also added the cases of input with a length of 2 and 1, which are stand-alone situations.

In this base cases, we tried to apply SLP, as shown in the snippet below (DFT of an input of length 4 in the array of complex `wave`, with CI memorization): the first block of instructions gathers data in a correct way, while the second one computes the additions/subtractions which give us the final result.

```
// ...
// _vecX are __m512d data
if(n == 4){
    // Load of data
    _vec0 = _mm512_broadcast_f64x2(_mm_load_pd((double *)&(wave[0])));
    _vec1 = _mm512_broadcast_f64x2(_mm_load_pd((double *)&(wave[1])));
    _vec1 = _mm512_permute_pd(_vec1, 0b01100110);
    _vec2 = _mm512_broadcast_f64x2(_mm_load_pd((double *)&(wave[2])));
    _vec3 = _mm512_broadcast_f64x2(_mm_load_pd((double *)&(wave[3])));
    _vec3 = _mm512_permute_pd(_vec3, 0b01100110);

    // Compute DFT
    _vec0 = _mm512_mask_sub_pd(_vec0, 0b01111000, _vec0, _vec1);
```

```
        _vec0 = _mm512_mask_add_pd(_vec0, 0b10000111,
_vec0, _vec1);
        _vec0 = _mm512_mask_sub_pd(_vec0, 0b11001100,
_vec0, _vec2);
        _vec0 = _mm512_mask_add_pd(_vec0, 0b00110011,
_vec0, _vec2);
        _vec0 = _mm512_mask_sub_pd(_vec0, 0b10110100,
_vec0, _vec3);
        _vec0 = _mm512_mask_add_pd(_vec0, 0b01001011,
_vec0, _vec3);

        // Store result
        _mm512_store_pd((void*)&wave[0], _vec0);

        return;
}
// ...
```

6 Experimental Results

In the next subsection we provide the obtained numerical results, while in the following we discuss why, in our implementation, block interleaved does not give any additional speedup.

6.1 Numerical Results

We measured the performance of six versions of the FFT:

- NO_AVX, which is a standard version of the FFT, optimized with the twiddle factor, compiled with 03 flag but without auto-vectorization;
- VE_AVX, which is the same as above, though auto-vectorization is enabled;
- CI_AVX, which is the version vectorized by hand with *complex interleaved* memorization, compiled with 03 flag.
- BI_AVX, which is the version vectorized by hand and *block interleaved* memorization, compiled with 03 flag.

In order to do that, we calculated how much time passed between the call of the function and its termination: after 2^{13} measures, we extracted the median of the data, which is more statistically stable than the arithmetic average.

Some of the results are shown in Table 1, while a complete overview for N between 2^3 and 2^{17} can be found in Fig. 3. As it is evident from the numbers, the vectorized versions are more efficient the the standard one, by the **33.78%**($\sim 1.51\times$).

6.2 Why Is *Block Interleaved* not Good Enough in our Setting?

As we immediately notice from the result, the *block interleaved* version is quite similar to the *complex interleaved* one; in some cases it is even slightly slower. That is not what we expected: since this memorization method is not common, we would need a major speedup to use it.

In [3] the authors use a *mixed* version of the two methods: they start from a CI input, then they use the computations of the algorithm itself to get a BI memorization (which makes some operations faster, such as the element-wise product, since it reduces the usages of slow instructions as permutations) and

Table 1. Execution time (μs) of FFT for some values of N

N	NO_AVX	VE_AVX	CI_AVX	BI_AVX
4096	976.0	952.0	717.0	730.0
8192	1955.0	1887.0	1415.0	1444.0
16384	2918.0	2784.0	2393.0	2312.0
32768	4506.0	4244.0	3168.0	3368.0
65536	8078.5	7778.5	5892.0	5489.5
131072	15409.0	13767.0	10203.5	10379.5

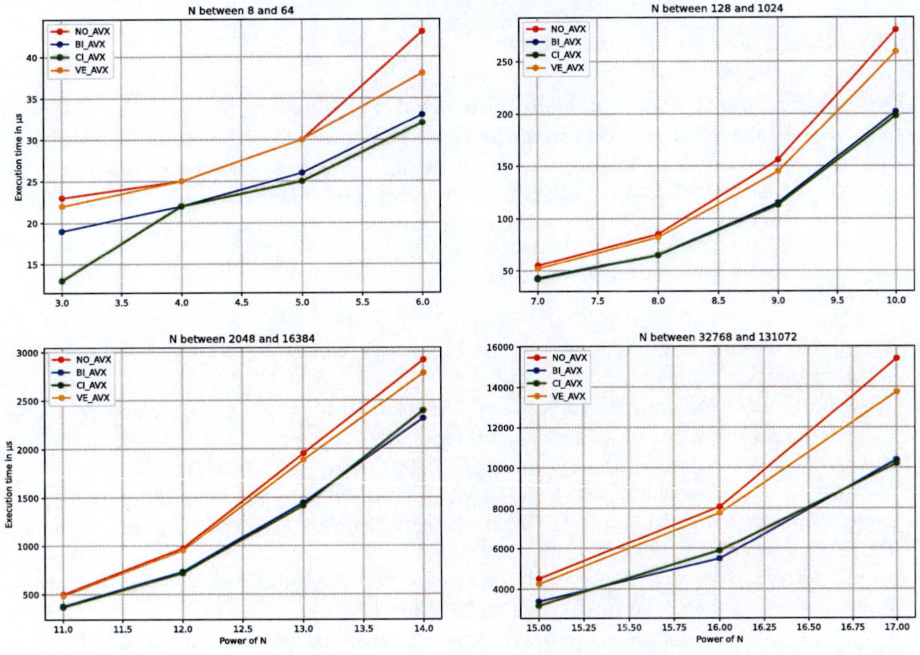

Fig. 3. Charts of the measures

they finally end up with a CI result. Instead, since we start directly from a block interleaved version, we are note able to replicate their speedups when using BI.

7 Conclusions

The final result of our experiments is a speedup of a **33.78%**(∼1.51×) between the first trivial version and the vectorized one. The automatic vectorization reaches a much lower speedup, which amounts to **10.66%**(∼1.12×). We have shown that BI memorization, while being not common and not compatible with standards like POSIX, does not provide any advantage over CI.

We would like to point out the importance of the AVX extension for programs which aim to achieve efficiency and speed. Clearly, writing our own vectorized code is not the best way to exploit this functionality, since we could make mistakes and it becomes difficult to maintain: the right approach should be to ask the compilers to introduce the functionality mentioned in the previous sections, suggesting some choices using `pragmas`.

In the end, the result could be not satisfying enough: in that case the programmer can disassemble the compiled code and try changing some instructions to speedup the program. And that is why it is important to be familiar with this extension. This is what we have learnt in this study. As a future work, we will:

- extend the code to exploit multi-threading, using the recently introduced C++ standard class for multi-threading;
- realize a port on CPU clusters [6];
- investigate how to optimize the impact on cache hierarchies [7].

Acknowledgments. Work partially supported by H2020 project TEXTAROSSA (grant no. 956831), https://textarossa.eu/) and partially by the Italian Ministry of Education and Research (MUR), CrossLab project (Departments of Excellence). We also want to thank Prof. Carlo Vallati for providing the machine used to run the experiments and Emanuele Ruffaldi for interesting discussions on the topic.

References

1. Frigo, M., Johnson, S.: The design and implementation of FFTW3. Proc. IEEE **93**(2), 216–231 (2005)
2. Vizcaino, P., Mantovani, F., Labarta, J.: Accelerating fft using nec sx-aurora vector engine. In: Chaves, R., et al. (eds.) Euro-Par 2021. LNCS, vol. 13098, pp. 179–190. Springer, Cham (2022). https://doi.org/10.1007/978-3-031-06156-1_15
3. Popovici, D.T., Franchetti, F., Low, T.M.: Mixed data layout kernels for vectorized complex arithmetic. In: 2017 IEEE High Performance Extreme Computing Conf. (HPEC), pp. 1–7 (2017)
4. Cooley, J.W., Tukey, J.W.: An algorithm for the machine calculation of complex Fourier series. Math. Comput. **19**(90), 297–301 (1965)
5. Gentleman, W.M., Sande, G.: Fast Fourier Transforms: for fun and profit. In: Proceedings of the November 7–10, 1966, Fall Joint Computer Conference, ser. AFIPS '66 (Fall), pp. 563–578. Association for Computing Machinery, New York (1966). https://doi.org/10.1145/1464291.1464352

6. Sharp, D., Stoyanov, M., Tomov, S., Dongarra, J.: A more portable HeFFTe: implementing a fallback algorithm for Scalable Fourier Transforms. In: 2021 IEEE High Performance Extreme Computing Conf. (HPEC), pp. 1–5 (2021)
7. Takahashi, D.: High-Performance FFT Algorithms, pp. 41–68. Springer, Singapore (2020). https://doi.org/10.1007/978-981-13-9965-7_6

An Image Processing Algorithm to Optimize the Output Configuration of a Photonic Integrated Circuit

Luca Gemma[1(✉)], Martino Bernard[2], and Davide Brunelli[1]

[1] Department of Industrial Engineering, University of Trento, 38123 Povo, Italy
luca.gemma@unitn.it

[2] Centre for Sensors and Devices, Fondazione Bruno Kessler, 38123 Povo, Italy

Abstract. The interest in silicon photonics as a quantum enabling technology is rapidly growing, and Photonic Integrated Chips (PICs) have been proven to be a robust and viable solution in such research fields. As this technology applied to the quantum world is relatively young, some areas of interest remain uninspected, especially the control and output optimization. In this work, we propose an image processing tool to control and optimize a PIC based solely on images captured by a camera and without invasive output detectors. We tested this architecture on a Silicon Oxynitride (SiON) PIC where several Mach-Zehnder interferometers can be voltage driven by Titanium-Titanium Nitride (TiTiN) thermistors. By comparing the results of the image processing algorithm with those retrieved by silicon photodetectors on the same chip, we have proven that our approach can match or even outperform the traditional approach of sensing outputs with silicon photodetectors.

Keywords: quantum circuits · PIC · image processing · optimization algorithm

1 Introduction

Quantum advantage has been claimed by different quantum technologies: in 2019, Sycamore, Google's quantum computer, solved in 200 s a problem that would have taken Summit, the fastest non-quantum computer, at least two days to resolve [16]. In 2020, Jiuzhang, a photonic-based quantum computer, solved the boson sampling problem, which would have taken a classical computer nearly 600 million years to solve [9]. Since then, many companies are inspecting different quantum technologies as enablers for quantum computing: IBM [8] and Rigetti [10] with superconducting circuits, QuiX Quantum [15] and PsiQuantum [7] with photonic circuits, DWave [4] with quantum annealing, IonQ [3] with trapped ions and Microsoft [18] with topological systems. Photonic Integrated Circuits (PICs) represent a viable solution to photonic quantum systems as they are based on well-known structures (used, for example, in telecommunications). They do not require bulky architectures for hard-cooling (as instead

R. Berta and A. De Gloria (Eds.): ApplePies 2022, LNEE 1036, pp. 264–272, 2023.
https://doi.org/10.1007/978-3-031-30333-3_35

superconducting circuits do) and can be embedded almost effortlessly with electronics as their fabrication process has many similarities with the Very Large Scale Integrated (VLSI) circuits. PICs usually require a chain of tunable channels (as shown for example, in [15]), and nowadays, such tuning is not straightforward but requires devices and a specific algorithm for the calibration. This paper proposes an automatic tool to achieve an optimal configuration without the need for bulk and invasive probes or fixed on-chip detectors and for arbitrary inspecting any region of the PIC and thus considering field intensity in arbitrary regions of the PIC as targets to be optimized. We tested the entire toolchain on PICs that we have fabricated in our facilities and compared the performances of the image processing algorithm with an analogue algorithm based on more classical feedback outputs, such as silicon photodetectors embedded in our PICs. This paper is organized as follows: in Sect. 2 we illustrate the state-of-the-art briefly for silicon photonics in quantum circuits. Section 3 summarizes our hardware and software architecture, focusing on our image processing tool. In Sect. 4 we illustrate our measurements and how they were achieved. Finally, in Sect. 5 conclusions are drawn.

2 Related Works

SiN has been proven to be a viable and robust solution for PIC architectures [13], particularly in the near-infrared, where our work is inserted. In addition, PICs are now considered one of the main architectures when dealing with basic quantum building blocks, such as the Mach-Zehnder Interferometers (MZIs) [2], e.g., an effective redundant alternative to beam splitters that allow for controlling both the amplitude and the phase-shift of the light. We based this work on previous studies we have conducted to characterize our output detectors [6], namely silicon photodiodes, and on our phase-shifters [5], i.e., metallic Ti-TiN thermistors. Concerning the layout, we have chosen to follow a Clements architecture, with which it is possible to achieve a compact structure with low propagation losses and a restricted optical depth [11]. Although being a better solution w.r.t. the simpler counterpart, called "Reck" architecture [12], the optimal calibration of such a structure is still an open issue. Even if image processing has already been used for integrated optics, from fiber alignment [14] to defects inspection [17], up to our knowledge, there are still no works proposing a solution based on image processing to calibrate and tune the PIC outputs based on scattering from the waveguide. This paper presents an optical tool to non-invasively retrieve the optical field intensity in arbitrary points within a configurable PIC, thus assessing its state by exploiting light scattering from its waveguides. This tool can be used to assess the state of a Quantum PIC by temporarily injecting laser light in the circuit to assess, map and set the state of the PIC, and then injecting few-photon states for a proper experiment.

3 Architecture

We fabricated our chips at the Bruno Kessler Foundation's (FBK) Sensor and Devices facility [1] on a 150 mm diameter crystalline wafer with a top epitaxially grown Silicon layer. The silicon photodiodes are realized through ion implantation, and an oxide cladding is realized over the substrate. On top of that, a patterned Silicon Oxynitride (SiON) layer constitutes the photonic layer. The photonic layer is then covered by silicon oxide, acting as protection and optical insulation. Finally, thermistors are realized on top, with a stacked layer of Titanium-Titanium Nitride and Titanium (Ti-TiN-Ti). After fabrication and dicing, the chips have all been connected to custom PCBs with wire bonding, each featuring a total of 10 MZIs with 27 thermistors. Thanks to the thermo-optic effect, by acting on our metallic thermistors, it is possible to change the optical path of the injected laser beam by effectively changing the refractive index of local regions of the waveguides. We controlled the thermistors using voltage, and thus, by Joule's effect, dissipating power and locally generating heat; such control is described in Fig. 1 where from a rest state of equally distributed output intensities, the system moves to a configuration where just a specific branch of an MZI (in (a)) or output of the entire system (in (b)) is active. The optical tool we built is based on images taken from the live stream of a Thorlabs camera which captures the image of a portion of the PIC from above. All the project is based on Python 3.9. We controlled our DoFs (i.e., the phase-shifters) through a Q8b driver model and its custom python module. For validating the approach, a 2450 Keithley digital multimeter was connected to the silicon photodiode of the target output to be optimized, using the PyVisa library to interface and control the instrument. The Opencv cv2 library was used for all the image processing operations. Throughout all our work, we aimed to find the best combination of drive voltages for the thermistors that better maximized the light intensity of a target output. The algorithm is briefly discussed as follows: first of all, a target output combination is specified through a vector with either "1" (in case of maximization) or "0" (for minimization). For our experiments, such a target vector was composed of a single "1" and all other elements set to "0" (i.e., we look for maximizing a single output and minimizing all the other ones). Then an image of the excited PIC is acquired. From the image, the field intensity in specific rectangular regions of identical dimensions is estimated as the sum of all the inner pixel values for each of such regions. An additional "dark" region, far from the waveguides, is acquired and used as a background correction to compensate for possible changes in the light conditions. Therefore, we compute the offset as the sum of the inner pixel values for a "dark" region, and then we subtract it from each region. The measured outputs (i.e., the computed pixel values for each region) are then assembled in a measured output vector. Finally, the measured output vector is normalized to unity to be comparable with the target output vector. A cost function is applied to the target and measured output vectors, which minimizes the norm of the difference between these two vectors. A similar approach is used to realize a twin set of data by using an integrated photodiode to estimate the state of the "1" output channel. To compare the optical tool we developed with the classical approach of on-chip sensing, we then developed an algorithm to retrieve the current measurements sensed by the digital multimeter connected to the output to be maximized and look for the combination of drive voltages that better maximized such current readings.

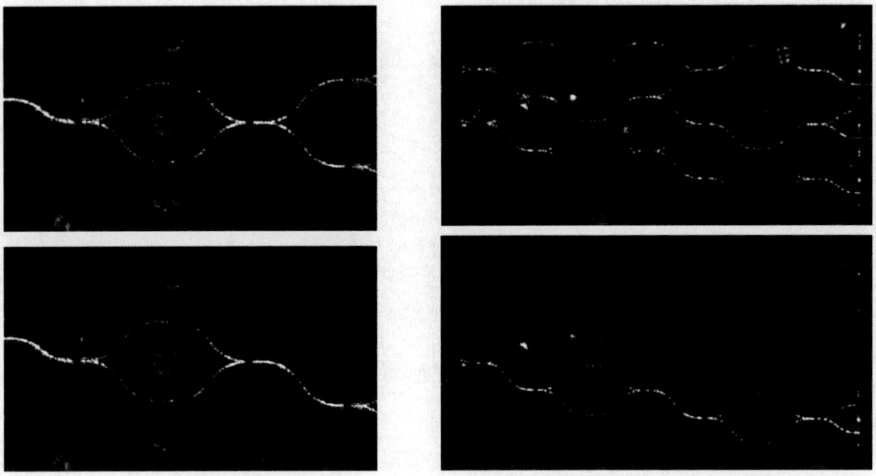

(a) single MZI (b) entire architecture

Fig. 1. Single MZI actuation (a): from an almost 50:50 split ratio (top) to a full transfer (bottom). Full architecture (b): from equally distributed output intensities (top) to a specific target output (bottom).

4 Results

We tested a Silicon Nitride PIC by controlling the 27 metallic thermistors, thus effectively achieving a 27 DoFs system and sensing one output at a time through 2450 Keithley digital multimeter connected to the respective silicon photodiode. The light beam was generated through a near-infrared 850 nm laser. We conducted two types of measurements: in the first set, the image processing tool was used to optimize (thus maximize) a target output, then a second set was produced as a reference and performance comparison by running the same optimization algorithm but based on sensed photocurrents from a single photodiode polarized in reverse bias at $-3\,\mathrm{V}$ [6], thus optimizing the same target output, which in turn means minimizing it (as the measurement is conducted in inverse polarization). In both cases, we controlled our thermistors by providing a 0 to 12 V voltage sweep with a step of 0.5 V. Each set of measurements was conducted for different input-output combinations, keeping the same configuration between the first and the second set (Fig. 2).

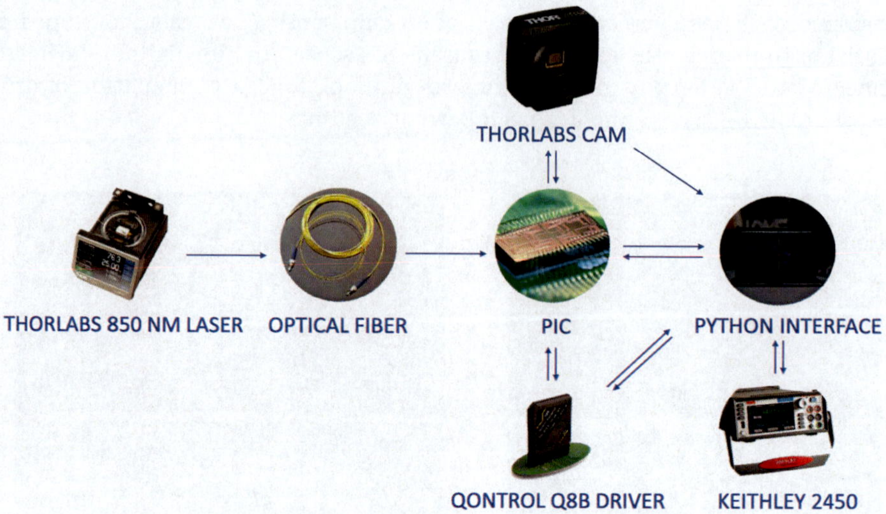

Fig. 2. System architecture.

For the sake of clarity for the reader, it is worth noting that we enumerate our inputs and outputs from bottom to top; thus, input (or output) 1 is the bottom one, and input (or output) 6 is the top one. We report the results obtained for one of the middle inputs (input 4) as this represents a scenario involving one of the highest combinations of degrees of freedom to be controlled and optimized. The starting configuration is shown in Fig. 3, where it can be noted that the majority of the incoming NIR beam is routed, due to the uncontrolled initial state of the system, into the last top couple branches (output 5 and 6). The image processing-based optimization algorithm was then run on such configuration specifying as target output configuration a maximization of output 4 and minimization for all the remaining outputs. The estimated pixel intensities evolution during optimization for each outputs are shown in section (a) of Fig. 4. On the y-axis, the computed pixel intensity sum is displayed as a function of the evolution coordinate (i.e., image number) as each voltage sweep is performed on all detected DoFs. It is worth noting that after each V-shaped curve (that represents a sweep for a single thermistor), the information about the optimization is preserved: after each maximization (red line), the signal jumps to the local maximum (i.e., the highest value reached during the single sweep) while after each minimization (all other colored lines) the signal jumps to the local minimum (i.e., the lowest value achieved). This means that the system is correctly set to the optimal local configuration reached during each sweep. The final maximization (i.e., the optimization of the output 4, in red) achieved reached more than twice the starting value and, generally, the minimization routines performed well, with two outputs (5 and 6, respectively in purple and brown) dramatically decreased, and two outputs (1 and 2, blue and orange) almost kept constant. Although the behavior of output 3 (in green) was counter-trend compared with the other minimizations, this trend can be simply explained by considering that such output is correlated to output 4 (i.e., the one to be maximized) as they lie on the same output branch. Further considerations can be made on the presence of negative values: this is an artifact due to the subtraction of a "dark" computed in a single region, which can reasonably have a different background level with respect to the probe regions, and thus it is possible for the background-subtracted value to fall below zero. This issue can be addressed in future implementations with region-wise dark subtraction. The same experiment was then repeated with the Keithley-based optimization algorithm (section (b) in Fig. 4), which aimed at maximizing output 4 by sensing the photocurrent in the reverse bias region of a silicon photodiode coupled with the waveguide. As previously stated, although this is a maximization task, the actual task is to maximize the modulus of the negative photocurrent as the experiment is conducted in reverse polarization, and therefore it appears as a minimization task in the image. Similarly to the image processing case, also in this scenario, the maximization trend is visible throughout the entire experiment, with the algorithm keeping the best (lowest) value reached during every single sweep, confirming the correct functioning of the optimization algorithm. Moreover, the result of the maximization procedure is really similar (about twice the starting value) to what we obtained with the optical tool, confirming the effectiveness of the optical tool as a viable and strong alternative to the more invasive and limited classical approach. The similarities of the final configurations can be seen in Fig. 3 where the image processing tool outcome (b) is notably similar to the Keithley-based one (c), and both achieve the goal of successfully maximizing output 4, thus again validating our developed optical tool.

(a) Starting Image (b) End image Image processing (c) End image Keithley

Fig. 3. Starting configuration for input 4 and output 4 (a). Final configuration after image processing tool (b) and Keithley-based tool (c).

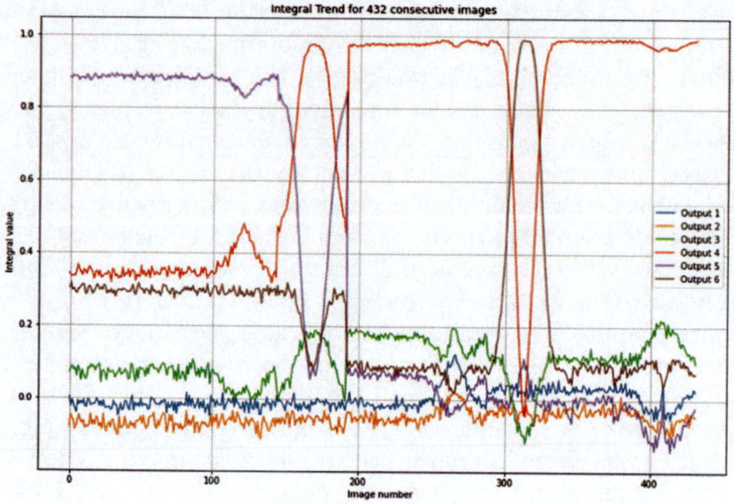

(a) Image Processing

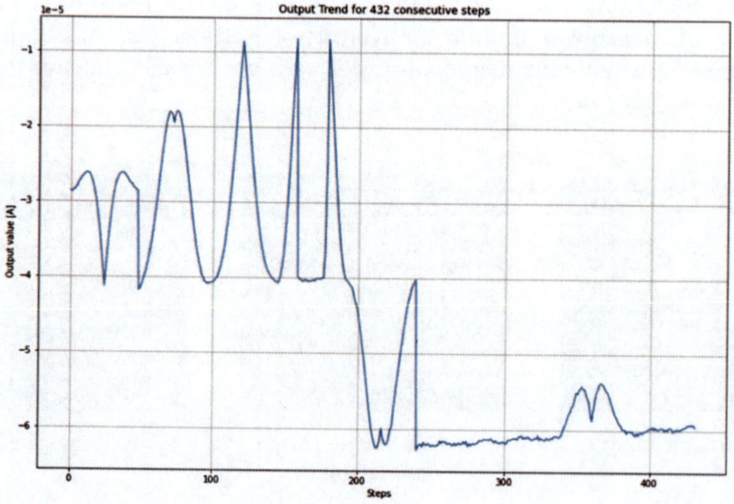

(b) Keithley

Fig. 4. Optimization plots for input 4 and output 4: image processing showing estimated pixel intensities for all outputs (a) and Keithley-based showing sensed photocurrent in reverse bias (b).

5 Conclusions and Future Work

This paper presents an optical tool based on image processing to automatically control and tune multiple outputs of a PIC architecture. This approach has been proven to achieve comparable results w.r.t. a classical sensor-based approach with silicon photodiodes. Moreover, a considerable advantage of this new solution is the ability to inspect and tune intermediate MZI outputs, which are, in general, not easily or arbitrarily accessible with monitor photodiodes. In conclusion, this tool can offer a valid alternative to invasive on-chip detectors. Future work can include the addition of further intelligence to the algorithm in order to automatically identify the best combination of degrees of freedom to use for each specific input/output combination (up to now, this is done by hand by the user as it involves non-trivial considerations of path choices). Moreover, an interesting work would be inspecting the impact of machine learning algorithms on the optimization algorithm itself, e.g., adding a Convolutional Neural Network (CNN) for the image processing tool.

Acknowledgment. We acknowledge the support of the MNF Laboratory staff of FBK during sample fabrication. We acknowledge financial support from the Autonomous Province of Trento, under the initiative *"Quantum at Trento - Q@TN"*, projects *Q-PIXPAD* and *CoSiQuP*. This project has received funding from the European Union's Horizon 2020 research and innovation programme under grant agreements No 777222, ATTRACT *INPEQuT and No 899368,* EPIQUS.

References

1. Sensor and devices, fondazione bruno kessler. https://sd.fbk.eu/en/about-us/
2. Annoni, A., et al.: Automated routing and control of silicon photonic switch fabrics. IEEE J. Sel. Top. Quantum Electron. **22**(6), 169–176 (2016)
3. Blümel, R., et al.: Efficient stabilized two-qubit gates on a trapped-ion quantum computer. Phys. Rev. Lett. **126**(22) (2021). https://doi.org/10.1103
4. Boothby, K., et al.: Architectural considerations in the design of a third-generation superconducting quantum annealing processor (2021). arxiv.org/abs/2108.02322
5. Gemma, L., Bernard, M., Ghulinyan, M., Brunelli, D.: Analysis of control and sensing interfaces in a photonic integrated chip solution for quantum computing. In: Proceedings of the 17th ACM International Conference on Computing Frontiers, CF 2020, pp. 245–248. Association for Computing Machinery, New York (2020). https://doi.org/10.1145/3387902.3394034
6. Gemma, L., Bernard, M., Ghulinyan, M., Brunelli, D.: Analysis of photodiode sensing devices in a photonic integrated chip solution for quantum computing. In: 2020 IEEE SENSORS, pp. 1–4. IEEE (2020)
7. Kim, I.H., Liu, Y.H., Pallister, S., Pol, W., Roberts, S., Lee, E.: Fault-tolerant resource estimate for quantum chemical simulations: case study on li-ion battery electrolyte molecules. Phys. Rev. Res. **4**(2), 023019 (2022)
8. Liu, Y., Minev, Z., McConkey, T., Gambetta, J.: Design of interacting superconducting quantum circuits with quasi-lumped models. Bulletin of the American Physical Society (2022)

9. Nature: Light on quantum advantage (2021). https://www.nature.com/articles/s41563-021-00953-0#ref-CR3
10. O'Brien, W., et al.: Superconducting caps for quantum integrated circuits (2017). https://arxiv.org/abs/1708.02219
11. Pérez, D., Gasulla, I., Mahapatra, P.D., Capmany, J.: Principles, fundamentals, and applications of programmable integrated photonics. Adv. Opt. Photon. **12**(3), 709–786 (2020). https://opg.optica.org/aop/abstract.cfm?URI=aop-12-3-709
12. Reck, M., Zeilinger, A., Bernstein, H.J., Bertani, P.: Experimental realization of any discrete unitary operator. Phys. Rev. Lett. **73**, 58–61 (1994). https://link.aps.org/doi/10.1103/PhysRevLett.73.58
13. Roeloffzen, C.G.H., et al.: Low-loss si3n4 triplex optical waveguides: technology and applications overview. IEEE J. Sel. Top. Quantum Electron. **24**(4), 1–21 (2018)
14. Samad, A.A., Unni, C.: Image processing based end-view alignment for symmetric specialty optical fibers. In: 2018 Second International Conference on Inventive Communication and Computational Technologies (ICICCT), pp. 1080–1082 (2018)
15. Taballione, C., et al.: 20-mode universal quantum photonic processor (2022). https://arxiv.org/abs/2203.01801
16. Times, N.Y.: Why google's quantum supremacy milestone matters (2019). https://www.nytimes.com/2019/10/30/opinion/google-quantum-computer-sycamore.html
17. Vahabi, N., Yang, D., Selviah, D.R.: Improving data transmission in fiber optics by detecting scratches on the fiber end face. In: 2018 IEEE British and Irish Conference on Optics and Photonics (BICOP), pp. 1–4 (2018)
18. Vaitiekenas, S., et al.: Fluxinduced topological superconductivity in full-shell nanowires. Science **367**(6485), eaav3392 (2020)

Multi-objective Framework for Training and Hardware Co-optimization in FPGAs

Mohammad Amir Mansoori(✉) and Mario R. Casu

Department of Electronics and Telecommunications,
Politecnico di Torino, Turin, Italy
mohammadamir.mansoori@polito.it

Abstract. Although several works have recently addressed the problem of performance co-optimization for hardware and network training for Convolutional Neural Networks, most of them considered either a fixed network or a given hardware architecture. In this work, we propose a new framework for joint optimization of network architecture and hardware configurations based on Bayesian Optimization (BO) on top of High Level Synthesis. The multi-objective nature of this framework allows for the definition of various hardware and network performance goals as well as multiple constraints, and the multi-objective BO allows to easily obtain a set of Pareto points. We evaluate our methodology on a network optimized for an FPGA target and show that the Pareto set obtained by the proposed joint-optimization outperforms other methods based on a separate optimization or random search.

Keywords: Machine Learning · FPGA · High-Level Synthesis · Bayesian Optimization · Co-optimization

1 Introduction

Like in other Machine Learning (ML) models, in Convolutional Neural Networks (CNNs) the number of layers, neurons, kernel size, number of filters, etc., can be considered as hyper-parameters to tune in order to maximize the accuracy achievable during training. When it comes to implementing a CNN in FPGA using a dedicated accelerator, however, the accuracy requirements and the corresponding network architecture that meets those requirements might be in contrast with hardware-related requirements and constraints, such as latency and FPGA resources utilization. Therefore, a trade-off must be found, and this can be done by solving a multi-objective optimization problem.

In recent years, several approaches based on Hardware-aware Neural Architecture Search (Hw-NAS) have addressed the problem of co-optimizing the network and its hardware accelerator [1,2]. To solve the multi-objective optimization problem, some works use a two-stage optimization, which we call *separate* method [3,4], in which the network parameters are first tuned to maximize the accuracy, and then the hardware configurations are tuned to meet the hardware constraints (e.g., data precision can be reduced). Other works try to reshape the problem into a single-objective one subject to some constraints on hardware

R. Berta and A. De Gloria (Eds.): ApplePies 2022, LNEE 1036, pp. 273–278, 2023.
https://doi.org/10.1007/978-3-031-30333-3_36

performance [5,6]. Another method combines multiple objectives in a single function and uses single-objective optimization approaches [7], even though merging multiple objectives can limit the performance of the optimization and degrade the final Pareto-optimal sets. The last methodology is to employ a *truly* multi-objective optimization approach to obtain the non-dominated Pareto solutions. This approach has attracted considerable attention in the evolutionary computation community [8]. However, the computational complexity in evolutionary algorithms is the main limitation of these approaches.

Not all Hw-NAS approaches co-optimize the network and its accelerator: most of them use a *fixed hardware configuration* (fixed-Hw) where the search space is limited to the model architecture [1]: if the hardware requirements are not met, the network architecture must change. Another category uses *multiple hardware configurations* (multi-Hw) in which the search space combines hardware configurations and network architectures, which is also our approach.

The optimization algorithms in Hw-NAS are usually divided into Reinforcement Learning (RL), Evolutionary algorithms (EAs), Gradient-based methods, and Bayesian Optimization (BO) methods [9]. RL and EA have been recently used for both *fixed-Hw* and *multi-Hw* categories (e.g., the Genetic Algorithm in [10]). Despite their proven effectiveness in several works, they have some drawbacks: EAs are computationally intensive due to their requirement for a large *population* size in each *generation*; in RL, customizing the *policies* and *reward function* for each optimization problem is a challenge.

Gradient-based methods fall in the *fixed-Hw* category. In these methods a super-network containing all the possible realizations in the search space is trained and a sub-network is sampled in each Hw-NAS iteration, guiding the search to the optimal network [11]. Although hardware metrics can be included in the loss function of the super-network, the search space includes only the network configurations, hence the fixed-Hw categorization.

BO approaches consider a Gaussian Process for each objective, which is a natural fit for the optimization problems in Hw-NAS, and no customization is required in contrast to RL methods [12,13]. In addition, BO enables a truly multi-objective optimization. Despite these advantage, to the best of our knowledge, BO methods in FPGAs have been used only for the fixed-Hw category.

In this paper, we propose a new framework based on Multi-Objective Bayesian Optimization with Constraints (MOBOC) on top of High Level Synthesis (HLS) to jointly optimize the network architecture and the HLS-based hardware configurations for FPGA devices. The search space supports multi-Hw configurations. In this multi-objective framework, we can assign multiple objectives and constraints related to the hardware and network performance, and use the truly multi-objective BO approach to find the optimal Pareto sets.

2 Proposed Methodology

Our method can be divided in the four parts shown in Fig. 1 (B-E) and connected inside the multi-objective BO framework (A). MOBOC will update the samples

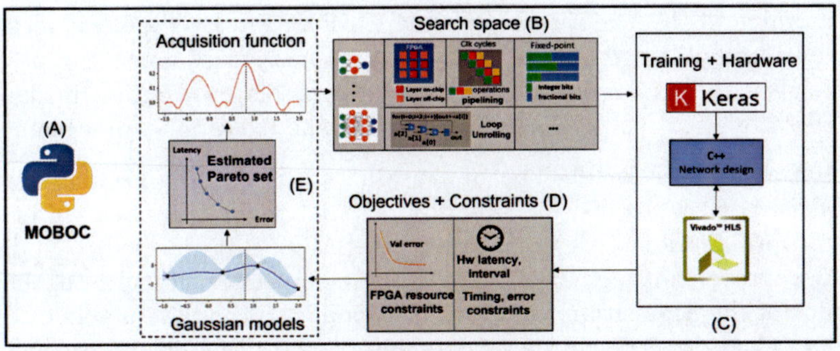

Fig. 1. Proposed methodology for training and hardware co-optimization in FPGAs.

from the search space based on the expected improvement in the Pareto sets to find the optimum configurations for network and hardware accelerator. In the following, we illustrate in detail MOBOC and each part of the framework.

Multi-objective BO with Constraints (MOBOC): BO is a method to optimize black-box functions that are expensive to evaluate. In MOBOC, a Gaussian model is initially built for each objective and constraint using few samples. An acquisition function is also built, based on the expected improvement of the estimated Pareto sets obtained from the Gaussian models. The maximum value of this function will suggest a new point in the search space. Objectives and constraints are evaluated for the new point and guide the search towards the optimum configurations by updating the Gaussian model at each iteration [14].

Search Space: It combines all the values of training hyper-parameters and HLS hardware configurations. Although the framework supports any ML model with an HLS code ready for FPGA implementation, in this work we focus on a CNN classifier for the MNIST dataset inspired from Lenet-5 [15] and consisting of two Convolutional, two max-pooling, and three dense layers. For each layer, the hyper-parameters include the number of neurons, filters, and kernel size.

Several HLS hardware knobs can configure the FPGA accelerator, such as the choice between on-chip or off-chip memory for storing the weights of a layer, specific HLS pragmas for loop pipelining and for loop unrolling with a configurable unroll factor, enabling or disabling the Dataflow pragma for task-level concurrency, the configuration of the fixed-point precision, and the value of the clock frequency. The optimal choice of these HLS *directives* and hardware parameters is achieved by MOBOC in conjunction with the optimal training parameters.

Function Evaluations: At each iteration, a new suggested point in the search space must be evaluated. For this evaluation, we used Keras for training and HLS C-simulation and Synthesis for hardware verification.

Objectives and Constraints: Even though we can assign any number of objectives and constraints, the larger the number, the longer the time to update the Bayesian model in each iteration. In this work, we selected to optimize the prediction error on the validation set, the hardware latency, and the throughput, with constrained FPGA resources (BRAM, DSP48, FF, LUT), clock period, and maximum admissible error (<10%). Note that instead of using the prediction error during training, we use the prediction error after execution in hardware: in this way the error includes also the effect of the fixed-point precision.

Update Bayesian Model: The new evaluation is used to update the Gaussian models. For the acquisition function, Predictive Entropy Search for Multi-objective Optimization with Constraints (PESMOC) [16] is used to update the expected improvement of the estimated Pareto set. After the update, the maximum value of the acquisition function corresponds to the parameters of the new suggestion. Convergence criteria or maximum iterations stop the procedure.

3 Results

The FPGA target is a Zynq-7000 SoC (XC7Z020-CSG484). For MOBOC, the latest *Spearmint* package is used [16]. The search space is described in Table 1, in which *On Chip* refers to the option of selecting on-chip BRAMs to store weights and bias values for each layer (as opposed to an off-chip external DRAM), hence 5 options are available for 5 layers. Total and Integer bits in Table 1 determine the fixed-point data precision that can be individually configured for weights, activations, and accumulations for the intermediate calculations (i.e., sum of products). Note that the search space size is on the order of 10^{16} total configurations—even without considering other parameters such as the learning rate, which prevents to do an exhaustive search of the best configurations. However, the BO convergence would be slower: in the current setting 100 BO iterations are sufficient to obtain a Pareto set, with more parameters more iterations would be needed.

Figure 2 shows the Pareto-optimal points in the space of prediction error evaluated in hardware (Hw error, x axis) and execution latency (Time, y axis) achieved by three different methods: Random search, Separate, and the proposed Joint method. The search space for all the methods is the same and the same limit of 100 iterations is used. In the Random case, we randomly choose in each iteration a point in the search space. In the Separate method, we first optimize the network hyper-parameters to obtain the best accuracy, and then BO optimizes the hardware configurations to maximize the hardware performance.

Configs	Range	Configs	Range
#Filters Conv1	[2,20]	Dataflow	active/inactive
		On Chip	active/inactive
#Filters Conv2	[2,20]	Clk (ns)	[8,20]
		Total bits	[10,20]
Kernel size	2*([1,4]) + 1	Integer bits	[1,9]
#Neurons Dense1/2	[50,150]	Conv1/2 UF	2^([0,4])

Table 1. Search space: network and hardware parameters (UF:Unroll Factor)

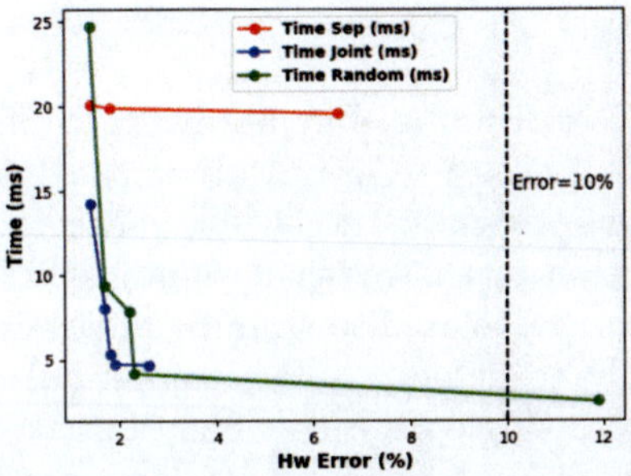

Random		Separate		Joint	
Error (%)	Time (ms)	Error (%)	Time (ms)	Error (%)	Time (ms)
1.4	24.67	1.4	20.1	1.4	14.22
1.7	9.36	1.8	19.89	1.7	8.06
2.2	7.83	6.5	19.54	1.8	5.31
2.3	4.19	-	-	1.9	4.74
11.9	2.53	-	-	2.6	4.68

Fig. 2. Pareto-points found by the joint method, random search, and separate method in the space of prediction error (Hw error) and execution latency (Time).

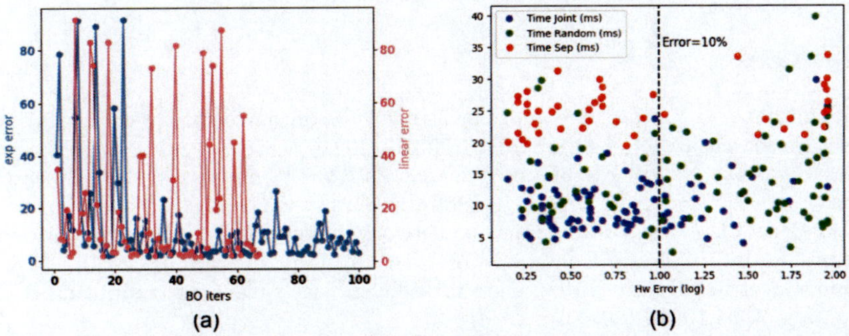

Fig. 3. (a) Hardware inference error in our joint method in each BO iteration (linear and exponential), (b) Total points suggested by the joint, separate, and random search methods; note the concentration of the joint method on low errors (<10%).

In the Joint method, BO jointly optimizes the hyper-parameters and the hardware configuration to maximize both prediction accuracy and performance. Note that since in Random search we cannot set an error constraint, there is one Pareto point with error larger than the admissible threshold (>10%). With the exception of one lucky point detected by the Random search, all the other points are Pareto-dominated by those found by the Joint method.

To help speed-up the BO convergence toward low prediction error, we modified the objective function with a non-linear exponential of the error. Figure 3(a) shows the error as a function of the BO iterations: the exponential definition of the error helps reach a quicker convergence than a linear definition does. The final results in this work are based on the exponential error.

Finally, Fig. 3(b) shows all the points found in 100 iterations. For better clarity, a log scale is used for the error. Notice how the error threshold concentrates the points below 10% for our joint MOBOC method, which helps in better optimization convergence by searching in the area of interest.

4 Conclusions

We proposed a joint optimization of CNN training and hardware configurations in FPGAs, within a multi-objective Bayesian Optimization (BO) framework. In contrast to more common BO approaches, the search space includes multiple hardware configurations in addition to the hyper-parameters used for CNN training. Moreover, conflicting objectives related to training and hardware performance can be considered separately to achieve a Pareto-optimal set of configurations without merging them into a single objective function. We compared our joint methodology with conventional approaches, random search and separate optimization method. The Pareto set achieved with our method outperforms those obtained with random and separate methods, with 1.7× and 1.4× improvement in execution time for the minimum error, respectively.

References

1. Benmeziane, H., et al.: A comprehensive survey on hardware-aware neural architecture search. CoRR, vol. abs/2101.09336 (2021)
2. Sekanina, L.: Neural architecture search and hardware accelerator co-search: a survey. IEEE Access **9**, 151337–151362 (2021)
3. Cai, H., et al.: Once-for-all: train one network and specialize it for efficient deployment. In: International Conference on Learning Representations (2020)
4. Han, S., et al.: Design automation for efficient deep learning computing. CoRR, vol. abs/1904.10616 (2019)
5. Tan, M., et al.: MnasNet: platform-aware neural architecture search for mobile. CoRR, vol. abs/1807.11626 (2018)
6. Wu, B., et al.: FBNet: hardware-aware efficient convnet design via differentiable neural architecture search. In: Proceedings of the IEEE/CVF Conference on Computer Vision and Pattern Recognition (CVPR) (2019)
7. Hsu, C., et al.: MONAS: multi-objective neural architecture search using reinforcement learning. CoRR, vol. abs/1806.10332 (2018)
8. Chu, X., et al.: Multi-objective reinforced evolution in mobile neural architecture search. CoRR, vol. abs/1901.01074 (2019)
9. Parsa, M., et al.: PABO: pseudo agent-based multi-objective Bayesian hyperparameter optimization for efficient neural accelerator design. In: ICCAD, pp. 1–8 (2019)
10. Yu, Y., et al.: Software-defined design space exploration for an efficient DNN accelerator architecture. IEEE Trans. Comput. **70**(1), 45–56 (2021)
11. Cai,H., Zhu, L., Han, S.: ProxylessNAS: direct neural architecture search on target task and hardware. CoRR, vol. abs/1812.00332 (2018)
12. Stamoulis, D., et al.: HyperPower: power- and memory-constrained hyper-parameter optimization for neural networks. In: DATE, pp. 19–24 (2018)
13. Liberis, E., et al.: μNAS: constrained neural architecture search for microcontrollers. In: 1st Workshop on Machine Learning and Systems, pp. 70-79 (2021)
14. Greenhill, S., et al.: Bayesian optimization for adaptive experimental design: a review. IEEE Access **8**, 13937–13948 (2020)
15. Lecun, Y., et al.: Gradient-based learning applied to document recognition. Proc. IEEE **86**(11), 2278–2324 (1998)
16. Garrido-Merchán, E., et al.: Predictive entropy search for multi-objective Bayesian optimization with constraints. Neurocomputing **361**, 50–68 (2019)

Transiently-Powered Batteryless Device-to-Device Communication Protocol Simulator

Alessandro Torrisi(✉), Federico Baggio, and Davide Brunelli

Department of Industrial Engineering, University of Trento,
Via Sommarive 14, 38123 Trento, Italy
{alessandro.torrisi,davide.brunellis}@unitn.it
https://www.dii.unitn.it/

Abstract. Energy harvesting batteryless wireless sensors survive in harsh environments with a minimal energy budget. Indeed, the ambient harvested energy may be sporadic and not constant. Thus, devices operate intermittently experiencing frequent power failures. Communication over these devices must guarantee efficiency and avoid loss of energy. TRAP (TRAnsiently-powered Protocol) can be a solution to avoid packet loss due to receivers' power failure. TRAP guarantees communication between devices thanks to the node's energy status awareness. Indeed, effective communication succeeds only if both sides of the communication channel involved in the data packet exchange process have enough energy to complete transmission and reception. TRAP relies on nodes' energy level information provided through an RF backscatter channel implemented as a self-sustainable and ultra-low-power demanding process. A possible TRAP implementation with RF backscatter has been proven in a laboratory setup for a few devices network. Although results were promising, deep validation and scale-up are necessary. For this purpose, we developed a simulator to verify how the protocol works in different conditions and how performance scales in large sensor networks. The simulation results show that the TRAP protocol reduces waste of energy avoiding failed transmissions due to power failures and increasing communication performances.

1 Introduction

One of the most rising challenges for the next-generation Internet of Things (IoT) devices is reducing energy requirements and operating without the need for batteries. Indeed, batteryless is nowadays a viable solution for environmentally friendly and cost-effective IoT applications. As a consequence of energy consumption minimization, operations such as computation and end-to-end communication become non-trivial. In general terms, a batteryless node comprises, outside the main structure of sensors and computation, an energy harvester and a small-sized energy buffer (e.g., a capacitor or a super-capacitor). The harvester charges the energy buffer up to an activation threshold. When active, the node

R. Berta and A. De Gloria (Eds.): ApplePies 2022, LNEE 1036, pp. 279–286, 2023.
https://doi.org/10.1007/978-3-031-30333-3_37

draws the stored energy until a lower threshold is reached and operations stop allowing for a new charging cycle.

One of the most energy-demanding processes for IoT devices is sending the information through the ether, especially if using active radios [6–8]. Indeed, a power failure or a power interruption due to a lack of energy from the ambient can affect the transmission process. Hence, during communication, both transmitter and receiver must have enough energy to complete the task. If either the transmitter or the receiver has a lower energy level, the communication fails and causes significant data packet loss and precious energy waste [1]. TRAP (TRAnsiently-powered Protocol) [2–4] aims to avoid this waste of energy and guarantee effective communication. It relies on the energy status of neighbouring nodes allowing communication between the transmitter and the receiver only after verifying that both of them have enough energy to complete the operation. A secondary communication channel, in addition to the main radio used for the data exchange, allows for sharing information about the energy level of each near node. Although integrating TRAP on batteryless nodes requires an additional communication channel, backscatter communication, due to ultra-low-power capability, enables self-sustainable energy status communication. RF backscatter operates by removing RF power-hungry components. Indeed, achieving quasi-zero energy cost for the transmitter. It bases on the reflection of an externally provided electromagnetic field. The impinging signal can be already environmentally present (e.g., Wi-Fi or TV signals) or created by a special external illuminator. The modulation of the energy status information is provided by a dedicated ultra-low-power circuit called "automodulator". In a context where energy is precious, the modulation must be simple and the energy required must be easily harvested. Indeed the automodulator consumes the lowest power of $10\ \mu W$, a power requirement that fits with the current status of the art in RF harvesting [8]. Indeed, the status update process runs even if the node is out of energy on board. Thus, bringing the complete energy status transmission a self-sustainable plug&play solution for any IoT batteryless device.

As TRAP protocol requires a physical setup including batteryless nodes and an RF backscatter infrastructure, testing a wide batteryless sensor network can be non-trivial and burdensome. Moreover, the testbed setup features fixed and constrained properties, providing results only for a specific application. Indeed, prior work [1] only provides results related to a very small network of two-three nodes. In this paper, we aim to analyze TRAP properties and operations in larger networks. We propose to simulate the batteryless network behaviour and include the TRAP policies. After building and validating the simulator we obtained remarkable results describing the protocol performance in different conditions and when used with large networks.

The rest of the paper is organized as follows: after a prior art revision (in Sect. 1.1) we present the simulator parameters (in Sect. 2) and we focus on the simulation results (in Sect. 3) before concluding the paper with final remarks (in Sect. 4).

1.1 Related Work

As aforementioned, due to energy discontinuity from ambient sources, batteryless devices operate intermittently bringing about several challenges. As an example, computation under frequent power failure suffers from a loss of computational state preventing flawless forward progress. Researchers proposed to instrument the firmware with checkpoints and save the memory content before a power failure happens [11]. When the node experience an energy shortage, it can recall the last saved state without the need for booting the system from scratch. Thus, saving precious energy.

A further challenging aspect involves communication over intermittently operated batteryless devices. Current research such as the one involving RF backscatter aims to reduce the power consumption of the radio links [10]. Furthermore, researchers proposed to combine ultra-low-power radio links with RF energy harvesting systems [8] [9]. In this scenario, the RF carrier is used not only for backscattering the information but also for collecting a small quantity of energy.

Anyway, the prior art considers communicating devices with enough energy and not experiencing power failures. On the contrary, in real batteryless applications, nodes can stop operation while data communication is ongoing. Thus, leading to a significant amount of packet losses and energy waste. Recently, Geissdoerfer at el. [5] considers nodes operating intermittently and proposes to maximize the probability that both sides of the communication channel have sufficient energy to perform a transmission-reception action successfully. However, data communication is not guaranteed. Moreover, an extension of the work presented in the Bonito protocol [12] share the knowledge of the nodes' charging times and uses this information to synchronize nodes' communication actions.

1.2 The Problem Statement and Contributions

Building large-scale IoT networks with many devices is costly. To this end, TRAP was tested in the laboratory with a network accounting for a few devices and with fixed parameters related to the main communication channel [1]. Here we present the results obtained with a simulator built to study how TRAP work with different energy configurations and a wide number of devices. By using the simulator, we compare the communication behaviour of the network with and without TRAP. Results provide highlights on the benefits of using TRAP in different conditions or on the protocol limitation. As a direct consequence of using TRAP, there is no loss of energy due to incomplete communications but it is interesting to study if the number of complete communications is smaller, comparable or larger concerning the results with no integration of TRAP. The system can simulate hundreds of nodes that communicate together through an energy-intensive channel and with TRAP through a less energy-demanding channel. All the parameters can be tuned to replicate the real energy costs of the main communication channel. All the nodes in the network charge themselves following a realistic and random curve of energy.

2 Simulator Design

TRAP protocol is a set of rules based on a hardware layer that allows for controlling communication between batteryless devices. Its main goal is energy awareness to regulate data transmission between devices on a batteryless sensor network to avoid incomplete communications. Before starting communication, TRAP ensures that transmitter and receiver nodes have enough energy to complete the task. This process allows for avoiding any power failure during communication. Thus, reducing the waste of energy and enhancing data packet throughput. TRAP relies on the energy information of each node provided through a dedicated RF backscatter channel. Thanks to backscatter, each node sends simple packets related to energy level to the entire network. The automodulator encodes the energy status of a node into a fixed modulation scheme. An ON-OFF keying (OOK) modulated burst with different lengths communicates the energy level: the longer the burst the higher the energy level. As a matter of low-power constraints, only a few energy levels can be transmitted and the energy status information is discrete. Indeed, a node can communicate its availability to transmit data, when it is fully charged, its availability to receive data, when an intermediate energy threshold is reached, and its unavailability to either transmit or receive data packet if the onboard energy is below the threshold.

The simulator describes the protocol operation from the hardware level and includes the energy behaviour of communicating nodes. Each node charges its energy buffer following a randomized energy increment profile. The backscatter channel provides the energy level knowledge to the entire network and the simulator integrates this process. The protocol, relying on this knowledge, regulates the communication on the main channel. Each communication discharge the battery of both transmitter and receiver by a different and fixed amount called energy transmission cost and energy reception cost. The simulator can be tuned to describe networks with different types of main communication channels controlling the parameters of discharge level during communication from 0% to 100%. It is possible also to turn off the TRAP operation to compare the results with and without the protocol using the same energy increment profile for the nodes.

3 Results

We started our evaluation by comparing the simulator with a physical setup experiment with two nodes that lasted 10 min and featured TRAP. The simulator run with the same energy increment profile. Figure 1 reports a snapshot of the behaviour of the experiment and the simulator with and without implementing TRAP. The experiment characteristics are a charging rate of about 1,33%/s, a cost for the data transmission of 100%, and a cost for the reception of 70% (the figure highlights the two thresholds). The results prove the simulator's capability to emulate a possible real-world scenario.

We exploit the simulator to test different communication scenarios, especially looking at scalability performance. The tests run for about 10 min simulated and

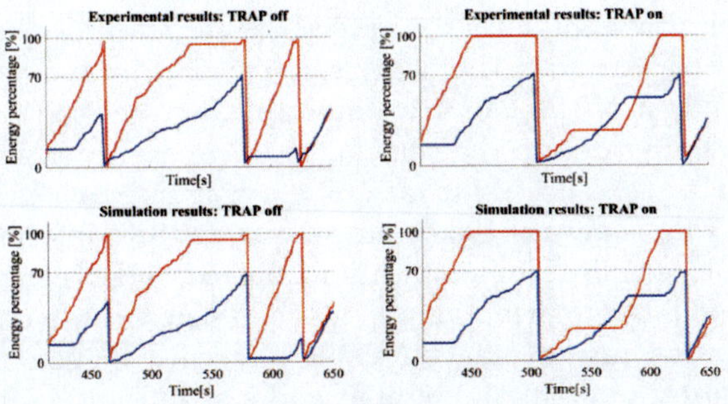

Fig. 1. Comparison between experimental results and simulation result by using the same energy profile both with (right) and without (left) TRAP protocol.

varying the required energy for completing a reception and the charging rate from 0.5 %/s to 3.5 %/s (i.e., reasonable values for batteryless IoT applications). It is worth mentioning that different reception energy costs can be associated with a different radio transceiver. Finally, we obtained two main resulting curves reporting the number of completed reception tasks per second (i.e., the throughput in Fig. 2), and the ratio between successful and failed communication (in Fig. 3).

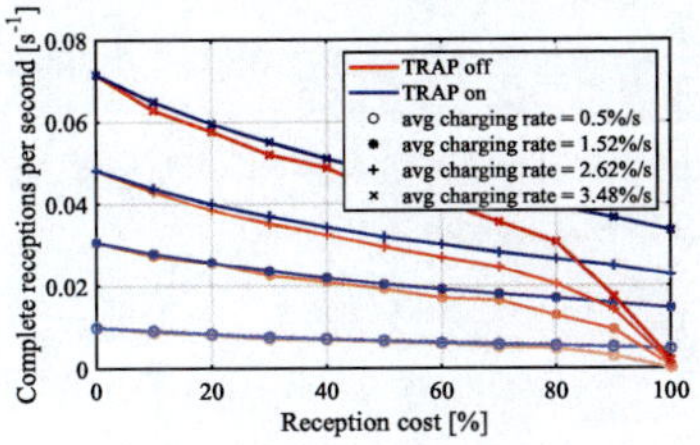

Fig. 2. Network throughput as a function of the energy reception cost with different energy increment averages.

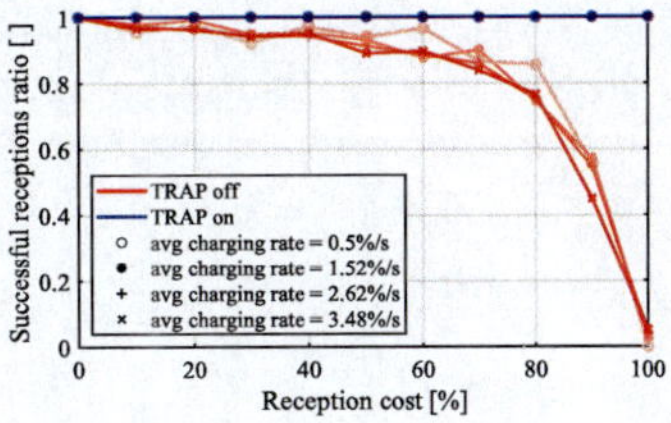

Fig. 3. Successful rate as a function of the energy reception cost with different energy increment averages. With TRAP is always 1.

Figure 2 represents the throughput by varying the energy reception cost, from 0% to 100%, while the transmission cost remains at 100 %. The most effective evidence is that in all cases when adopting TRAP, the network throughput is higher than when not using the protocol. Each couple of curves, with and without TRAP, is obtained at different charging rates. For lower reception cost values, the curves at equal charging rates are very close. Indeed, the probability that the receiver node has sufficient energy to complete the communication task is higher. On the other hand, when the reception cost is higher the curves start differing and, in the case of without TRAP, the throughput drastically drops due to the lower probability that a receiver node features the proper energy level when required. Anyway, the decreasing trend with and without TRAP is the natural behaviour of the network. Indeed if the reception cost is higher the network evolves slower due to the constant average energy inlet.

Figure 3 reports the successful rate of the communication. It is clear that with TRAP the rate is always 100 % but we can retrieve information about the energy waste due to failed communication. Indeed, when communication fails both the energy and the information are lost and a new communication attempt is required to spend further energy. Especially with higher energy reception costs, we have both the worst throughput and a significant waste of energy and data packet loss. Moreover, by looking at different energy increments, we can understand that the energy rate does not affect the success rate. Thus, both the complete and non-complete communications per second increase.

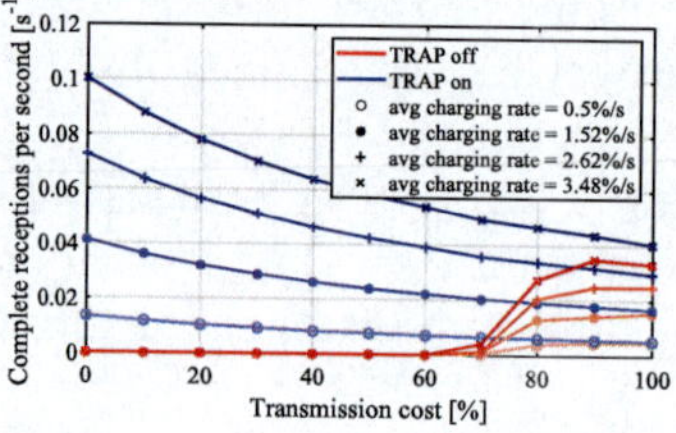

Fig. 4. Network throughput as a function of the energy transmission cost with different energy increment averages.

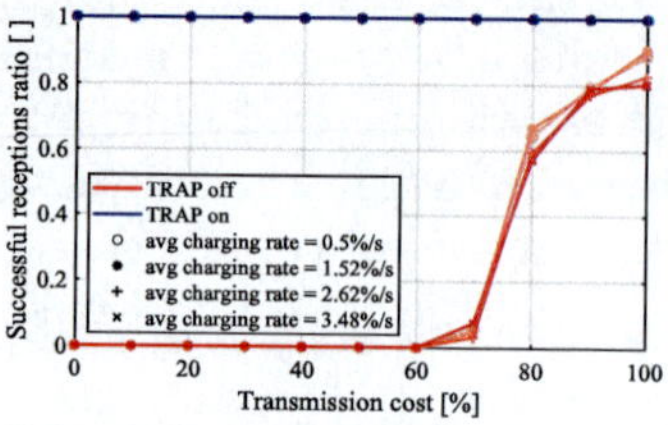

Fig. 5. Successful rate as a function of the energy transmission cost with different energy increment averages. With TRAP is always 1.

Figure 4 and Fig. 5 report simulation output curves obtained by changing the energy transmission cost from 0% to 100% at different charging rates by keeping the reception cost fixed at 70 %. Figure 4 reveals the throughput drops to zero when the transmission cost is lower than the reception cost. In this case, only a few complete receptions lead to high energy waste. In this condition, nodes become quickly ready for transmission while they need more time to have enough energy to complete the reception. As a consequence, the communication does not allow receivers to charge enough by holding the energy level below the reception threshold.

In Fig. 4 we can see the different behaviour with and without TRAP. Similarly to Fig. 2, the number of complete receptions per second decreases while the transmission cost grows. Indeed, if the energy cost increases also the time to reach the proper energy level increases and, in turn, the throughput decreases.

Finally, one of the main limitations of a laboratory test concerns the limited number of nodes that can be tested. Indeed scaling up requires a large number of nodes that may not fit with available space and manufacturing costs. Thanks to the simulator, scale-up validation is easily possible with hundreds of devices.

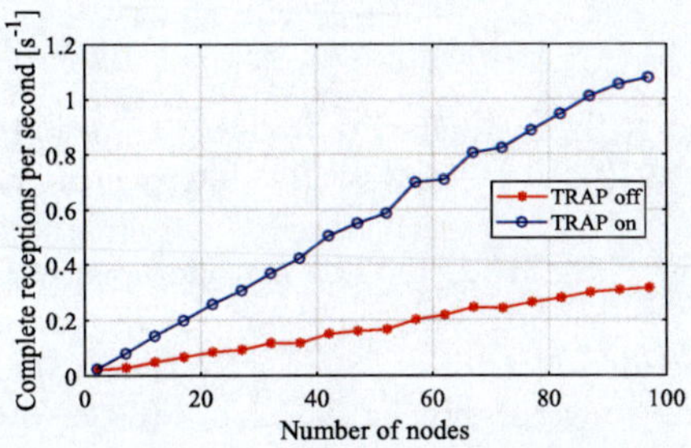

Fig. 6. Network throughput as a function of the network number of nodes with a fixed energy increment average.

Figure 6 we report simulation results of the network throughput as a function of nodes in the network. The parameters are transmission cost 100%, reception cost 70% and an average energy increment of 1.9%/s. The number of nodes varies from 2 to 102. Both curves with and without TRAP reveal almost a linear trend with two different slopes. Indeed, as a consequence of using TRAP, the throughput is enhanced. In conclusion, the simulation results highlight that in addition to the elimination of waste of energy due to incomplete communication, TRAP increases the communication performance of a network with a large number of nodes.

4 Conclusion

In this article, we presented the challenging scenario of the so-called intermittent communication over batteryless IoT devices. We evaluated TRAP, an energy-aware protocol allowing for secure data packet communication when dealing with sporadic and limited energy harvesting resources in combination with a limited energy buffer. We developed a simulator able to reproduce energy increment profiles and upon those emulating the behaviour of intermittent IoT device operations. For validation, we tested a two-nodes setup and compare the experimental result with simulated results by using the same energy increment profile. After validation, we exploit the simulator to extract performance information in different scenarios proving the protocol's effectiveness and scalability. As a final remark, an energy-aware protocol is a fundamental building block to create secure networking over batteryless devices, indeed the data exchange success can be obtained only if the transmitter is aware of possible active receivers.

References

1. Torrisi, A., Yıldırım, K.S., Brunelli, D.: Reliable transiently-powered communication. IEEE Sens. J. **22**(9), 9124–9134 (2022). https://doi.org/10.1109/JSEN.2022.3158736
2. Torrisi, A., Yıldırım, K.S., Brunelli, D.: Autonomous energy status sharing and synchronization for battery-less sensor networks. In: Proceedings of the 19th ACM Conference on Embedded Networked Sensor Systems (SenSys 2021), pp. 569-571. Association for Computing Machinery, New York (2021). https://doi.org/10.1145/3485730.3493360
3. Torrisi, A., Brunelli, D., Yıldırım, K.S.: Zero power energy-aware communication for transiently-powered sensing systems. In: Proceedings of the 8th International Workshop on Energy Harvesting and Energy-Neutral Sensing Systems (ENSsys 2020), pp. 43-49. Association for Computing Machinery, New York (2020). https://doi.org/10.1145/3417308.3430269
4. Torrisi, A., Yıldırım, K.S., Brunelli, D.: Enabling transiently-powered communication via backscattering energy state information. In: Saponara, S., De Gloria, A. (eds.) ApplePies 2020. LNEE, vol. 738, pp. 192–201. Springer, Cham (2021). https://doi.org/10.1007/978-3-030-66729-0_22
5. Geissdoerfer, K., Zimmerling, M.: Bootstrapping battery-free wireless networks: efficient neighbor discovery and synchronization in the face of intermittency. In: NSDI (2021)
6. Nardello, M., et al.: Camaroptera: a batteryless long-range remote visual sensing system. In: Proceedings of the 7th International Workshop on Energy Harvesting and Energy-Neutral Sensing Systems, pp. 8-14. Association for Computing Machinery, New York (2019)
7. Sartori, D., Brunelli, D.: A smart sensor for precision agriculture powered by microbial fuel cells. In: IEEE Sensors Applications Symposium (SAS), vol. 2016, pp. 1–6 (2016)
8. Assimonis, S.D., Daskalakis, S., Bletsas, A.: Sensitive and efficient RF harvesting supply for batteryless backscatter sensor networks. IEEE Trans. Microw. Theory Tech. **64**(4), 1327–1338 (2016). https://doi.org/10.1109/TMTT.2016.2533619
9. Ryoo, J., Jian, J., Athalye, A., Das, S.R., Stanaćević, M.: Design and evaluation of "BTTN" a backscattering tag-to-tag network. IEEE Internet Things J. **5**(4), 2844–2855 (2018). https://doi.org/10.1109/JIOT.2018.2840144
10. Majid, A.Y., Jansen, M., Delgado, G.O., Yildirim, K.S., Pawełzak, P.: Multi-hop backscatter tag-to-tag networks. In: IEEE INFOCOM 2019 - IEEE Conference on Computer Communications, pp. 721–729 (2019). https://doi.org/10.1109/INFOCOM.2019.8737551
11. Balsamo, D., et al.: Hibernus++: a self-calibrating and adaptive system for transiently-powered embedded devices. IEEE Trans. Comput. Aided Des. Integr. Circuits Syst. **35**(12), 1968–1980 (2016). https://doi.org/10.1109/TCAD.2016.2547919
12. Geissdoerfer, K., et al.: Learning to communicate effectively between battery-free devices. In: 19th USENIX Symposium on Networked Systems Design and Implementation (NSDI 22), pp. 419–435 (2022). isbn 978-1-939133-27-4, https://www.usenix.org/conference/nsdi22/presentation/geissdoerfer, USENIXAssociation, Renton, WA

Digital Modulation Recognition Method Based on High-Order Cumulant Feature Learning

Hao Li[1], Hua Wu[1,3](✉), Qinghe Zhen[2](✉), Yang Liu[1,3], Abdussalam Elhanash[4], and Sergio Saponara[4]

[1] School of Information Science and Electrical Engingeering (School of Artificial Intelligentce), ShanDong JiaoTong University, Jinan 250357, China
wuhua@sdjtu.edu.cn

[2] School of Intelligence Engineering, Shandong Management University, Jinan 250357, China
15005414319@163.com

[3] CAAC Key Laboratory of General Aviation Operation, Civil Aviation Management Institute of China, Beijing, China

[4] Department of Information Engineering, University of Pisa, 56122 Pisa, Italy

Abstract. Automatic modulation recognition (AMR) of communication signals is an important research topic in the processing of intercepted signals. In this paper, aiming for the automatic recognition of modulated signals, we propose a feature learning and classification method based on the high-order cumulants, which effectively suppresses Gaussian white noise. Six digital modulation schemes including BPSK, QPSK, 8PSK, 8QAM, 16QAM, and 64QAM can be recognized by comparing the feature with the threshold. During the experiments, we plot the confusion matrix under different conditions. Moreover, it is derived and verified with simulation experiments and actual data acquisition.

Keywords: Cognitive radio · modulation recognition · high-order cumulant

1 Introduction

Automatic recognition of signal modulation schemes [9] is a key technology in the wireless communications. Its purpose is to recognize the modulation types of various intercepted signals, and to provide modulation parameters to provide subsequent prior information.

According to the difference between feature extraction and recognition methods, there are mainly the following two common signal modulation recognition techniques. One is a method based on the likelihood ratio judgment theory. The other is a pattern recognition way based on the feature extraction [10], and in this way, the high-order cumulant has strong anti-noise ability and thus is suitable for practical applications. Kim et al. [1] correctly distinguishes BPSK, QPSK and eight-phase phase shift keying signals by using the combination of fourth-order cumulants. In this paper, the high-order cumulants of 6 kinds of PSK and QAM modulated signals are experimentally analyzed, new feature parameters are proposed, and different signal classifications are achieved by setting different judgment thresholds.

R. Berta and A. De Gloria (Eds.): ApplePies 2022, LNEE 1036, pp. 287–293, 2023.
https://doi.org/10.1007/978-3-031-30333-3_38

2 Modulation Recognition Based on High-Order Cumulant

The process of extracting feature parameters of the modulated signal is as follows: first, the signal is preprocessed to reduce the interference effect of noise on the signal, the extracted signal features should reflect the detailed information of the modulated signal, then calculate the higher-order accumulation and process the signal.

2.1 Principle of Digital Modulation

In this section, we first introduce the parameters and the time and frequency domain waveforms of different signals.

Assuming that the received digital modulated signal sampling complex sequence in the Gaussian environment [2], it can be expressed as:

$$\mathrm{S(t)} = \sqrt{E}\sum_{n} a_n g(t - nT_s)e^{j(w_c+\theta_c)} + n(t) \tag{1}$$

where E represents the energy of the symbol waveform transmitted in the channel; a_n represents the symbol sequence of signal; $g(t)$ represents symbol waveform; W_s represents carrier frequency; T_s represents symbol width; θ_c represents the carrier phase; $n(t)$ represents a complex Gaussian white with zero mean noise, the signal S_t is subjected to down-conversion, intermediate frequency filtering, demodulation, and symbol synchronization, and the symbol synchronization sampling sequence obtained by the matched filter output is:

$$S_{k=}\sqrt{E}e^{j\theta}a_k + n_{k,}k = 1, 2, ..., N \tag{2}$$

where N is the length of the observation data. According to the feature of digital modulated signals, the symbol synchronous sampling complex signal of MPSK and MQAM signals can be expressed as:

$$MPSK : S_k = \sqrt{E}e^{j\theta}a_k + n_k,\ a_k \in \{\exp(j2\pi(m-1)/M),\ m = 1, 2, ..., M\} \tag{3}$$

$$MQAM : S_k = \sqrt{E}e^{j\theta}a_k + n_k,\ a_k = (a_{1,k}, a_{1,k}, a_{Q,k}) \in \{2m - 1 - \sqrt{M}d, m = 1, 2, ..., \sqrt{M}\} \tag{4}$$

2.2 High-Order Cumulant Learning

In this part, we set the transmitted symbols to satisfy the condition of independent and identical distribution, In this case, high-order cumulants are calculated respectively. Assuming that the signal energy is E. In order to carry out the experiment, the symbol sequence [3] is normalized according to Table 1.

Table 1. High-order cumulants of wireless signals with different modulations.

Signal	C_{21}	C_{40}	C_{41}	C_{42}	C_{60}	C_{63}
BPSK	E	$2E^2$	$2E^2$	$2E^2$	$16E^3$	$13E^3$
QPSK	0	E^2	0	E^2	0	$4E^3$
8PSK	0	0	0	E^2	0	$4E^3$
4QAM	0	E^2	0	E^2	0	$4E^3$
16QAM	0	$0.68E^2$	0	$0.68E^2$	0	$2.08E^3$
64QAM	0	$0.61E^2$	0	$0.68E^3$	0	$1.79E^3$

2.3 Classifier Design

Recognize signal types, use cumulant combinations to extract characteristic parameters [4], calculate the high-order cumulant, analyze icon, and give following characteristic parameters:

$$M_1 = \frac{|C_{40}|}{|C_{21}^2|} M_2 = \frac{|C_{40}|}{|C_{42}|} M_3 = \frac{|C_{20}^2|}{|C_{41}|} Q = M_1 \tag{5}$$

$$A_1 = \frac{|C_{60}^2|}{|C_{40}^3|} A_2 = \frac{|C_{63}^2|}{|C_{40}^3|} P = \frac{|C_{40}|}{|C_{42}|} + \frac{|C_{41}|}{|C_{42}|} \tag{6}$$

The corresponding features of the inter-class signals are shown in the Table 2. In an ideal case, MPSK envelope is a constant, and its variance is approximately to 0. The envelope of MQAM signal is a multi-order function, and the envelope variance is constant, Different high-order cumulant characteristic parameters are used to i-th type of signal [5].

(1) The received modulation signal is processed by down-conversion to obtain a baseband modulation signal.
(2) Calculate the higher order cumulants of baseband modulated signals and their feature parameters.
(3) The characteristic parameter p is used to recognize the PSK, QAM signal. The PSK is (0, 2), and the QAM is (0, 1), and the two signals are identified according to [6].
(4) The baseband modulated signal is differentiated and then passed through a median filter to calculate the transformed high-order cumulant, and use the difference of characteristic parameters to realize signal recognition.

Table 2. Feature values of different modulated signals

Features	BPSK	QPSK	8PSK	parameter	4QAM	16QAM	64QAM
p	2	1	0	Q	1	0.68	0.61

2.4 Modulation Recognition Process

(1) By calculating the cumulant f of the PSK signal and setting the threshold $flag_1 = 0.5$ [7], we recognize the signal type as PSK according to

$$\begin{aligned} f_1 &= abs(C_{40}(signal))/abs(C_{42}(signal)) \\ f_2 &= abs(C_{41}(signal))/abs(C_{42}(signal)) \end{aligned} \tag{7}$$

(2) If f1 > flag1, and f2 > flag1, the signal is recognized as BPSK
(3) If f1 > flag1, and f2 < flag1, the signal is recognized as QPSK.
(4) If f1 < flag1, and f2 < flag1, the signal is recognized as 8PSK.
(5) By calculating the subtraction clustering value c of QAM and setting the cluster radius and cluster radius as $ra = 0.25$ and $flag_2 = 5$, respectively [8]. Then we classify the signal modulation type as QAM according to

$$c = subclust(abs(signal'), ra) \tag{8}$$

$$n = length(c) \tag{9}$$

(6) If n < flag2, the signal is recognized as 16QAM
(7) If n > flag2, and f2 < flag1, the signal is recognized as 64QAM.

3 Experiments

3.1 Experimental Settings

According to the above analysis, Matlab software is used to simulate, respectively, to classify and recognize 6 kinds of digital modulation signals of four types: BPSK, QPSK, 8PSK, 4QAM, 16QAM 64QAM. The initial parameters of the experiment are set as: the number of symbols equals to 400, the symbol rate $R = 3.2 \times 10^9$ bit/s, and the sampling $f_s = 10$ kHz. In the experiment, 200 independent trials are carried out for the signal to be recognized, and carry out experimental analysis.

3.2 Experimental Results

We first present the classification accuracy curves under different sample sizes, as shown in Fig. 1. In the range of 0–10 dB SNR, the accuracy of PSK signal recognition gradually increased, while the modulation recognition accuracy between 16QAM and 64QAM drops first and then tends to be stable.

Then we show the confusion matrix for 50 db SNR and 90 db SNR in Fig. 2. The number of experiments was 200 times, and the SNR was 50 and 80 db respectively. The confusion matrix recognition was basically accurate, close to 100%

In Fig. 3, we present the feature distribution of higher-order cumulants. The graph shows that the distribution of C_{40} cumulants, QPSK is around the threshold of 1, 8QAM is limited around the threshold of 1.3, 16QAM and 64QAM fluctuate from 0.3 to 0.6.

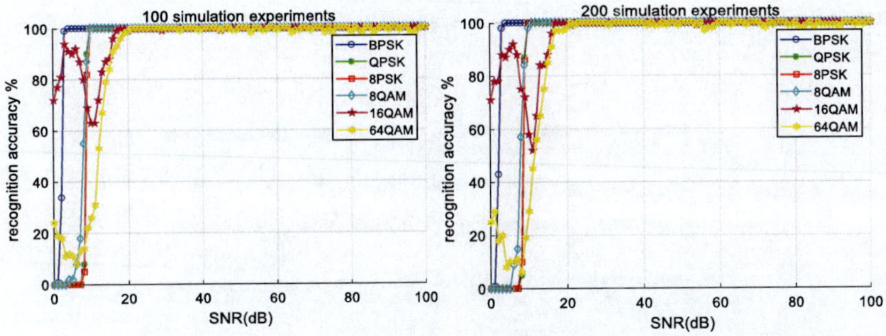

Fig. 1. Classification accuracy curves for 50 and 100 simulation experiments.

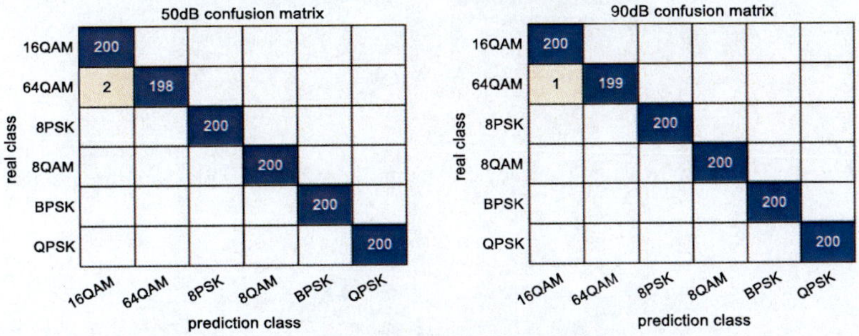

Fig. 2. Confusion matrix for 50db SNR and 90db SNR.

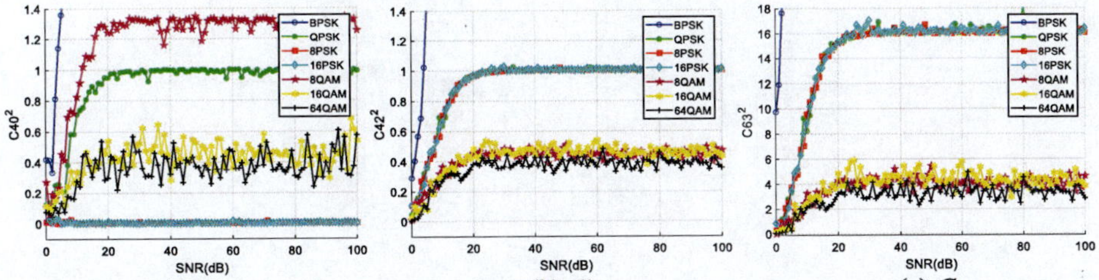

(a) C_{40} (b) C_{42} (c) C_{63}

Fig. 3. Distribution of higher-order cumulants.

Then we report the classification accuracy and confusion matrix under different decision thresholds in Fig. 4 and Fig. 5. During the 200 simulation experiments, the BPSK signal recognition accuracy is 100% when the threshold is 0.5 ~ 3, and the accuracy is 0 when the threshold is 5. As for QPSK signal and 8 PSK signal, when the threshold is 0.5, the recognition rate is 100%; when the threshold is 1, it gradually increases to 100%, and the recognition rate is 0 when the threshold is 3 to 5.

According to the confusion matrix, the classification is basically accurate when the threshold is 0.5 during 200 simulation experiments. When the threshold is 1, QPSK and 8PSK are all wrongly judged, and the accuracy of 16QAM and 64QAM is 90%. When the threshold is 3, QPSK and 8PSK are all wrongly judged, and the accuracy of 16QAM and 64QAM is 90%.

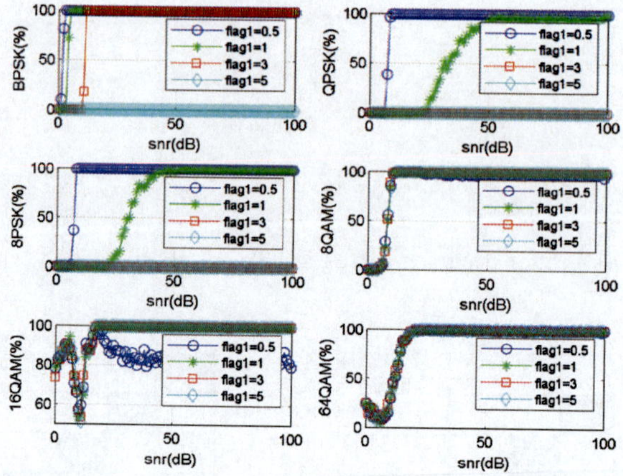

Fig. 4. The classification accuracy curves under different SNRs and thresholds.

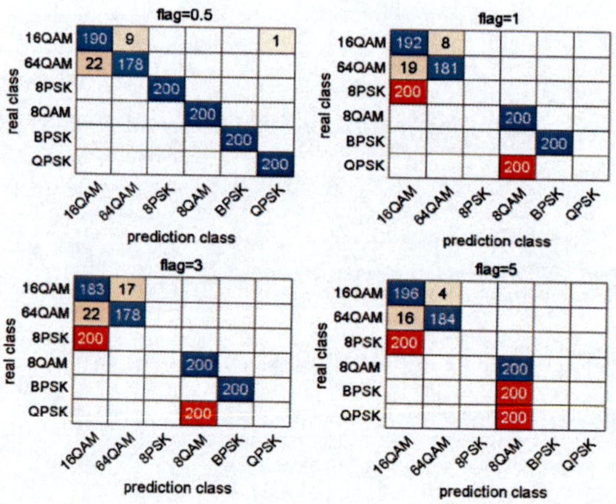

Fig. 5. TThe confusion matrix under different thresholds

4 Conclusion

Automatic modulation recognition (AMR) is an important research topic in wireless communications. In this paper, the calculation and judgment of high-order cumulants are used to realize the recognition of various signals including BPSK, QPSK, 8PSK, 4QAM, 16QAM, and 64QAM. Firstly, PSK signal is identified by threshold judgment, then QAM signals are identified by subtractive clustering algorithm, and finally experimental comparison is carried out. We draw the recognition accuracy curve and confusion matrix under different quantization conditions, different SNRs and different judgment thresholds to prove the effectiveness of the algorithm. The method has stable recognition accuracy, effectively resists noise interference, low computational complexity, and it is easy to implement in engineering.

Acknowledgements. This research was supported by Shandong Provincial Natural Science Foundation (Grant ZR2020MF151), CAAC Key Laboratory of General Aviation Operation (Civil Aviation Management Institute of China) (Grant CAMICKFJJ-2020–2), National Natural Science Foundation of China (Grant U1933130 and 71731001), and Research and Demonstration of Key Technologies for the Air-Ground Collaborative and Smart Operation of General Aviation (Grant 2022C01055).

References

1. Swami, A., Sadler, B.M.: Hierarchical digital modulation classification using cumulants. IEEE Trans. Commun. **48**(3), 416–429 (2000)
2. Sreekantamurthy, M., Popescu, D.C ., Joshi, R.P.: Classfication of digital modulation schemes in multipath environment u-sing higher order statistics. IEEE SoutheastCon **12**(4) (2016)
3. Guo, J., Yi, H., Jiang, L.: Recognition of digitally modulated signals using high-order cumulants. Commun. Technol. **47**(11), 1255–1260 (2014)
4. He, J.A., Du, P.P.: Recognition method of digital modulation signal based on high-order cumulant. Measurem. Control Technol. **36**(40) (2017)
5. Chen, W.D., Yang, S.Q.: Classification of MPSK signals using cumulative invariants. J. Xidian Univ. (Nat. Sci. Ed.) **29**(12), 229–232 (2002)
6. Li, S.P., Chen, F.C.: Digital modulation recognition algorithm based on wavelet and higher-order cumulant. Comput. Appli. 31(11) (2011)
7. Yu, Y., Li, X., Zhang, X.: Signal modulation recognition of psk. Electronic Sci. Technol. **28**(9) (2015)
8. Li, Y.L., Li, B.B., Yin, C.Y.: MQAM signal recognition based on particle swarm and subtractive clustering extraction of classification features. J. Northwestern Univ. (Nat. Sci. Ed.), **41**(3) (2011)
9. Zheng, Q., Zhao, P., Wang, H., Elhanashi, A., Saponara, S.: Fine-grained modulation classification using multi-scale radio transformer with dual-channel representation," IEEE Commun. Lett. **26**(6), 1298–1302 (2022)
10. Zheng, Q., Zhao, P., Zhang, D., Wang, H.: MR-DCAE: Manifold regularization-based deep convolutional autoencoder for unauthorized broadcasting identification. Int. J. Intell. Syst. **36**(12) 7204–7238 (2021)

Integrated Photonics for *NewSpace*

G. Brunetti, N. Saha, G. Campiti, A. di Toma, N. Sasanelli, F. Hassan, M. N. Armenise, and C. Ciminelli(✉)

Optoelectronics Laboratory, Politecnico di Bari, Via Orabona 4, 70125 Bari, Italy
caterina.ciminelli@poliba.it

Abstract. In the last years, an innovative approach in conceiving Space systems has been proposed, known as *NewSpace*, aiming at developing less expensive satellites in short periods of time, also saving costs, reducing time to access Space and enabling constellation flights. The miniaturization of on-board systems, even preserving reconfigurability and reliability, is required to fulfil the *NewSpace* goals. These features match with the photonics development trends, increasingly focusing on photonic integrated circuits that show EMI immunity, transparency, low propagation-induced loss, wide bandwidth, and radiation hardness. Here, the recent advances in the field of micro- and nano-photonic devices and systems, with potential applications mainly in the field of satellite technologies, are overviewed, with reference to materials, performance, reliability aspects, and technical bottlenecks, also reporting the development directions and perspectives.

1 Introduction

During the last decades, the Space industry has experienced a deep transformation, with the advent of the paradigm *NewSpace,* also called *democratized Space* [1]. It is often used to describe the global trend to develop systems for a faster and cheaper access to Space. The *NewSpace* revolution in the Space sector allows to new players/commercial entrepreneurs entering in a domain traditionally dominated by institutional players [2, 3]. The revolution aims to investigate and offer new opportunities and services, supported by the migration from huge satellites to constellations of small- and medium-sized satellites. These constellations will add much more complexity to Space operations, by shifting the functionalities of a single large satellite to dozens of smaller ones. The launch of over 20,000 small satellites in the next 10 years has been predicted [4]. However, to support this transition, new technologies need to be investigated.

Photonics ensures the fulfilment of all *NewSpace* tasks, such as compactness, low costs and robustness [5, 6]. Moreover, photonics technologies in Space applications bring several advantages such as unlimited bandwidth, very low propagation losses, immunity against electromagnetic interferences (EMI), potentially low power consumption, excellent mechanical properties such as flexibility, light weight, reduced volume and mass, good resistance to vibrations/shocks, and large immunity to radiations [7, 8].

In recent years, several photonic subsystems have already been developed for both transport and ground segment, while photonics solutions for the Space segment, including payload and satellite platform, are currently under investigation (see Table 1).

R. Berta and A. De Gloria (Eds.): ApplePies 2022, LNEE 1036, pp. 294–299, 2023.
https://doi.org/10.1007/978-3-031-30333-3_39

Table 1. Photonic systems for satellite sub-systems (TRL: *Technology Readiness Level*).

Satellite sub-system	Photonic system	TRL
Structure and mechanism	Optopyrotechnics	Low
Thermal subsystem	FBG-based distributed temperature sensing systems	In-orbit demonstration
Data processing	High bit rate fiber data links for onboard data handling	High
Energy supply	III-V multi-junction solar cells	On the market
Communications	Satellite-to-satellite and satellite-to-ground free-space optical links	In-orbit demonstration
Attitude Regulation	He-Ne laser and fiber gyro Integrated microphotonic gyro Star tracker	On the market Experimental study On the market
Propulsion	Thrust generation via photon–matter interactions	Low
Payload	Optical payloads for EO and planetary rovers	In-orbit demonstration

2 Integrated Optoelectronic Platforms

Photonic Integrated Circuits (PICs) enable the transmission of data at high speeds using optical carriers with a remarkable improvement of bandwidth, data rate and significant savings in terms of mass (>50%), power (>50%) and size (100 X) with respect to the electronic counterparts.

The optoelectronics platforms include silicon (Si), silicon-on-insulator (SOI), silicon nitride (SiN), silica-on-silicon (SOS), indium phosphide (InP), polymers, mainly devoted to passive devices, while gallium arsenide (GaAs) and indium gallium arsenide phosphide (InGaAsP) are used for active devices [9]. A comparison of the aforementioned technologies in terms of refractive index contrast, propagation losses at 1550 nm, thermo-optical coefficient, CMOS compatibility and reliability is summarized in Table 2.

PIC integration approaches have been explored to meet Size, Weight, and Power (SWaP) goal, as well as performance goals for many applications. Monolithic and hybrid integration approaches use a single technological platform, i.e. InGaAsP, which guarantees the manufacturing on the same chip of both active and passive devices, or multi platforms to choose the best technologies to perform a single functionality, i.e. Si- or Si-derived combined with InP, respectively [10, 11]. In 2020, Infinera demonstrated the monolithic integration of several InP-based optoelectronic devices, such as amplifiers, lasers, modulators, detectors, passive waveguides and so on [12]. Although a data rate of 1000 Gb/s has been demonstrated, the PIC suffers from very high propagation losses, strictly correlated to InP platform.

Table 2. Overview of integrated optoelectronic platforms (5 = most favourable; 0 = less favourable) (ΔRI: refractive index contrast; α: propagation losses; TOC: Thermo-Optical Coefficient; CCMOS: Compatibility with CMOS; REL: RELiability).

Material	ΔRI	α	TOC	CCMOS	REL
III/V (InP, GaAs)	4	2	4	No	5
Si	5	2	5	Yes	5
Si_3N_4	3	5	1	Yes	5
SOS	2	4	2	Yes	5
Polymer	3	3	5	Yes	2

To overcome this limitation, the integration of InP lasers and amplifiers with the silicon-derived platform paves the way to PIC with low power consumption, low propagation losses, at expense of high cost. The hybrid integration is currently a hot topic for several research groups.

3 Space Applications of Integrated Photonics

3.1 Microwave Photonics

Microwave Photonics (MWP) represents a disruptive technology to improve the performance of RF systems, by merging the benefits of photonics and microwave domain [13–16]. In the Space context, microwave photonics systems can be used in payloads for Earth Observation (EO) and telecom (TLC) purposes, by pushing towards miniaturized satellites with excellent performance, especially in terms of bandwidth [17–25]. The most interesting recent advances of integrated MWP include low phase noise oscillators, RF waveform generation, analog-to-digital conversion (ADC), RF filtering, and wideband beamforming.

For telecom payload, a high-quality and ultra-compact microwave local oscillator has been reported in [18], with an operating frequency of about 9 GHz, a phase noise of –92 dBc/Hz @ 1 MHz from the carrier, and a chip footprint of 5×6 cm^2.

Moreover, a reconfigurable bandpass filter plays a crucial role to safeguard the in-orbit payload reconfigurability [19, 20]. A promising reconfigurable bandpass filter with bandwidth of 10 GHz, large extinction ratio (>15 dB), central frequency reconfigurability of about 15 GHz is reported in [19].

In the context of SAR payload, a linearly chirped microwave waveform generator is needed to ensure a high range resolution [21, 22]. Very large figures of merit, as the time-bandwidth product, have been reported in literature [21] for linearly chirped microwave generator in Ka-band, overcoming the noise problems of the electronic counterparts.

Several examples of optical beamforming networks have been proposed in literature [23–25], where the fast and wideband control of the optical path phases guarantee the high resolved shaping of the transmitted beams from the antenna array.

3.2 Attitude and Orbit Control Systems (AOCS)

PICs could represent a disruptive technology in several satellite subsystems, as Attitude and Orbit Control Systems (AOCS). For example, integrated photonic gyroscopes could help to get over the intrinsic bottlenecks of the strategic/navigation grade in-orbit solutions, such as Fiber Optical Gyroscopes (FOG), Ring Laser Gyroscopes (RLG), and Hemisperichal Resonant Gyroscopes (HRG), in terms of weight, size and power consumption. A Resonant Micro Optical Gyroscope (RMOG), whose operation is based on the Sagnac effect within a ring resonator (RR), represents a potential solution, combining high performance within a small footprint. Since the resolution is strictly correlated to the Q-factor of the RR, in the last years a great research effort has been spent to engineer RRs by exploring different configurations and technological platforms, aiming at achieving resolution <1°/hr. The latest advances in the RMOG context are reported in Table 3.

Table 3. Overview of Resonant Micro Optical Gyroscopes [26–30] (MD: MicroDisk; WR: Wedge Resonator; RR: Ring Resonator).

RMOG based on	Technology platform	Volume (cm^{-3})	Theoretical Resolution(°/hr)	Experimental bias drift (°/hr)	Experimental resolution (°/hr)
Coil [26]	Si3N4	10	<7	59	66
MD [27]	SiO_2	3	<0.7	-	22
WR [28]	SiO_2	68	<0.05	<3.6	5
RR [29]	InP	10	<10	-	-
RR [30]	Si_3N_4	1	0.01	-	-

4 Prospects

The use of photonics in the Space field could be enlarged by overcoming the following technical bottlenecks:

- *High reliability chip packaging.* A reliable and stable packaging of optoelectronic devices should be ensured to safeguard operations in harsh environment and reduce the losses of fiber or grating – chip butt coupling (3–4 dB/facet).
- *High on-orbit reliability of lasers.* For Space optical communications, the laser sources should be able to transmit light with large operational lifetimes despite atmospheric/environmental disturbances (e.g., cloud cover, wind, radiation, etc.).
- *Regulations for testing rad-hard photonic devices.* Space Agencies should publish basic requirements for testing photonic devices in Space environment.
- *Reduction of power consumption.* Continuous process optimization and innovative device design are needed to reduce the power consumption, which should be almost about 1 order of magnitude less than the electrical counterparts (100 fJ/bit).

- *Industrialization of the optoelectronic integration process, mass production capacity and reduction of the processing costs.* In order to preserve the compactness of the chip, a great research effort should be spent to increase the TRL of the hybrid integration approach.
- *Increase data rates.* For optical communications, the data rates should be improved (i.e., 10 – 100 X) without increasing mass, volume, power, and/or spectrum.

5 Conclusions

Integrated optoelectronic technology represents an efficient solution to support the *NewSpace* revolution, with significant savings in terms of size, weight, and power (SWaP) with respect to electronic counterparts. The main Space Agencies, like *National Aeronautics and Space Administration* (NASA), *European Space Agency* (ESA), and *Japan Aerospace eXploration Agency* (JAXA), have invested research funds in the topic, pushing the rapid development of rad-hard photonic devices and systems for frontier exploration and technology demonstration. The current scientific research is mainly focusing on miniaturization, integration, and increase of reliability of optical payload systems for several applications, including laser communication, microwave photonics, optical sensing, medical health, navigation and imaging.

References

1. Logue, T.J., Pelton, J.: Overview of commercial small satellite systems in the "New Space" age. In: Handbook of Small Satellites: Technology, Design, Manufacture, Applications, Economics and Regulation, pp. 1–18 (2019)
2. Muelhaupt, T.J., Sorge, M.E., Morin, Wilson, R.S.: Space traffic management in the new space era. J. Space Safety Eng., 80–87 (2019)
3. Aglietti, G.S.: Current challenges and opportunities for space technologies. Front. Space Technol, 1 (2020)
4. Curzi, G., Modenini, D., Tortora, P.: Large constellations of small satellites: A survey of near future challenges and missions. Aerospace, 133 (2020)
5. Bozovich, A.N.: Photonic integrated circuits (PICs) for next generation space applications. In: Virtual 2020 NEPP Electronics Technology Workshop (ETW), Greenbelt, Massachusetts (2020)
6. Krainak, M., et al.: Integrated photonics for NASA applications. In: Components and Packaging for Laser Systems V. SPIE (March 2019)
7. Brunetti, G., McKenzie, I., Dell'Olio, F., Armenise, M.N., Ciminelli, C.: Measured radiation effects on InGaAsP/InP ring resonators for space applications. Optics Express, 24434–24444 (2019)
8. Brasch, V., Chen, Q. F., Schiller, S., Kippenberg, T.J.: Radiation hardness of high-Q silicon nitride microresonators for space compatible integrated optics. Optics Express, 30786–30794 (2014)
9. Shi, W. H., Lv, W.M., Sun, T.Y., Zhang, B.S.: Optoelectronic platform and technology. Front. Inf. Technol. Electronic Eng., 439–457 (2019)
10. Hao, T., et al.: Toward monolithic integration of OEOs: from systems to chips. J. Lightwave Technol., 4565–4582 (2018)

11. Briles, T. C., et al.: Hybrid InP and SiN integration of an octave-spanning frequency comb. APL Photonics, 026102 (2021)
12. Sun, H., et al.: 800G DSP ASIC design using probabilistic shaping and digital sub-carrier multiplexing. J. Lightwave Technol., 4744–4756 (2020)
13. Capmany, J., Novak, D.: Microwave photonics combines two worlds. Nat. Photonics, 319 (2007)
14. Marpaung, D., Yao, J., Capmany, J.: Integrated microwave photonics. Nat. Photonics, 80–90 (2019)
15. Tzintzarov, G.N., Rao, S.G., Cressler, J.D.: Integrated silicon photonics for enabling next-generation space systems. Photonics, 131 (2021)
16. Seeds, A. J., Williams, K.J.: Microwave photonics. J. Lightwave Technol., 4628–4641 (2006)
17. Tu, X., Song, C., Huang, T., Chen, Z., Fu, H.: State of the art and perspectives on silicon photonic switches. Micromach., 51 (2019)
18. Tang, J., et al.: Integrated optoelectronic oscillator. Optics Express, 12257–12265 (2018)
19. Ciminelli, C., Dell'Olio, F., Brunetti, G., Conteduca, D., Armenise, M.N.: New microwave photonic filter based on a ring resonator including a photonic crystal structure. In 2017 19th International Conference on Transparent Optical Networks (ICTON). IEEE (July 2017)
20. Fandiño, J.S., Muñoz, P., Doménech, D., Capmany, J.: A monolithic integrated photonic microwave filter. Nat. Photonics, 124–129 (2017)
21. Brunetti, G., Armenise, M.N., Ciminelli, C.: Chip-Scaled Ka-Band Photonic Linearly Chirped Microwave Waveform Generator. Front. Phys., 158 (2022)
22. Zhang, K., Zhao, S., Yin, Y., Lin, T., Li, X., Jiang, W., Wang, G.: Photonic generation and transmission of linearly chirped microwave waveform with increased time-bandwidth product. IEEE Access, 47461–47471 (2019)
23. Liu, Y., et al.: Tuning optimization of ring resonator delays for integrated optical beam forming networks. J. Lightwave Technol., 4954–4960 (2017),
24. Seeds, A. J.: Optical technologies for phased array antennas, IEICE Trans. Electron., 198–206 (1993)
25. Vidal, B., Mengual, T., Martí, J.: Fast optical beamforming architectures for satellite-based applications. Adv. Optical Technol., 1–5 (2012),
26. Gundavarapu, S., et al.: Interferometric optical gyroscope based on an integrated Si3N4 low-loss waveguide coil. J. Lightwave Technol., 1185–1191 (2017)
27. Li, J., Suh, M. G., Vahala, K.: Microresonator brillouin gyroscope. Optica, 346–348 (2017)
28. Lai, Y. H., et al.: Earth rotation measured by a chip-scale ring laser gyroscope. Nat. Photonics, 345–349 (2020)
29. Ciminelli, C., et al.: A high-Q InP resonant angular velocity sensor for a monolithically integrated optical gyroscope. IEEE Photonics J., 1–19 (2015)
30. Ciminelli, C., et al.: Rigorous model for the design of ultra-high Q-factor resonant cavities. In: 2016 18th International Conference on Transparent Optical Networks (ICTON). IEEE (July 2016)

Implementation of Dynamic Acceleration Unit Exchange on a RISC-V Soft-Processor

Saeid Jamili(✉), Abdallah Cheikh, Antonio Mastrandrea, Marcello Barbirotta, Francesco Menichelli, Marco Angioli, and Mauro Olivieri

Department of Information Engineering, Electronics and Telecommunications, Sapienza University of Rome, Rome, Italy
Jamili.1911484@studenti.uniroma1.it

Abstract. Using Artificial Intelligence (AI) techniques has become the best solution in many applications. By the end of Moore's Law, implementing a platform capable of such massive processing for edge-IoT applications has become a significant challenge. However, using static hardware accelerators can be an excellent solution; even so, they typically require a great deal of silicon area and are not optimized for all operation modes. Reconfigurable computing lets parts of the hardware change proportionally to the task during operation, allowing for optimized operation and the use of many hardware accelerators without requiring a large area. In this study, we present a dynamic acceleration unit exchange on a RISC-V soft-processor based on the open-source Klessydra-T13 RISC-V core. We show how reconfiguration can be used to make the hardware accelerator more flexible and improve its performance. As a case study, we show how reconfiguration techniques can be used to speed up AI architectures by reconfiguration of vector accelerator units.

1 Introduction

We live in an era where intelligence IoT devices are developing widely, and a wide range of appliances are facing the possibility of taking an integral part in the IoT world. Intelligence edge devices in a cloud architecture can decentralize the computation and transmit lower amounts of data to the cloud centers [1, 2] This is particularly desired in edge nodes because of the typical bandwidth limitations imposed on IoT physical communication networks. Moreover, safety-critical systems, such as self-driving cars and health care devices, require real-time response and should be able to work autonomously even when losing communication with the central nodes. Integrating AI and migrating the process from the cloud to the edge makes the design of high-performance and low-power processor units for edge devices the next critical challenge to overcome [3]. Reconfigurable computing considers some hardware units to have the flexibility to dynamically change their behavior, implementing different logic functions. The scope is to increase the performance/power consumption tradeoff in applications by customizing the hardware to the needs of the specific tasks, reusing the same logic circuits for different computations, and in general, reducing the complexity of the final design. In fact, a classical static architecture containing an integrated hardware accelerator is not going to utilize

R. Berta and A. De Gloria (Eds.): ApplePies 2022, LNEE 1036, pp. 300–306, 2023.
https://doi.org/10.1007/978-3-031-30333-3_40

that part at full time, leaving the circuit inactive for some portions of time while the costs for its fabrication and standby power consumption are still present. By using hardware reconfiguration in Intelligence IoT devices, it is possible to put many hardware accelerators in a single physical logic circuit. This makes the device less complicated overall and makes better use of the hardware.

2 Background

In this section, we briefly describe the Dynamic Function Exchange (DFX) concept and the Klessydra-T13 core, showing how they interact with each other.

2.1 DFX

DFX is the ability to exchange one or more logical parts dynamically without stopping the operation of the other running parts. FPGAs are widely known for their low-to-medium volume production, and, when compared to ASICs, FPGAs exhibit lower speed, higher cost, and higher power consumption. However, ASICs have considerably higher NRE costs, which is why they are generally used in mass production devices but are not cost-effective in low or medium-volume production. On the other hand, the advantages of FPGAs lie in their flexibility and their runtime reconfiguration. The use of DFX can help in reducing the gap between ASICs and FPGAs regarding power and cost. It can be even considered a technique for reducing the gap between programmable architectures (CPU) and hardwired logic in terms of flexibility [4]. In some cases, it is possible to assimilate both the advantages of ASICs and FPGAs, making it possible to design high-speed hybrid systems with extremely high flexibility [5].

2.2 Klessydra-T13

Klessydra-T13 is an open-source 32-bit Interleaved Multi-threaded (IMT) processor that fully supports the RISC-V RV32IM ISA and one instruction from the Atomic "A" extension, along with supporting custom instruction extensions for accelerating vector operations [6, 7]. This architecture, by exploiting the benefits of IMT architectures and integrated hardware accelerators, can significantly decrease idle time (stalls) and consequently mitigate the total energy consumption per operation [8, 9].

3 Klessydra-T13-DFX

3.1 Architecture

The Klessydra-T13-DFX is a soft processor based on the Klessydra-T13 with the ability to dynamically change the Vector Coprocessor Unit (VCU) accelerator unit. As shown in the Vector Coprocessor Unit (VCU) accelerator is tightly coupled with the core, having a direct interface for both control and data signals. This can be regarded as an efficient way of implementing accelerators accessed very frequently and requiring low latency responses. However, a device like the VCU might demand a large on-chip area depending on the depth of both data-level parallelism and thread-level parallelism, which in turn translates to increasing costs, limiting the ability to implement a wide set of operations. However, by using DFX for functional reconfiguration, this limitation can be mitigated by reconfiguring the VCU and changing some of its internal functionality, thus reducing the overall complexity of the design and still having the ability to implement a wide set of operations. An example of this will be illustrated later in the article (Fig. 1).

The process of reconfiguration includes some steps, as shown in Fig. 2. The process can start with a trigger via software (writing on registers of the DFX controller) or via hardware, and when it is done, an interrupt is sent to the core to let it know the reconfiguration status. During the reconfiguration, all signals of the module that is being reconfigured are disconnected, and the output of the interface port can be set to default values. This is done to avoid sending unknown values to the input ports of other modules, thus allowing a safe reconfiguration operation, and preventing the system from behaving in an unpredictable way. The speed of reconfiguration depends on the interface port speed (for the Xilinx 7-series, the bandwidth for PCAP is about 1 Gb/s and for ICAP it is about 3.2 Gb/s), memory latency, and the configuration file size.

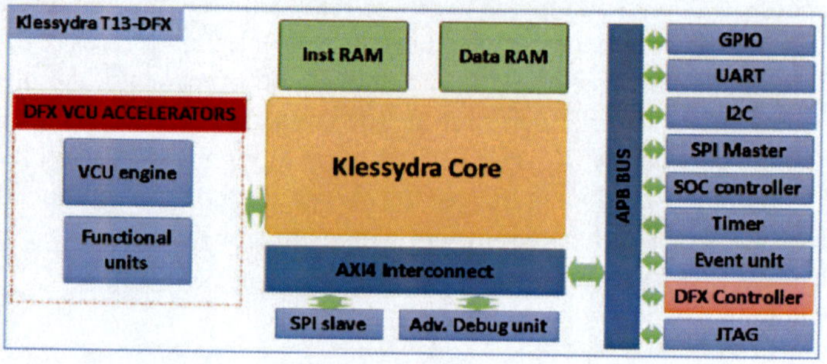

Fig. 1. Architecture of the Klessydra-T13-DFX

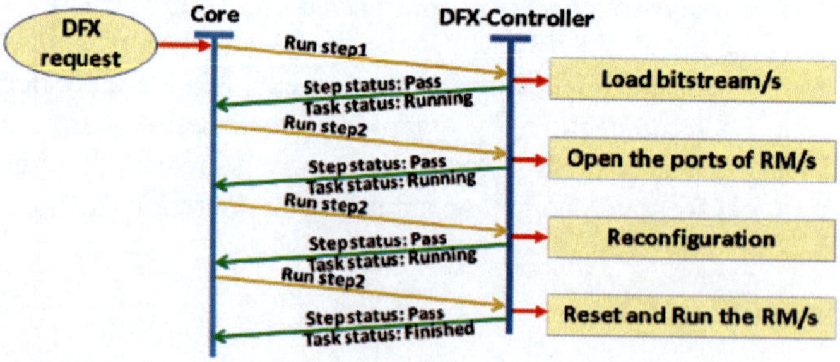

Fig. 2. Reconfiguration process in Klessydra-T13-DFX

3.2 Dynamic VCU

Figure 3 shows the Klessydra-T13-DFX core with the integrated dynamic VCU that connects directly to the pipeline. The VCU is made up of an interface port for safe connection reconfiguration, a VCU engine (hardware loop, scratchpad memory, and mapper), and a functional unit with the most frequent vector operations [7] and optional ones for implementing any other vector operations. The VCU engine and its embedded functional units can be reconfigured in two ways: by exchanging the whole VCU for a different configuration with a different level of SIMD/MIMD parallelism, or by exchanging a functional unit that isn't being used at the moment for one that the application needs.

For the optional functional unit, we proposed an example architecture as shown in Fig. 4 that is coupled with the basic operation function units for better utilization. Also, the state machine is designed for the ability to work with a variable number of stages. The input vector elements are read and processed in order (the width depends on the SIMD parameters), and the process will keep reading all the vector elements until the vector length is reached. After all the vector's parts have been processed, the VCU can store the result/s in local scratchpad memory. Some operations, like the reduction for subtracting, multiplying, and exponential, searching for a value and finding the minimum and maximum values in the vector, are implemented and tested as an example.

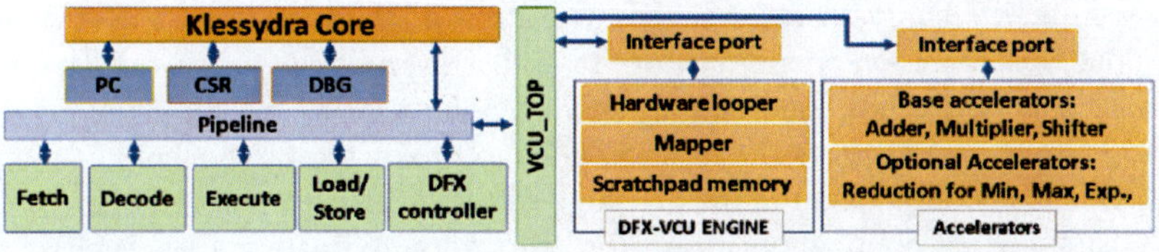

Fig. 3. Architecture of Klessydra-T13-DFX Core and the Interface port (decoupler) for reconfiguration regions

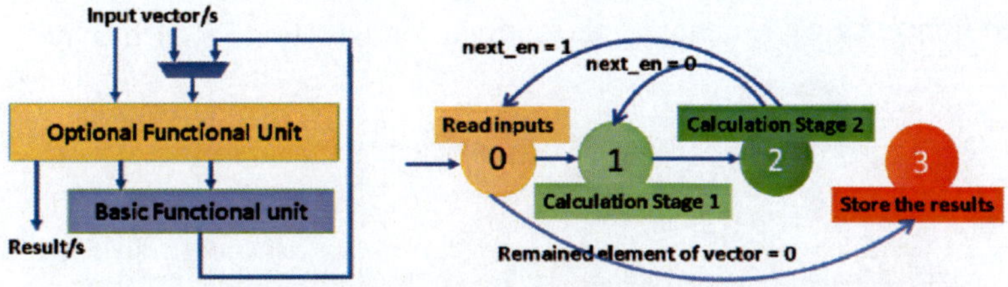

Fig. 4. Block diagram optional functional unit and the state machine.

3.3 Simulation Environment

In the DFX architecture, the reconfigurable regions may be updated with reconfiguration modules (RMs). Simulation software like Modelsim do not support DFX features by default. However, some tricks can be put in place to perform RM switching in a simulation environment. Figure 5 shows a simple architecture containing some switches that multiplex the chosen RMs. With this method, we can simulate pseudo-DFX designs and change the RMs while the simulation is running.

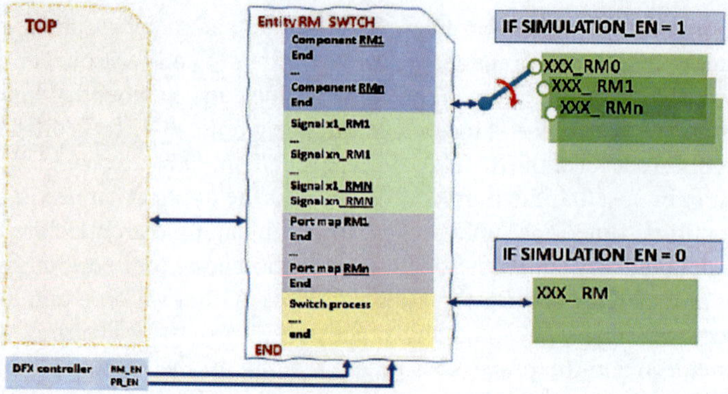

Fig. 5. Simulation model imitating the DFX implementation

4 Performance Evaluation of the DFX

In this section, we study a performance evaluation of the Klessydra-T13-dfx core by running the most common operations for pre-and post-processing such as filters, feature extraction, data converting, and neural networks like VGG16 [2, 6]. Figure 6 depicts the block diagram and the results of adding an approximation exponential function to the optional accelerator function over a range of 0 to 10 with a maximum error of 9%, which is suitable for many applications that allow approximate values, such as neural networks. Figure 7 shows how reconfiguration of the VCU can speed up the process by choosing the best accelerator functions and configuration for some common operations in AI applications. Also, Fig. 8 shows the results of running VGG16 with architectures containing classical static implementations, each of which is running at the maximum supported frequency while choosing the optimal configuration for each layer.

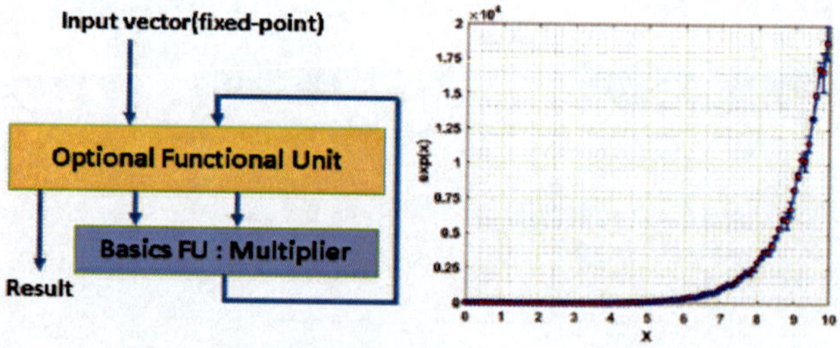

Fig. 6. Block diagram of an implementing exponential function in optional accelerator part and the output result for inputs between 0 and 10.

The overhead of reconfiguration was set to 10 ms for each reconfiguration by considering the PCAP interface. As can be seen from Fig. 6, dynamically switching the VCU configuration with faster RMs for each layer gives us a top-speed improvement over the optimal static configuration as shown in Table 1.

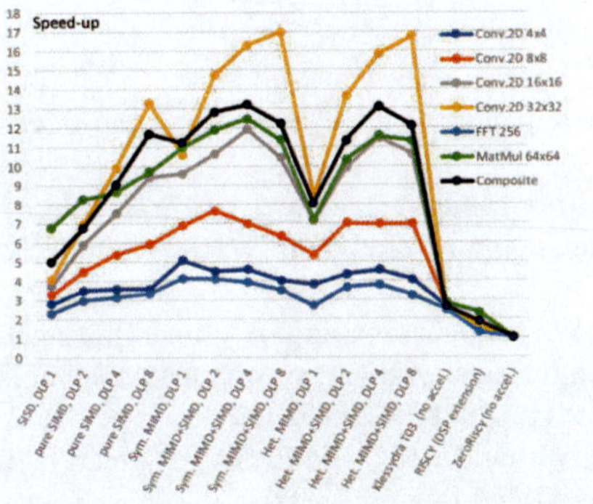

Fig. 7. Acceleration of execution time relative to the Zero-riscy core, as a reference for different configurations and some common operations. [6]

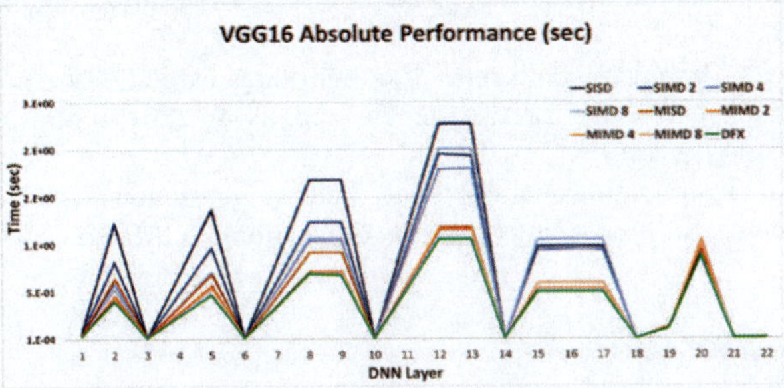

Fig. 8. Time per layer in VGG16 DNN for SIMD, MIMD, and DFX on layers

Table 1. Total time to run the VGG16 DNN for SIMD, MIMD, and DFX configs

VCU Config	SISD	SIMD 2	SIMD 4	SIMD 8	MISD	MIMD 2	MIMD 4	MIMD 8	DFX on layers
VGG 16 Time (s)	17.31	14.08	12.56	13.44	9.81	8.57	8.30	8.89	8.05

5 Conclusion

We have shown a case study on using DFX in both functional reconfiguration, which enhances the versatility of the processor without statically added area overhead, and full vector unit (VCU) replacement, which can be beneficial when the process needs many different accelerators for complex/different neural network algorithms and the switching time must be very small compared to the total time of the running task. In the future, more accelerator variations will be implemented in the VCU unit, and the performance of more complex AI systems will be studied, targeting the optimization of neural network execution via re-configuration at run-time.

References

1. Wang, X., et al.: Convergence of edge computing and deep learning: a comprehensive survey (2020)
2. Sordillo, S., et al.: Customizable vector acceleration in extreme-edge computing: a RISC-V software/hardware architecture study on VGG-16 implementation. Electronics **10**(4), 518 (2021)
3. Cao, J., Zhang, Q., Shi, W.: Edge Computing: A Primer. Springer (2018)
4. Koch, D.: Partial Reconfiguration on FPGAs; Architectures, Tools and Applications. Springer (2013). https://doi.org/10.1007/978-3-030-02083-5
5. Zuchowski, P.S., et al.: A hybrid ASIC and FPGA architecture. In: IEEE/ACM International Conference on Computer Aided Design (2002)
6. Cheikh, A., et al.: Klessydra-T: Designing vector coprocessors for multithreaded edge-computing cores. IEEE Micro **41**(2) (2021)
7. Cheikh, A., et al.: Efficient mathematical accelerator design coupled with an interleaved multi-threading RISC-V microprocessor. In: ApplePies 2019 (2019). https://doi.org/10.1007/978-3-030-37277-4_62
8. Cheikh, A., et al.: The microarchitecture of a multi-threaded RISC-V compliant processing core family for IoT end-nodes. LNEE, vol. 512, pp. 89–97 (2019). https://doi.org/10.1007/978-3-319-93082-4_12
9. Olivieri, M., et al.: Investigation on the optimal pipeline organization in RISC-V multi-threaded soft processor cores. In: Proceedings 1st New Generation of CAS, NGCAS 2017, pp. 45–48 (2017)

Investigating High-Level Decision Making for Automated Driving

Alessio Capello(✉), Luca Forneris, Alessandro Pighetti, Francesco Bellotti, Luca Lazzaroni, Marianna Cossu, Alessandro De Gloria, and Riccardo Berta

Department of Electrical, Electronic and Telecommunication Engineering (DITEN), University of Genoa, Via Opera Pia 11a, 16145 Genoa, Italy
{alessio.capello,luca.lazzaroni,marianna.cossu}@edu.unige.it,
{francesco.bellotti,alessandro.degloria,riccardo.berta}@unige.it

Abstract. As the quality of perception systems available for automated driving (AD) increases, we investigate the development of an AD agent based on Reinforcement Learning which exploits underlying systems for longitudinal and lateral control. The goal is addressed by designing high-level actions, trying to imitate the commands of a real driver. The proposed agent is trained in a simulated motorway environment and compared to an agent which outputs low-level actions. Our preliminary results show similar performance results, a more pronounced human-like behaviour and a huge reduction in needed training time because of the higher-level of the available actions.

1 Introduction

Availability of ever more accurate perception systems in vehicles has paved the way to higher-level decision making (DM). This paper explores the development of a DM for automated driving (AD), based on Reinforcement Learning (RL), in an end-to-end architecture. Our goal is to make the decisional process of the agent more human-like and also more suitable to posterior explanation [1]. In our proposal, the DM exploits context information and outputs high-level actions to control Mid-Level systems that are already present in vehicles, such as the Adaptive Cruise Control (ACC). This approach also abstracts the model of the vehicle to the DM, so that it can be applied to different vehicles without further training or constraints on physical parameters (e.g., acceleration curve, steering ratio, etc.).

2 The Leurent Environment for Automated Driving Based on RL

RL is one of the three main paradigms of Machine Learning, alongside with Supervised and Unsupervised Learning. The goal of RL is to train an Agent that will learn a model to estimate the outcome of all possible actions applied on an uncertain dynamic system. The Agent will use that model to act on that system to maximize the estimated outcome. Since the true outcome between actions and state of the dynamic system is unknown, the Agent is trained with an empirically approximated outcome function called Reward

R. Berta and A. De Gloria (Eds.): ApplePies 2022, LNEE 1036, pp. 307–311, 2023.
https://doi.org/10.1007/978-3-031-30333-3_41

Function (RF). The training consists in a MonteCarlo trial-and-error approach, usually in a simulated environment, where the Agent can explore different strategies and get the corresponding outcome with the RF. When the internal model is learned using a Neural Network, this paradigm is called Deep Reinforcement Learning (DRL). Due to its model learning capability, DRL has become the most important paradigm of RL to perform real-world tasks [2].

The environment chosen for training and testing our agent comes from the PhD thesis Eduard Leurent, who developed a framework for decision-making in autonomous driving [3]. This framework consists of several Python modules [4], that can be divided in three categories:

- Environments: ready to use automotive environments such as Highway (Fig. 1), Merge, Intersection and Parking
- Vehicle Physical Model: the framework can work with kinematic or dynamic model for vehicles. The dynamic one is more complex as it models physical laws such as friction between tires and road, car drifting, etc
- Road: this module contains road definition that can be used as building blocks when defining custom environments.

Our work focused on the Highway environment, where the Ego Vehicle (EV) driven by the DM agent should travel in a multi lane scenario avoiding collisions with other Non-Playing-Vehicles (NPV). To model the physical behaviour of the vehicle we modified the kinematic model to manage several vehicle parameters that could differentiate one vehicle from another, like max acceleration, brake intensity, etc.

The behaviour of our Agent will be described in detail in the next section; NPVs follow a heuristic behaviour with some randomness applied to achieve a non-deterministic behaviour, which is required to avoid over-fitting of the Agent to a particular behaviour of NPVs.

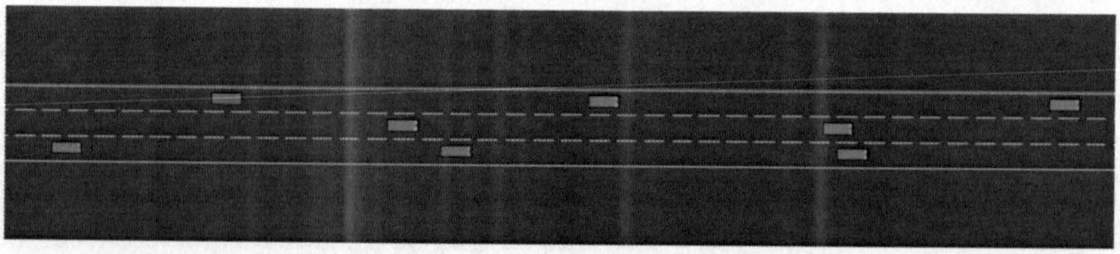

Fig. 1. View of the Highway environment. Ego Vehicle is coloured in green; Non-Playing-Vehicles are coloured in blue

3 High-Level Decision-Making Model

The High-Level Decision-Making model is the module that controls the EV in a human-like way, maintaining a safe behaviour in the highway environment.

The module lies over state of the art systems already present in vehicles (e.g., ACC). In order to reflect a human-like control, we defined that the decisor could output the following actions:

- Keep lane: the ACC is activated to keep a safe distance from the front vehicle, if present, otherwise reach the max speed allowed
- Overtake: change lane to the left if a front vehicle is present, otherwise use the Cruise Control to reach the max speed allowed
- Go to right most lane: change lane to the right, wait few seconds and repeat, if possible.

Figure 2 depicts the behaviours enacted by these actions.

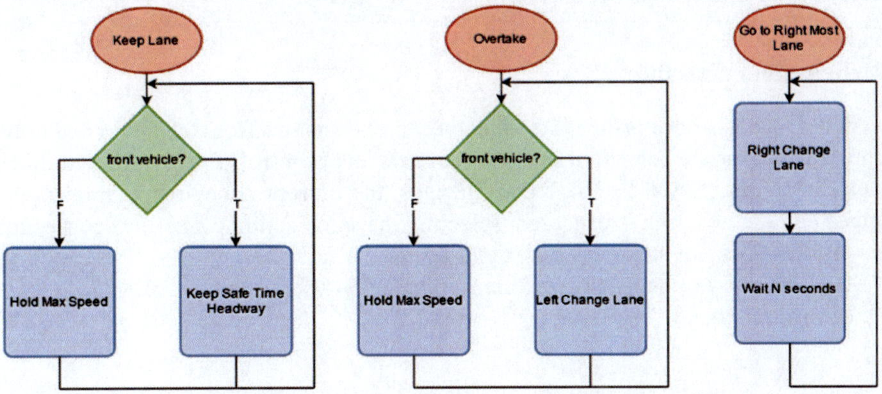

Fig. 2. Flowcharts of high-level actions

The High-Level Decision-Making Model is composed by two subsystems stacked together, as in [5]: Driver Decision Maker (DDM) and Behaviour Executor (BE). The DDM controls the BE using the above-mentioned high-level actions, while the BE implement those actions into a series of low-level controls (steering and throttle) and sends them to the simulated environment. Some implementation details on these two modules are provided in the next two sub-sections (Fig. 3).

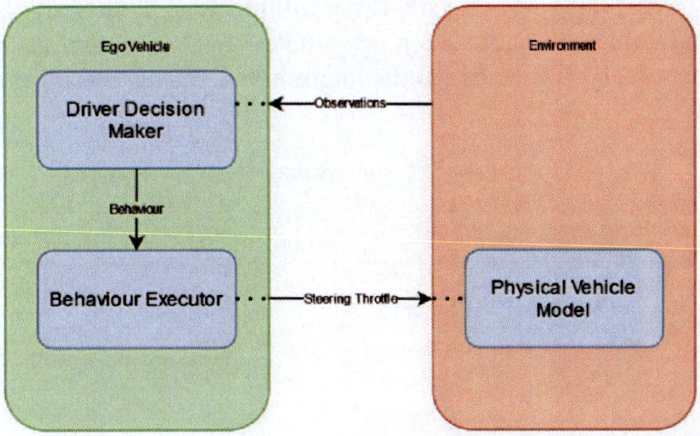

Fig. 3. Ego Vehicle and Environment schema with sub-systems and connections

3.1 Driver Decision Maker

The top layer of our architecture is the DDM, which is trained in the DRL framework, currently using the Proximal Policy Optimization (PPO) Algorithm [6]. As the designer can specify different types of rewards and penalties, DRL allows also training AD agents implementing different driving styles (e.g., optimize fuel consumption, minimize overtake, etc.) [7]. DDM gets observations from the surrounding environment and selects the appropriate high-level action. In our experiments, the DDM receives observations and selects an action once per second.

3.2 Behaviour Executor

The Behaviour Executor is the module that receives decisions from the DDM and implements the appropriate behaviour in a series of low-level controls fed to the environment. When the action chosen by the DDM changes, the current behaviour is immediately stopped and the next one is executed; selecting the same action for several consecutive time instants does not stop the current execution.

State of the art onboard devices are implemented with PID controllers in a closed loop control schema.

4 Experimental Results

The agent presented in this article was trained on a PC with Intel Core i7-10750H CPU,16 GB of ram, NVIDIA RTX 2060 mobile in 60 min. Table 1 compares the resulting policy with that of an agent [8] which outputs five low-level actions: faster, slower, left, right, idle. The comparison considers the number of collisions and kilometres travelled in 60 episodes of 60 s each. Results show that the high-level agent is a bit slower but eliminates all the collisions. It is important to highlight that the training of the higher-level agent took more than one order of magnitude less steps than the lower-level agent, which can be well explained because of the possibility of outputting much higher-level decisions exploiting powerful subsystems controlling longitudinal and lateral motion. An expert assessment also noticed a more human-like behaviour on the part of the high-level agent, which also facilitates the interpretation of the neural model outputs.

Table 1. Performance comparison over 60 60-s episodes

Algorithm	Collisions	Km travelled	Training steps
High-level agent	0	99	60K
Low-level agent [8]	2	105.7	1.75M

5 Conclusions and Future Work

The paper has presented an architecture that can be used to implement a high-level decision maker for motorway driving. We tested our approach in a simulation environment, and obtained promising preliminary results, particularly in terms of training time reduction, arguably because of the reduction in the action space dimensionality. We observed a human-like behaviour of the agent, while also keeping good performance (distance travelled and number of collisions). Future works should focus on further improving the training of the agent (e.g., considering different network architectures and training algorithms), improving performance and expected acceptability by possible human passengers. Further work should also concern explainability of the model.

References

1. Bibal, A., Lognoul, M., de Streel, A., Frénay, B.: Legal requirements on explainability in machine learning. Artifi. Intell. Law **29**(2), 149–169 (2020). https://doi.org/10.1007/s10506-020-09270-4
2. Mnih, V., et al.: Human-level control through deep reinforcement learning. Nature **518**, 529–533 (2015). https://doi.org/10.1038/nature14236
3. Leurent, E.: An Environment for Autonomous Driving Decision-Making (2018)
4. Leurent, E., Mercat, J.: Social Attention for Autonomous Decision-Making in Dense Traffic, http://arxiv.org/abs/1911.12250, (2019). https://doi.org/10.48550/arXiv.1911.12250
5. Mirchevska, B., Pek, C., Werling, M., Althoff, M., Boedecker, J.: High-level decision making for safe and reasonable autonomous lane changing using reinforcement learning. In: 2018 21st International Conference on Intelligent Transportation Systems (ITSC), pp. 2156–2162 (2018). https://doi.org/10.1109/ITSC.2018.8569448
6. Schulman, J., Wolski, F., Dhariwal, P., Radford, A., Klimov, O.: Proximal Policy Optimization Algorithms, http://arxiv.org/abs/1707.06347, (2017). https://doi.org/10.48550/arXiv.1707.06347
7. Massoud, R., Berta, R., Poslad, S., De Gloria, A., Bellotti, F.: IoT sensing for reality-enhanced serious games, a fuel-efficient drive use case. Sensors. **21**, 3559 (2021). https://doi.org/10.3390/s21103559
8. Campodonico, G., et al.: Adapting autonomous agents for automotive driving games. In: de Rosa, F., Marfisi Schottman, I., Baalsrud Hauge, J., Bellotti, F., Dondio, P., Romero, M. (eds.) GALA 2021. LNCS, vol. 13134, pp. 101–110. Springer, Cham (2021). https://doi.org/10.1007/978-3-030-92182-8_10

A Blind Modulation Classification Method Based on Decision Tree and High Order Cumulants

Yulai He[1], Hua Wu[1,3](✉), Qinghe Zheng[2](✉), Yang Liu[1,3], Abdussalam Elhanashi[4], and Sergio Saponara[4]

[1] School of Information Science and Electrical Engingeering (School of Artificial Intelligentce), ShanDong JiaoTong University, Jinan 250357, China
wuhua@sdjtu.edu.cn

[2] School of Intelligence Engineering, Shandong Management University, Jinan 250357, China
15005414319@163.com

[3] CAAC Key Laboratory of General Aviation Operation, Civil Aviation Management Institute of China, Beijing, China

[4] Department of Information Engineering, University of Pisa, 56122 Pisa, Italy

Abstract. The classification of modulation schemes will be widely used in future communication systems, among which the classification accuracy and speed of modulation classification methods have always been the focus of the research. Our goal is to simplify the complexity of the classification model and improve the modulation recognition accuracy. In this paper, we propose the blind modulation classification method based on the decision tree and high order cumulants. During the experiments, the results are visualized to observe the effect. By comparing the probability of accurate classification at different signal-to-noise ratios at each stage, it is used as an overall measure of the performance of the entire classifier. Moreover, experiments show that the performance of the independent block level and the whole system has excellent accuracy and stability under the optimal threshold, signal-to-noise ratio, and the number of samples.

Keywords: Wireless communication · cognitive radio · modulation classification · feature learning

1 Introduction

Nowadays, wireless communication plays an important role in the communication systems in both civil and military fields. This requires us to make full use of channel resources and improve the effectiveness and accuracy of wireless communication [7]. In order to meet the different needs of various customers, the communication signals often use the method of matching modulation. In the development of wireless communication, the modulation scheme has also experienced many changes: from simple analog to complex digital. At the same time, the environment of wireless communication is becoming more and more complicated, and a variety of modulation schemes are included in a certain communication frequency band [8]. To effectively monitor and identify wireless

R. Berta and A. De Gloria (Eds.): ApplePies 2022, LNEE 1036, pp. 312–319, 2023.
https://doi.org/10.1007/978-3-031-30333-3_42

signals, the research on modulation classification has attracted numerous concerns by the academic community to make a breakthrough.

After repeated replication and modulation of traditional analog signals, long-distance transmission is more vulnerable to the interference of channel random noise [1]. In order to obtain more stable communication effects, digital communication application scenarios based on digital signals are gradually increasing [2]. In the digital communication process, the choice of modulation schemes will seriously affect the effect of information transmission and communication security, so it is important to choose scientific and secure signal modulation schemes. When the transmitter can freely set the modulation scheme of the signal, the receiver must get the information to demodulate the signal in order to transmit successfully.

Automatic modulation classification (AMC) methods are mainly divided into two categories: likelihood-based methods [3, 4] and feature-based classification methods. The former is based on the likelihood function of received signals and compares the likelihood ratio with the threshold value, which can be easy to implement; The latter typically uses multiple features and makes decisions based on the observations. Specifically, Zhang et al. [5] proposed a modulation classification method based on high-order cyclic cumulative feature which can realize the modulation classification of multi-channel signals.

This paper is organized as follows. In Sect. 2, we first introduce five kinds of modulation schemes and three sub blocks of the classification system. The proposed method focuses primarily on multi-carrier (MC), and single-carrier (SC) classification. Five modulation schemes include MC in orthogonal frequency division multiplexing (OFDM) signal and 4-quadrature amplitude modulation (4-QAM), 16-quadrature amplitude modulation(16-QAM), and 2-pulse amplitude modulation (2-PAM), 4-pulse amplitude modulation (4-PAM). In Sect. 3, extensive experimental results are intuitively reflected through confusion matrices. It can be found that performance is acceptable when the threshold was set to −0.2 and SNR equals to 20 dB.

2 Modulation Classification Method Based on Decision Tree and High Order Cumulants

Through a series of modulation techniques, the higher communication rates and more efficient spectrum utilization can be obtained, thus spectrum shifting can be carried out, and the modulated symbols can be converted into a modulated signal suitable for channel

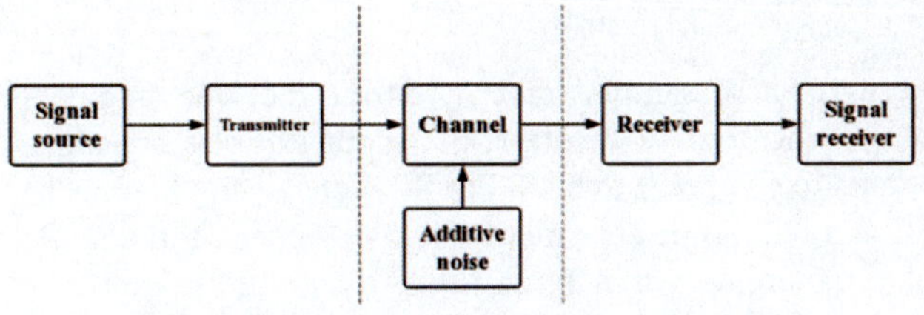

Fig. 1. Wireless communication system model.

transmission or channel multiplexing, which can improve the transmission effectiveness and transmission reliability of the system. The wireless communication system can be demonstrated by the following communication model in Fig. 1.

2.1 System Model Structure

The proposed system model is based on the decision tree, and the complete system can effectively identify the modulation scheme. As shown in Fig. 2, the modulation classification system is mainly divided into 3 blocks: (1) MC-SC block classifies between multi-carrier and single carrier signals; (2) MTC block distinguishes between M-QAM and M-PAM signals if the input to the block is a SC signal; (3) MLC block behaves as a modulation level classifier for classifying 4-QAM/16-QAM or 2-PAM/ 4-PAM.

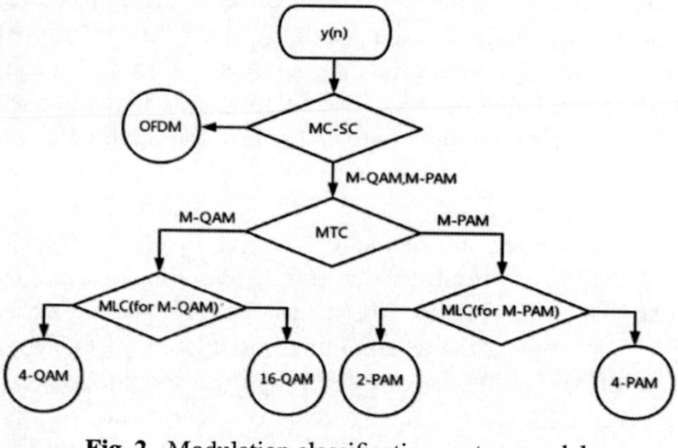

Fig. 2. Modulation classification system model.

2.2 MC-SC Block

The MC-SC block identifies MC and SC by judging the fourth-order cumulants of different signals by performing a Gaussian test. The cumulunt C_{42} can be computed as follows:

$$C_{42}=\frac{1}{N_{\mathrm{m}}}\sum_{n=1}^{N_m}|y[n]|^4-|C_{20}|^2-2C_{21}^2 \quad (1)$$

where N_{m} is the number of samples and $y[.]$ is the received signal. Since OFDM is a mixture of many subcarrier waveforms, the input samples are close to a Gaussian distribution, its statistic C_{42} tends to be 0. For SC signals, C_{42} approaches to a non-zero value so that MC and SC can be classified. The C_{42} values for the modulation schemes with a threshold of 0.5 are shown in Table 1.

Table 1. The C42 values for the different modulation schemes.

Modulation scheme	C_{42}
OFDM	0
4-QAM	−0.82
16-QAM	−0.52
2-PAM	−1.69
4-PAM	−1.1

2.3 MTC Block

For received signals, MTC block uses cyclic-auto-correlation (CAC) function to determine their cyclostationary features. The conjugate and non-conjugate CACs can be computed as

$$\begin{aligned} R_{x^*}(\alpha, v) &= \frac{1}{N}\sum_{n=0}^{N-1} x[n]x^*[n-v]e^{-j2\pi\alpha n T_s} \\ R_x(\alpha, v) &= \frac{1}{N}\sum_{n=0}^{N-1} x[n]x[n-v]e^{-j2\pi\alpha n T_s} \end{aligned} \tag{2}$$

where α is the cyclic frequency, v is the lag variable, N is a finite number of samples, T_S is the sampling period. Based on the CAC has peaks at different locations of cyclic frequencies α which is a function of symbol rate (1/T) and carrier frequency (fc), modulation schemes can be distinguished. The peaks of the CAC of M-QAM in terms of T and fc is (1/T, 0)*, M-PAM are (1/T, 0)*, (2fc, 0) and (2fc/T, 0). Combining these theoretical values, a feature vector of length 6 can be obtained as

$$F = \begin{bmatrix} |R_{x^*}(1/T, 0)|, |R_{x^*}(2fc - 1/T, 0)|, |R_{x^*}(2fc - 1/2T, 0)|, \\ |R_{x^*}(2fc, 0)|, |R_{x^*}(2fc + 1/2T, 0)|, |R_{x^*}(2fc + 1/T, 0)| \end{bmatrix} \tag{3}$$

where the element is proportional to the signal power and the feature vector has been normalized to unit power. The normalized asymptotic feature vector for M-QAM is and M-PAM are $V_1 = [1, 0, 0, 0, 0, 0]$ and $V_2 = [0.5, 0.5, 0, 0.5, 0, 0.5]$, respectively. By comparing the signal with feature vector, MTC can recognize M-QAM or M-PAM.

2.4 MLC Block

MLC block calculates the communication data field (CDF) of the received signals and compares them with theoretical CDFs of available modulation schemes and then chooses the one with the smallest distance from them. The CDF is calculated as

$$F_N(t) = \frac{1}{N}\sum_{n=1}^{N} I(z(n) \le t) \tag{4}$$

where N is the number of samples, t is a threshold range, $z(n)$ is the received signal, and $I(.)$ is 1 if the condition true, 0 otherwise. By calculating CDFs by replacing $z(n)$ with the corresponding input signal $x(n)$, MLC block further breaks down the signal into its modulation level: whether 4-QAM, 16-QAM or 2-PAM, 4-PAM.

3 Experimental Results and Analysis

3.1 Experiment Process

The performance of MC-SC block can be measured by evaluating the classification probability By changing SNR, the classification probability of MLC for M-QAM and M-PAM is tested respectively.

3.2 Signal Waveform Plot

The initial signal provided was set before the experiments begin. Experiments are performed by setting up a transmitter that randomly transmits one signal in the MC and SC. The MC signal is an OFDM signal with 64 sub-carriers, 16 samples, and a sub-carrier spacing of 15.625 kHz. The SC signal applied by raised cosine filter is randomly composed of one of 4-QAM, 16-QAM, 2-PAM, and 4-PAM. The received signal is given as

$$y(n) = x(n)e^{j2\pi fcnTs} + w(n) \tag{5}$$

where $x(n)$ is 1 MHz bandwidth transmitted signal, carrier frequency $fc = 2$ MHz, sampling duration $Ts = 0.25$ µs, the Gaussian noise $w(n) \sim CN(0, \sigma^2)$. The number of MC symbols and SC symbols are 25 and 2000 respectively. The frequency spectrum can then be generated by fast Fourier transforming (FFT) the signal. In Fig. 3, we show the amplitude spectrum and frequency spectrum of various modulation schemes.

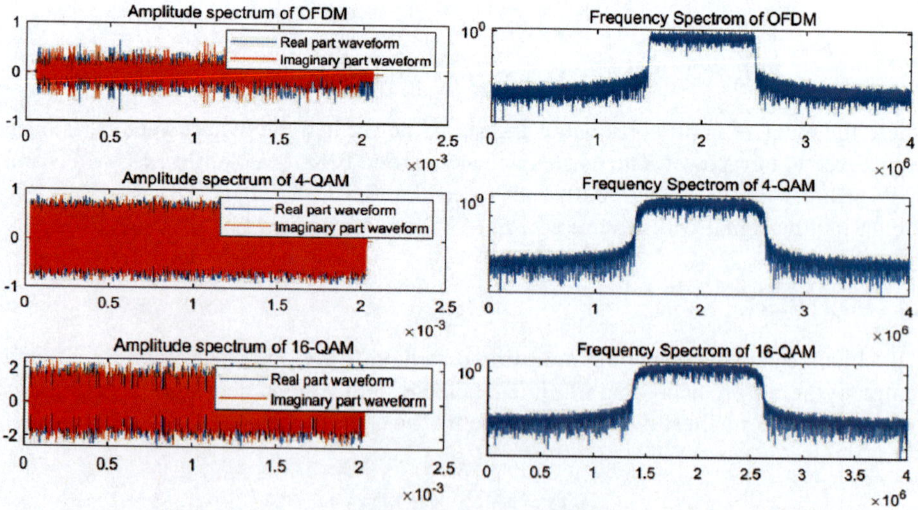

Fig. 3. Amplitude spectrum and frequency spectrum of various modulation schemes.

3.3 Experiment Results and Analysis

Probability of detection (Pd) is the probability that the received signal classified as MC. Probability of false alarm is the probability that the received signal classified as MC though it is not MC. As shown in Fig. 4, Nm determines the performance since a higher value of Nm can obviously improve the performance of MC-SC block. At the same time, SNR can make an impact on performance as well. We can obviously discover that SNR reaches the best state at 20 dB, which is better than 10 dB or 0 dB.

In Fig. 5, we depict the performance of MTC blocks in terms of classification probability. Here, the performance also increases as the number of samples to perform the calculation increases.

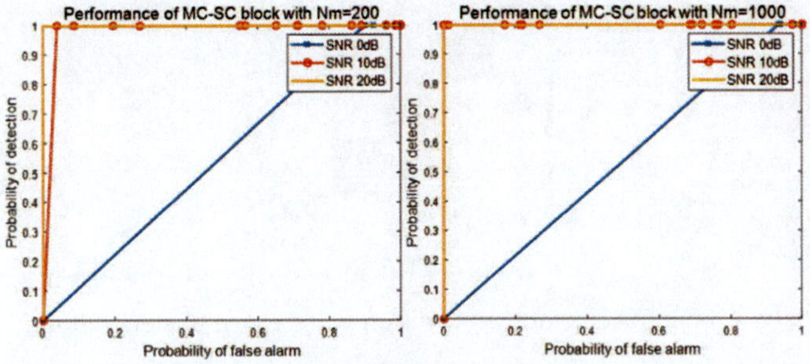

Fig. 4. Performance of MC-SC block (Nm = 200 and Nm = 1000).

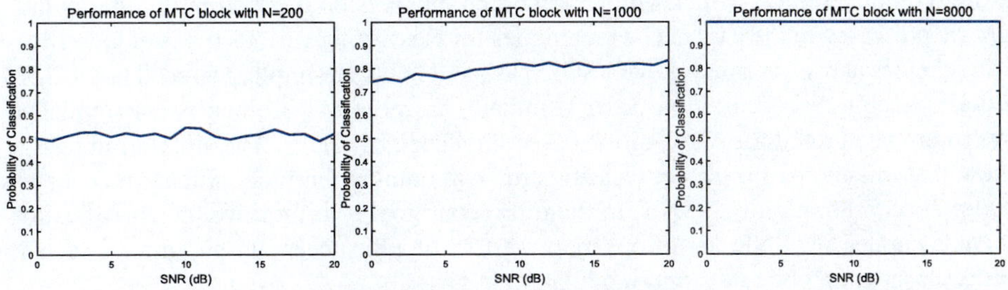

Fig. 5. Performance of MTC block (N = 200, N = 1000, and N = 8000).

Figure 6 shows the performance of MLC block. It can be seen that PAM classifier performs better than QAM classifier, which is due to that transmitted QAM signals' empirical CDF lies between the theoretical CDF of both level's signal. As shown in Fig. 7, the model achieves best performance (almost 100%) when the threshold γ was -0.2 and SNR = 20 dB.

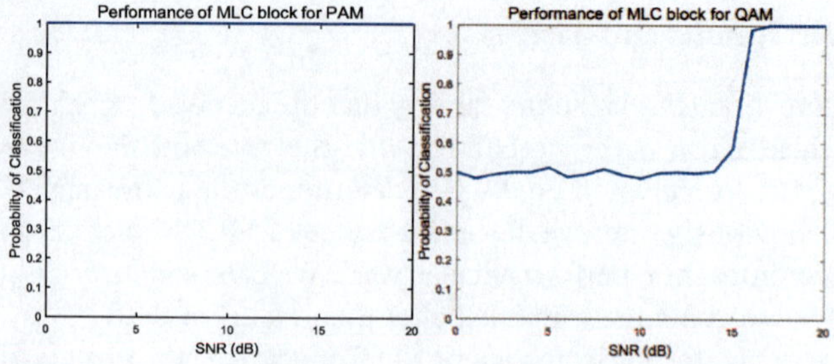

Fig. 6. Performance of MLC block for PAM and MLC block for QAM.

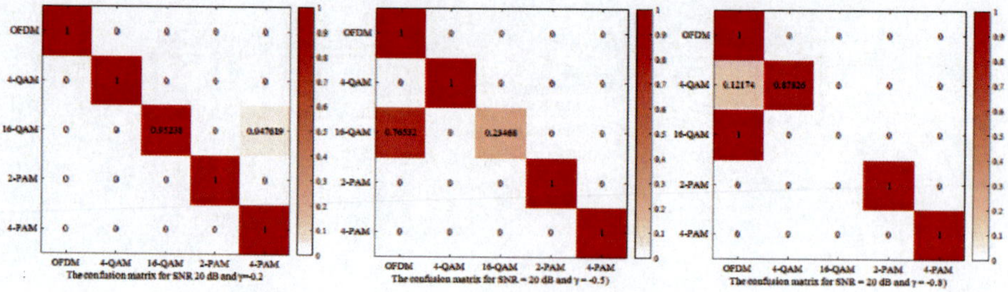

Fig. 7. Confusion matrices of different hyper-parameters setting.

4 Conclusion

A blind modulation classification method based on decision tree and high order cumulants is presented in the paper. The techniques for classifying various types of classification schemes at every stage viz. MC-SC, MTC, and MLC are implemented. The MC-SC block is accomplished by high-order cumulant, then the MTC block is accomplished by feature vector and the MLC block is accomplished by CDFs. The simulation results show that this method of direct classification can obtain a high classification accuracy under the condition of high SNR. In the future, our goal is to improve the classification accuracy under low SNR conditions by pre-estimating the necessary parameters such as carrier frequency, code rate, and modulation index.

Acknowledgements. This research was supported by Shandong Provincial Natural Science Foundation (Grant ZR2020MF151), CAAC Key Laboratory of General Aviation Operation (Civil Aviation Management Institute of China) (Grant CAMICKFJJ-2020-2), National Natural Science Foundation of China (Grant U1933130 and 71731001), and Research and Demonstration of Key Technologies for the Air-Ground Collaborative and Smart Operation of General Aviation (Grant 2022C01055).

References

1. Kim, D.H., Choi, J.H.: Analysis of the transmission performance of control pilot signal lines for charging communication in electric vehiclesa. Int. J. Automot. Technol. **21**(2), 519–525 (2020)

2. Al-Salehi, A.R., Qureshi, I.M., Malik, A.N., et al.: Dual-function radar-communications: information transmission during FDA radar listening mode. Int. J. Microwave Wirel. Technol. **12**(1), 1–12 (2020)
3. Dobre, O., Zarzoso, J., Bar-Ness, Y., Su, W.: On the classification of linearly modulated signals in fading channel. In: Proceedings CISS Conference, Princeton, March 2004
4. Zheng, Q., Zhao, P., Wang, H., Elhanashi, A., Saponara, S.: Fine-grained modulation classification using multi-scale radio transformer with dual-channel representation. IEEE Commun. Lett. **26**(6), 1298–1302 (2022)
5. Zhang, C.Y., Yue, H.T., Wang, B.W., et al.: Deep learning pattern recognition method of partial discharge ultrasonic signal based on similar matrix blind source separation and convolutional neural network. Power Syst. Technol. **43**(6), 7 (2019)
6. Zheng, Q., Zhao, P., Zhang, D., Wang, H.: MR-DCAE: Manifold regularization-based deep convolutional autoencoder for unauthorized broadcasting identification. Int. J. Intell. Syst. **36**(12), 7204–7238 (2021)
7. Zheng, Q., Yang, M., Wang, D., et al.: An intelligent wireless communication model based on multi-feature fusion and quantile regression neural network. J. Intell. Fuzzy Syst. (2021). Early Access
8. Zheng, Q., Zhao, P., Li, Y., Wang, H., Yang, Y.: Spectrum interference-based two-level data augmentation method in deep learning for automatic modulation classification. Neural Comput. Appl. **33**(13), 7723–7745 (2020). https://doi.org/10.1007/s00521-020-05514-1

Design and FPGA Synthesis of BAN Processing Unit for Non-Archimedean Number Crunching

Federico Rossi, Lorenzo Fiaschi(✉), Marco Cococcioni, and Sergio Saponara

University of Pisa, Largo Lucio Lazzarino 1, 56122 Pisa, Italy
{federico.rossi,marco.cococcioni,sergio.saponara}@unipi.it,
lorenzo.fiaschi@phd.unipi.it

Abstract. This work presents the design and synthesis of a processing unit for numbers encoded according to the recently introduced BAN format. Such an encoding allows one to represent numbers which are not only finite (as the reals) but also infinitely large or infinitely small, i.e., non-Archimedean. The motivation behind this study is the significant burst the non-Archimedean numerical computations have received in the last 20 years and the applications that have been found. With a hardware support, this operations would significantly increase in speed, enlarging the spectrum of possible applications to industrial and real-time ones.

Keywords: Non-Archimedean fields · Alpha Theory · Bounded Algorithmic Number (BAN) · Arithmetic Unit · FPGA

1 Introduction

A non-Archimedean number set is an ordered set containing numbers which can be either infinitely large or infinitely small other than finite. Even if the theoretical study of these sets is quite dated, only in recent days researchers started investigating the numerical calculus using such numbers, pioneered by Sergeyev and his Grossone Methodology [1]. The introduction of numerical encodings allowed researchers to implement software simulators for algebraic operations between non-Archimedean numbers, paving the way for their use in numerical computations. Some of the applications of these numbers regard multi-objective optimization problems combined with linear programming [2–5], quadratic programming [6], evolutionary algorithms [7], game theory [8], artificial intelligence [9,10], etc. A computer architecture able to manipulate Grossone-based numbers was patented in 2009 (see [11]).

This work considers a different non-Archimedean encoding, the so-called Bounded Algorithmic Number (BAN) format. The latter has been proposed to represent a particular subset of non-Archimedean numbers, namely the Euclidean ones, which have been proposed by Benci as a variant of Robinson's hyperreal numbers. Notice that BAN format is a fixed length encoding, which allows to obtain a result of an operation between two BAN of degree L, to be of the same degree, exactly as happens with operations between 32-bit IEEE 754

R. Berta and A. De Gloria (Eds.): ApplePies 2022, LNEE 1036, pp. 320–325, 2023.
https://doi.org/10.1007/978-3-031-30333-3_43

floats, where the result of addition, subtraction, addition and division is again a 32-bit float (the same happens for 64-bit precision). Having a fixed length will also allow us to extend a RISC-V CPU by adding specific registers for loading/storing BAN numbers to/from the BAN Processing Unit (BPU) and the main memory (RAM).

This work represents the first successful hardware implementation of the four arithmetic operations $(+, -, \times, /)$ between non-Archimedean numbers a fixed length format (the BAN one). We have also implemented a few more operations, such as the square root, the comparison operations, etc., as explained in the following sections.

2 Euclidean Numbers and the Alpha Theory

The first appearance of the set of Euclidean numbers $\mathbb{E}$ dates back to 1995 within Benci's Alpha Theory seminal work [12], a mathematical theory which has grown and continues to grow still nowadays [13]. The peculiar name of this theory comes from the fact that it fixes a value infinitely large and calls it α. Then, it represents any number in $\mathbb{E}$ as a function of α (and only functions of α are numbers in $\mathbb{E}$). For instance, the following are all Euclidean numbers

$$\alpha^2, \qquad \frac{1}{\alpha}, \qquad 1 - \frac{\alpha}{2}, \qquad e^{\alpha} - \frac{1}{\alpha^2}, \qquad -\frac{1}{2^{\alpha}}, \qquad \alpha^{-\pi} - \ln\left(\frac{1}{\alpha}\right) \tag{1}$$

Such a construction guarantees that the numbers in $\mathbb{E}$ and the mathematical operations among them behave according to their counterparts in $\mathbb{R}$. Therefore, sum and multiplications are still commutative and the latter distributes over the former, there always exists the inverse of a Euclidean number but the zero, $f(x) = x^2$, $x \in \mathbb{E}$, is still a quadratic, continuous, twice differentiable, non-negative function, etc. Notice that $\mathbb{R} \subset \mathbb{E}$ since $\alpha^0 = 1$ and $x \in \mathbb{R} \implies x = x\alpha^0 \in \mathbb{E}$ by definition.

As opposed to Archimedean Mathematics, in this context, the concepts of infinite and infinitesimal numbers are sharply defined rather than vague concepts.

Definition 1. *Given $\xi \in \mathbb{E}$, then*

- *ξ is infinite* $\iff \forall n \in \mathbb{N}, |\xi| > n$
- *ξ is finite* $\iff \exists n \in \mathbb{N}, \frac{1}{n} < |\xi| < n$
- *ξ is infinitesimal* $\iff \forall n \in \mathbb{N}, |\xi| < \frac{1}{n}$

Therefore, in (1) the first, the second, the third and the fifth numbers are infinite, while the remaining one is infinitesimal. The interested reader can find a more detailed presentation of Alpha Theory and the set $\mathbb{E}$ in [14].

3 The BAN Format

The BAN encoding was introduced for the first time in [15]. It consists of a finite length representation for Euclidean numbers. As for IEEE 754 floating point numbers, BANs cannot represent the whole set $\mathbb{E}$, since this is in contrast with the finite length of the representation and the finiteness of any computer. A BAN is any Euclidean number which fits with the following representation:

$$\xi = \sum_{i=1}^{L} r_i \alpha^{p-i},$$

where $L \in \mathbb{N}$ is the encoding length, $r_i \in \mathbb{R}$ and $p \in \mathbb{Z}$. Paraphrasing, a BAN is a Euclidean number which can be represented by a linear combination of L subsequent integer powers of α.

The main consequence of such an encoding is that the representation of a Euclidean number is very similar to the one of a polynomial with α as variable. This also implies that algebraic operations between them must have complexity at least proportional to the ones among polynomials. The next lines show an example of addition and multiplication between BANs with $L = 3$:

$$(2.3\alpha^2 - \alpha + 1) + (2\alpha - 3 - 0.5\alpha^{-1}) = 2.3\alpha^2 + \alpha - 2 - 0.5\alpha^{-1}$$

$$(2.3\alpha^2 - \alpha + 1) \times (2\alpha - 3 - 0.5\alpha^{-1}) = 4.6\alpha^3 - 8.9\alpha^2 + 3.85\alpha - 2.5\alpha^{-1} - 0.5\alpha^{-2}$$

Notice that, in both computations, the result is not a BAN since it requires more than three consecutive powers of α to be represented. Therefore, there is a need to approximate them considering only the first three highest powers of α, that is executing a truncation of the result. The next lines show the numerical execution of the two operations above, along with the division, by a software simulator of the BPU. The latter manipulates and outputs BANs in the normal form [14,15], i.e., in the standardized format which guarantees the uniqueness of the representation.

```
First operand: α^2(2.3 - 1η^1 + 1η^2)
Second operand: α^1(2 - 3η^1 + 0.5η^2)

Sum: α^2(2.3 + 1η^1 - 2η^2)
Product: α^3(4.6 - 8.9η^1 + 6.15η^2)
Division: α^1(1.15 + 1.225η^1 + 2.05η^2)
```

4 BAN Processing Unit (BPU) Design

Considering a hardware implementation of a BAN processing unit we wanted to support the following operations (both between BANs and between a BAN and a 32-bit float, except for the square root):

- Addition/Subtraction
- Multiplication
- Division
- Square root
- Comparisons

We designed the unit with a configuration of BANs that has 3 floats as coefficients and one integer as reference power. This resulted in a module that has two BANs as input (128-bit each), an optional 32-bit float input and a 32-bit integer input for the opcode. As for the output, we got a result on 128-bit containing either:

- a BAN if the operation was a sum, subtraction, multiplication or division or square root
- a boolean if the operation was a comparison between BANs

We have implemented all the mentioned operations using the C language, and then we have used an automatic synthesis in HDL to obtain the hardware description of the BPU, ready to be loaded and tested into an FPGA. Figure 1 shows an overview of the internal organization of the BPU (BAN Processing Unit) module with the implemented operations (the circuital part involving the optional 32-bit float input is omitted for a better clarity).

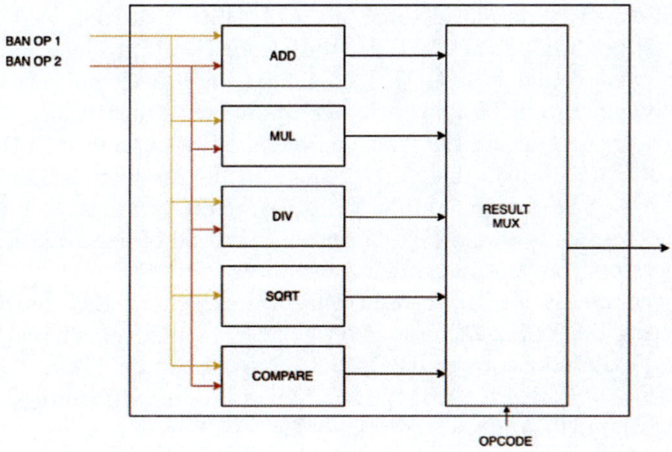

Fig. 1. BAN Processing Unit (BPU) module: internal illustration of the circuit implementing the operations involving BANs.

Listing 1.1 shows a possible implementation for a BAN binary operation. In particular, we show the sum between two BAN numbers with a reference power difference of *diff_p*. Notice that, to accommodate the automatic synthesis for this piece of code, we needed to insert the `HLS_UNROLL` pragma, to avoid branches inside the for loop. Indeed, thanks to the pragma and to the adopted size of 3 for the polynomial, the for-loop is unrolled in 3 different *if* blocks, each computing a different polynomial coefficient for the sum.

Listing 1.1. Example of implementation of the sum between two BAN

```
Ban Ban::_sum(const Ban &a, const Ban &b, int diff_p){
        Ban c(a);
        T tmp;
        for(unsigned i=0; i<3; ++i){
            #pragma HLS_UNROLL
                if(i>=diff_p){
                        tmp = c.num[i] + b.num[i-diff_p];
                        c.num[i] = tmp;
                }
        }
        if(!diff_p) // diff_p == 0
                c.to_normal_form();
        return c;
}
```

5 BPU Synthesis and Results

The BAN Processing Unit designed in the previous section was then synthesized using the Xilinx Vitis high-level synthesis tool. With this tool, we were able to generate an IP containing the HDL version of our BPU interface.

Moreover, we leveraged the C/RTL co-simulation functionality to thoroughly test the newly generated RTL hardware component. In particular, we tested all the operations by comparing the resulting output by the unit with the one provided by a software library, therefore demonstrating the functional equivalence of the software and hardware module. We instrumented the test to raise an error whenever the two values would be different. At the end of the evaluation, we did not get any errors from the tests.

Since the co-simulation test passed, we synthesized the RTL hardware component targeting the Xilinx ZCU106 FPGA platform and reporting the following three metrics: utilization of the FPGA resources (Look-up tables - LUT - and Configurable Logic Blocks - CLB), power consumption, and timing.

Regarding the utilization, we obtained the results reported in this box:

Total LUT used:
11656 out of 230400 (5.06%).
In particular:
- 11582 for logic functions
- 74 for shift registers

Total CLB used:
8147 out of 460800 (1.77%).
In particular:
- 8147 for flip-flops (i.e., *all*)

For the power report, we only focused on dynamic power consumption we obtained a CLB logic power consumption of $0.029W$, a signal power consumption of $0.037W$ for an overall consumption of $0.093W$. For the timing report, we targeted a $10ns$ period clock, obtaining a worst negative slack of $5.343ns$ and a maximum achievable frequency of $0.2GHz$ ($5ns$ period).

6 Conclusions

This work showed the design and synthesis of a processing unit for the fixed-length BAN encoding. As future work, we will integrate the designed BPU within a RISC-V processor, following the methodology adopted in [16].

Acknowledgments. Work partially supported by H2020 project TEXTAROSSA (grant no. 956831), https://textarossa.eu/) and partially by the Italian Ministry of Education and Research (MUR), CrossLab project (Departments of Excellence).

References

1. Sergeyev, Y.D.: Numerical infinities and infinitesimals: methodology, applications, and repercussions on two Hilbert problems. EMS Surv. Math. Sci. **4**(2), 219–320 (2017)
2. De Cosmis, S., De Leone, R.: The use of grossone in mathematical programming and operations research. Appl. Math. Comp. **218**, 8029–8038 (2012)
3. Sergeyev, Y.: The olympic medals ranks, lexicographic ordering and numerical infinities. Math. Intell. **37**(2), 4–8 (2015)
4. Cococcioni, M., Pappalardo, M., Sergeyev, Y.D.: Lexicographic multi-objective linear programming using grossone methodology: theory and algorithm. Appl. Math. Comput. **318**, 298–311 (2018)
5. Cococcioni, M., Cudazzo, A., Pappalardo, M., Sergeyev, Y.D.: Solving the lexicographic multi-objective mixed-integer linear programming problem using branch-and-bound and grossone methodology. Comm. Nonlinear Sci. Numer. Simul. **84**, 105177 (2020)
6. Fiaschi, L., Cococcioni, M.: A non-archimedean interior point method and its application to the lexicographic multi-objective quadratic programming. Mathematics **10**(23), 4536 (2022). https://doi.org/10.3390/math10234536
7. Lai, L., Fiaschi, L., Cococcioni, M., Deb, K.: Solving mixed pareto-lexicographic many-objective optimization problems: the case of priority levels. IEEE Trans. Evol. Comput. **25**, 971–985 (2021)
8. Cococcioni, M., Fiaschi, L., Lambertini, L.: Non-archimedean zero sum games. J. Comput. Appl. Math. **393**, 113483 (2021)
9. Astorino, A., Fuduli, A.: Spherical separation with infinitely far center. Soft Comput. **24**(23), 17751–17759 (2020)
10. Cavoretto, R., De Rossi, A., Mukhametzhanov, M.S., Sergeyev, Y.D.: On the search of the shape parameter in radial basis functions using univariate global optimization methods. J. Global Optim. **79**(2), 305–327 (2021)
11. Sergeyev, Y.: Computer system for storing infinite, infinitesimal, and finite quantities and executing arithmetical operations with them. USA patent 7,860,914 (2010), EU patent 1728149 (2009), RF patent 2395111 (2010)
12. Benci, V.: A construction of a nonstandard universe. In: Albeverio, S., et al. (eds.) Advances of Dynamical Systems and Quantum Physics, pp. 11–21. World Scientific, Singapore (1995)
13. Benci, V., Di Nasso, M.: How to Measure the Infinite: Mathematics with Infinite and Infinitesimal Numbers. World Scientific, Singapore (2018)
14. Benci, V., Cococcioni, M., Fiaschi, L.: Non-standard analysis revisited: an easy axiomatic presentation oriented towards numerical applications. Appl. Math. Comput. **32**(1), 65–80 (2022)
15. Benci, V., Cococcioni, M.: The algorithmic numbers in non-archimedean numerical computing environments. Disc. Contin. Dyn. Syst. - Ser. S **14**(5), 1673–1692 (2021)
16. Cococcioni, M., Rossi, F., Ruffaldi, E., Saponara, S.: A lightweight posit processing unit for RISC-V processors in deep neural network applications. IEEE Trans. Emerg. Topics Comput. **10**, 1898–1908 (2021)

Check for updates

Prototyping and Preliminary Testing of a Revamped Electric Bus with a Fast Recharge System

Adriano Alessandrini[1], Lorenzo Berzi[2], Fabio Cignini[3], Tommaso Favilli[1], Adelmo Niccolai[1], Fernando Ortenzi[3], and Luca Pugi[2](✉)

[1] Department of Civil and Enviromental Engineering, University of Florence, Florence, Italy
[2] Department Industrial Engineering, University of Florence, Florence, Italy
luca.pugi@unifi.it
[3] ENEA, Research Center, Casaccia, Rome, Italy

Abstract. A fast and reliable recharge of the electric public transportation system is a fundamental prerequisite for extended electrification and decarbonization of public transportation systems. As a part of continuous research cooperation, authors have proposed and tested different fast recharge technologies that have been tested on the fleet of electrical Minibus from TecnoBus. In this way, different recharge technologies are compared on identical vehicles with power management systems that are customized and revamped for the testing of a specific technology that must be tested on the vehicle. In this work, the authors focus their attention on the testing of highpower hybrid lithium Titanate cells that can support repeated fast recharge cycles that are performed using a recharge station powered by a flywheel storage system. In this preliminary work, it's presented the design and integration of the on-board storage system on the benchmark vehicle and its preliminary testing at ENEA Casaccia research center in Rome, Italy.

1 Introduction

Tecnobus Gulliver ESP 520 is an electric minibus which has been adopted and used in different Italian towns such as Rome, Florence etc. Main features of the first series model are described in Table 1: the original version of the bus was equipped with a lead battery pack mounted on a modular frame that was designed for a fast battery swap. This mature electric minibus was chosen as benchmark vehicle for the testing of different storage technologies for the following reasons:

- Widely diffused vehicle for which large amounts of data from real public service are available; this is a very important feature to properly simulate and size the system respect to realistic mission profiles
- Compact Dimensions and reduced weight, an important feature to reduce maintenance costs of a fleet for research purposes.
- Availability of low cost used vehicles and commercially available components for repair, customization, and maintenance of the experimental fleet of vehicles owned by University of Florence and Managed by ENEA.

R. Berta and A. De Gloria (Eds.): ApplePies 2022, LNEE 1036, pp. 326–332, 2023.
https://doi.org/10.1007/978-3-031-30333-3_44

This fleet of vehicles have been successfully employed for the testing of a fast recharge systems with supercapacitors that have been successfully tested in previous research activities [1, 2] where the bus has been also modified to install a pantograph dedicated for the fast current recharge (original bus was originally designed for battery swap). In this work it's verified the possibility of equipping the bus with a hybrid lithium Titanate battery assembled by XBW with last generation of Kokam cells described in Table 2 [3]. Cells are recharged using a fast recharge station whose main features are described in Table 3: the recharge station is feed by a flywheel system which assures high peak power (20kWwhich is high for the scale of the proposed benchmark which is a minibus) and relatively longer life respect to other storage technologies [4]. A feature that is highly desirable for the high number of recharge cycles that should be foreseen for a public transportation system.

Table 1. Tecnobus ESP 520, main features [1, 2]

Vehicle Parameters	
Weight (Tare/Maximum)	4270/6045[kg]
Batteries (kind/capacity/weight)	Lead Acid/ @72[V]/1500[kg] 595[Ah]
Inst.Traction System (nom. peak power)	21[kW]/ 25[kW]
Lights and on board instr. (power)	200 [W]
Air Conditioning System	2[kW]

Table 2. Elementary Parameters of Kokam Cells SLPB120216216HR2 used for installed XBW battery

Parameter	Value	Parameter	Value
Capacity	46 [Ah] (ref. to 0.2C at 25°)	Voltage (min./nom./max)	2.7 [V]/3.7 [V]/4.7 [V]
Energy Density (Grav.& Vol.)	150 [Wh/kg] 348 [Wh/dm3]	Max & Peak Charge Currents	8C (continuous), 12C (10 s.)
Max Int. Resistance	0.5 [mΩ]	Expected Life	>10000 90% cycles at 1C
Dimensions	226x227 × 12.2 [mm]	Battery Cap./Voltage/Energy	46[Ah]/355[V]/16.3[kWh]
Batt. Dimensions	1074x476x354 [mm]	Batt. Mass	200 [kg]

Table 3. Main features of the Flywheel Recharge Station

Parameter	Value	Parameter	Value
Rot. Speed	2000–4000 [rpm]	Motor/Alt. Efficiency	0.93
Max Torque	60 [Nm]	Power	20.32 [kW]
Ext. Diameterr	742 [mm]	Max Recharge Current	180 [A]
height	750 [mm]	Max Voltage	450[V]
Mass	2500 [kg]	Energy	9754[kJ]-/2.704 [kWh]
Inertia	171.41 [Kgm^2]		

2 Revamping of the Benchmark Minibus

As visible in the scheme of Fig. 1 a Gullliver Tecnobus was revamped to install the new 350 V lithium storage system that is used for a fast vehicle recharge. Since the original system was designed for a 72 V lead battery, new storage must be integrated in the original traction system through a reversible DC-DC converter able to assure the bi-directional power flow between traction motors and installed storage. The whole powerpack is installed in mobile frame originally designed to perform battery swap. For what concern DC-DC converter authors adopted two parallel connected power converters configured to keep stable the input voltage of traction system at 72 V. To reduce the risk of chattering of the voltage loops a master-slave configuration is adopted. Data of adopted converters are visible in Table 4. All the system is managed through a CAN network controlled by Barth Mini Lococube Mini PLC and acquired data are automatically stored on a cloud data repository using a commercial CSS data logger. All the system is designed and calibrated respect to the same mission profile produced from experimental field data to manage design and testing of super cap storage on the same bus [1, 2].

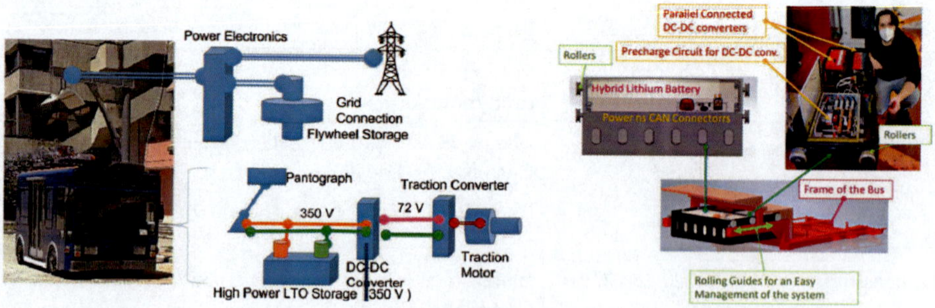

Fig. 1. Revamping of Gulliver Tecnobus to test the proposed storage system.

Table 4. Main Feature of adopted DC/DC RedPrime™ converters from Zekalabs™

Parameter	Value	Parameter	Value
Power (nominal)	40 [kW]	Max/Nom. Voltage side B	450/350 [V]
Max Current Side B	115 [A]	Max/Nom. Voltage side A	410/72 [V]
Max Current Side A	200 [A]	Number of Parallel Connected DC-DC	2

3 Preliminary Testing of the System at ENEA Labs

The whole system, designed at University of Florence was transferred to ENEA Labs for testing on Load Simulators that are able to perform an HIL (Hardware In the Loop) [5, 6] according the scheme of Fig. 2/a. An example of typical mission profile is shown in Fig. 2/b. However, HIL testing on the rig was slowed by severe stability issues due high

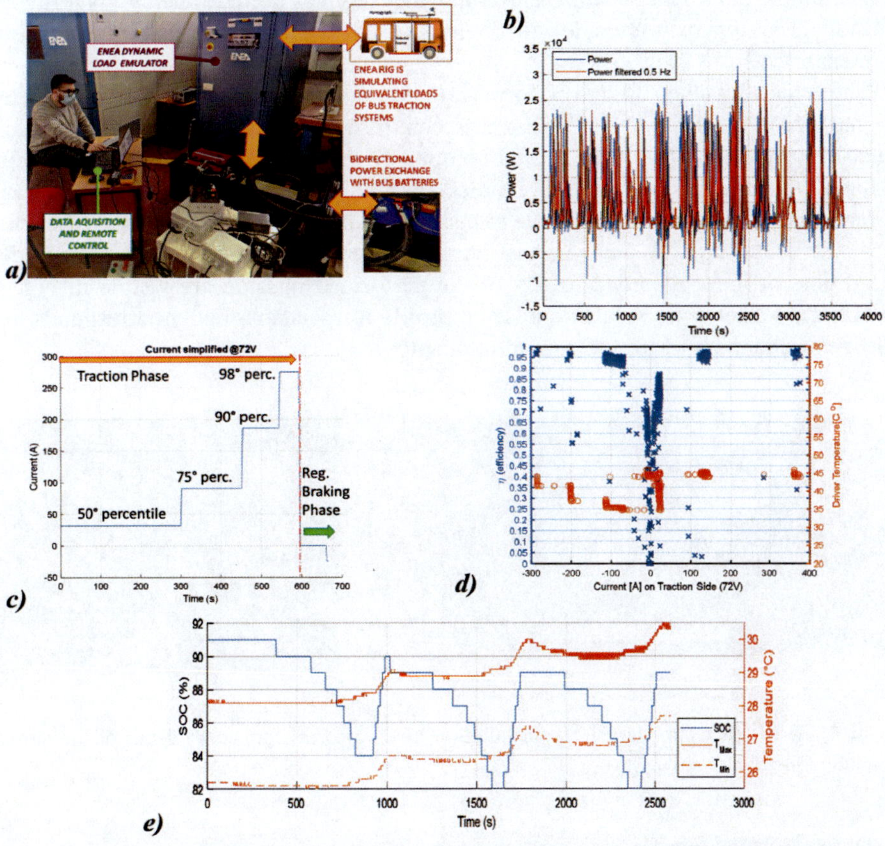

Fig. 2. Testing at ENEA labs (a), Reference Mission Profile (b), aggregated equivalent profile (c), Measured efficiency and temperature of DC-DC Converters (d), thermal and SOC profiles of batteries during testing

frequency interactions between tested drives: ENEA load simulator is connected to the traction bus to simulate traction loads so three DC-DC converters (two parallel connected of the bus and the ENEA load simulator) are acting in closed loop on a shared node with a limited, near to null capacity (which is limited to the output capacity of interfaced converters). For this reason, authors preferred to perform tests with equivalent power profiles in which cumulated statistical distribution of mission profile is used to produce a power-equivalent load profile visible in Fig. 2/c that was used to evaluate temperature and efficiency profiles of DC-DC converters and batteries over repeated load cycles as visible in Fig. 2/d/e.

4 Preliminary Testing of the System on an Internal Circuit at ENEA Casaccia Research Center

As visible in Fig. 3, proposed powerpack was installed on the bus using the same simple platform that was originally employed by the designers of the vehicle to perform battery swap: a simple trans pallet/ manual elevator maneuvered by a single operator is sufficient to install the system in the vehicle with obvious positive consequences in terms of system maintenance.

The vehicle was then used to perform various simulated missions inside the protected internal circuit available at the ENEA Casaccia research center in Rome. The vehicle is not currently equipped with an autonomous drive system, so it was impossible to perform the same mission profile recorded on real mission in the city of Rome and Florence. However, authors were able to instruct the human driver of the bus to perform a realistic sequence of maneuvers that were at least similar or equivalent to recorded data. Some of these recorded results of the performed mission are visible in Fig. 4: it's noticeable that with a realistic mission profile foreseen autonomy corresponds to a mileage of 20 km and 150–180 min of total autonomy.

a) b)

Fig. 3. UNIFI-ENEA Bus during the installation of the power storage (a) detail of the simple installation procedure (b)

Duration of the mission is important since a significant part of the energy is not consumed by the traction system (vehicle speed is modest and the vehicle is often stopped) but by the auxiliaries (pneumatic systems, steering and HVAC). During performed tests

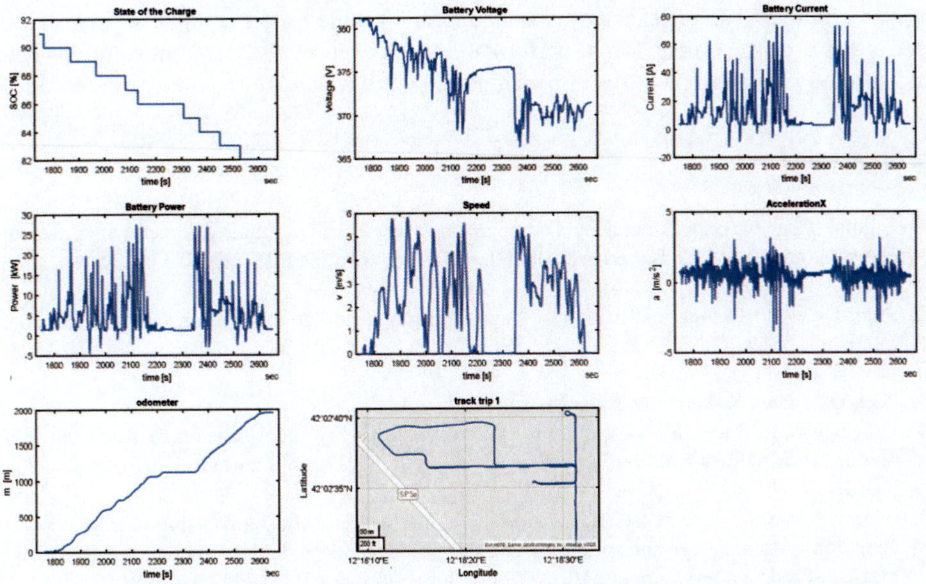

Fig. 4. Example of experimental results on the internal exp. Circuit available ENEA Research center of Casaccia, Rome, Italy

also installed DC-DC converters proved their functionality on multiquadrant operations. This autonomy was considered sufficient considering that the typical distance between to bus stops corresponds to a travel duration of about 300–500 m and a duration of 3–5 min. Considering a minimal recharge time of 2–3 min at each bus stop to assure the transit of passengers, the battery can be substantially recharged at bus stop assuring a continuous and release service. Expected life of proposed battery with proposed battery is estimated in more than 10000 cycles corresponding to cautiously evaluated mileage of at least 200000250000 km. This assured mileage is not very high; however, it should be considered the small size of the battery which make its substitution relatively cheap and the proposed installation of the storage that has been derived from a battery swap application. So, it should be concluded that the cost of these maintenance operations is relatively affordable. Finally, it should be noticed that these are only preliminary test runs of the proposed system in which most of the calibrations of energy management should be reasonably improved so authors are quite confident that final performances of the system should be better than expected ones.

5 Conclusions and Future Developments

Authors have preliminary tested the assembled power storage of the revamped bus, so the system is ready for the testing of the flywheel storage system. This second phase of performed research is on going and it will be the object of more detailed future publications. Authors are working to an autonomous drive system to be installed on the same fleet of benchmark buses. This system which will be the object of future publications, will be very useful to reproduce exactly the same mission profile on different buses in order to perform a precise comparison of performances achieved by the different solutions that are currently installed on buses of benchmark fleet of buses.

References

1. Cignini, F., et al.: Experimental data comparison of an electric minibus equipped with different energy storage systems. Batteries **6**(2) (2020). Art. no. 26. https://doi.org/10.3390/batteries6020026
2. Pugi, L., et al.: Design and testing of a supercapacitor storage system for the flash recharge of electric buses. Int. J. Electr. Hybrid Vehicles **13**(1), 57–80 (2021). https://doi.org/10.1504/IJEHV.2021.115197
3. Tech Data from Kokam site. https://kokam.com/cell
4. Dragičević, T., Sučić, S., Vasquez, J.C., Guerrero, J.M.: Flywheel-based distributed bus signalling strategy for the public fast charging station. IEEE Trans. Smart Grid. **5**(6), 2825–2835 (2014)
5. Pugi, L., Galardi, E., Carcasci, C., Lucchesi, N.: Hardware-in-the-loop testing of bypass valve actuation system: design and validation of a simplified real time model. Proce. Inst. Mech. Eng. Part E J. Process Mech. Eng. **231**(2), 212-235 https://doi.org/10.1177/0954408915589513
6. Allotta, B., Pugi, L., Malvezzi, M., Bartolini, F., Cangioli, F. A scaled roller test rig for high-speed vehicles. Veh. Syst. Dyn. **48**(SUPPL. 1), 3–18 (2010). https://doi.org/10.1080/00423111003663576

On the Deployment of Low-Cost Sensors to Enable Context-Aware Smart Classrooms

Edgar Batista(✉), Oriol Villanova, Joan Rosell-Llompart, F. J. Huera-Huarte, Antoni Martínez-Ballesté, and Agusti Solanas

School of Engineering, Universitat Rovira i Virgili, 43007 Tarragona, Spain
{edgar.batista,oriol.villanova,joan.rosell,francisco.huera, antoni.martinez,agusti.solanas}@urv.cat

Abstract. Smart classrooms augment traditional classrooms with sensors, IoT devices and data networks, and have become a promising arena for context-awareness studies. The collection of contextual data opens the door to numerous applications like environmental sensing, one of the most demanded abilities to face the COVID-19 emergency. To this end, this article presents a flexible and autonomous IoT-based platform for environmental sensing using inexpensive sensors in classrooms. Preliminary results prove the feasibility of the proposed platform and allow us to draw several conclusions on aspects to be considered when deploying these sensors in real settings.

Keywords: Smart classroom · Context awareness · Embedded Systems · Sensors · Ubiquitous Computing · Internet of Things · Education

1 Introduction

The widespread use of IoT devices along with the deployment of faster data networks have propelled environments augmented with contextual information in real-time. Smart classrooms [1,2] are emerging context-aware environments equipped with sensors, embedded systems, IoT devices and communication technologies able to capture rich amounts of contextual data. These data are extremely valuable to enhance the teaching and learning experiences as well as to achieve more efficient and sustainable schools. Among the many use cases [3,4], environmental monitoring is a promising application to be researched, *e.g.,* for determining its influence on the students' academic performance or on the spread of respiratory viruses via aerosol transmission. The latter has special relevance in the current context of the COVID-19 pandemic so as to make classrooms a safer space for children. This is the ultimate goal of our ACTUA project [5], developed under the smart healthcare paradigm [6,7].

This article presents the technical aspects of an IoT-based platform that augments classrooms with sensing capabilities for environmental sensing using low-cost devices. The rest of the article is organised as follows: Sect. 2 explains

R. Berta and A. De Gloria (Eds.): ApplePies 2022, LNEE 1036, pp. 333–338, 2023.
https://doi.org/10.1007/978-3-031-30333-3_45

the physical and logical architecture of our platform, Sect. 3 provides details on the prototype implementation, Sect. 4 presents preliminary results of some sensing measurements from different locations, and Sect. 5 closes the article.

2 Architecture

This section describes the architecture of the proposed platform (see Fig. 1).

2.1 Sensing Kit

The *sensing kit* (or *kit*, in short) is the core enabler of context-aware smart classrooms. It comprises a small single-board computer (SBC) acting as a controller node, and a number of independent sensors that capture relevant contextual variables. These variables may include, but are not limited to, air temperature, relative humidity, barometric pressure, CO_2 concentration, air pollution, room's ventilation, noise level and people's movement. The SBC and the sensors can be connected through wires or wirelessly (*e.g.,* BLE, ZigBee,...).

The functioning of the sensing kit is governed by the SBC, which implements the entire logical architecture of the kit. This architecture is modular by design: each part (*i.e.,* module) is well-isolated and able to operate autonomously. Hence, each module runs as an independent process. We distinguish four kinds of modules. First, the *sensor module* implements the specific functionalities to interact with a particular sensor. The architecture comprises as many sensor modules as physical sensors within the kit. Therefore, the sensing kit could easily be scaled up by adding new sensors without interfering with the already working sensors. Also, malfunctioning sensors do not disrupt the operation of the rest of the kit (only the specific sensors modules), thus minimising the risk of failure of the kit as a whole. Second, the *sensor manager module* orchestrates the logical functioning of the sensing kit. This module verifies that the sensor modules are running, terminates those sensor modules that present an abnormal behaviour, and communicates with the sensor modules and the back-end server. Third, the *database module* provides read and write operations to the local database installed in the

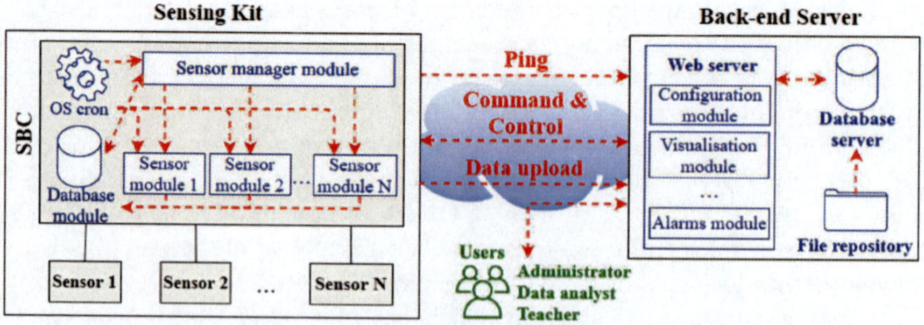

Fig. 1. High-level architecture of the platform (the physical components in black, the logical components in blue, the communications in red, and the users in green). (Color figure online)

SBC. Each sensor module writes the sensor's values to the database, which are then retrieved by the sensor manager module and sent to the back-end server. And fourth, the *OS's cron* periodically verifies whether the processes associated to those modules are running and, if not, initiates them. This process, also triggered at startup, makes the sensing kit a "plug and play" solution.

2.2 Back-End Server

The *back-end server* manages the functioning of the proposed platform, stores the information, communicates with the sensing kits, and provides interaction capabilities with the users. It comprises three main components. First, a *web server* hosting a web application able to set up the platform (*e.g.,* management of kits and classrooms where they are installed, users profiles and their roles), represents the data collected from the kits using charts, and alerts users in case of unexpected behaviours, among others. Second, it also hosts a *database server* as a central data repository. This datastore is fed from the sensing kits' data and the platform's management configurations. And third, it contains a *file repository* that kits use to upload their sensors data.

2.3 Communications

In order to have access to the Internet, each kit is connected to the school's LAN, using Ethernet or WiFi. The three main communication channels are highlighted as follows: First, kits periodically ping (*e.g.,* every ten minutes) the web application (through an HTTP request) to inform that they are up and running. Prolonged time periods without receiving this signal indicates a problem in the kit's functioning (*e.g.,* not powered on, SBC broken,...) and, then, alarms are raised. Second, to endorse the system with a more flexibility, a command & control functionality is implemented so that users could remotely control the kits with predefined commands, such as turning on/off a specific sensor or setting the sensing frequency. To achieve this, kits periodically request (*e.g.,* every hour) the web application (through an HTTP call) for new commands. Last but not least, kits establish FTP connections to the server's file repository to upload their sensors data in CSV format.

2.4 Users and Roles

The platform distinguishes three kinds of users: The *administrator* is responsible for the management of the platform (*e.g.,* associates kits to classrooms, sends commands to the kits, sets up other users,...). The *data analyst* is able to visualise the overall status of the platform and to download the data of the sensing kits. Last, the *teacher* is a school's worker who can visualise the data collected from the kits installed in the school as well as to receive alarms via email in case of abnormal functioning of the kit.

3 Implementation

To demonstrate the design feasibility of the proposed platform, a prototype has been developed. This prototype is going to be installed in hundreds of classrooms of preschools and primary schools in Catalonia (Spain) for environmental sensing. To avoid distracting students, the kits must be compact, discrete and minimally intrusive. Moreover, the used physical components are generally low-cost in order to keep down the cost of the whole kit.

A Raspberry Pi 4 has been used as SBC, and seven types of sensors have been connected to it through wires. These sensors, detailed in Table 1, capture the most relevant contextual variables for environmental sensing. Some variables are sensed by more than one sensor, which allows for more accuracy and fault tolerance. The kit has been mounted on a wooden base of 20 × 20 cm and covered by a 0.8 mm thick aluminium mesh. Reed switch sensors, embedded in auxiliary plastic cases of 11 × 7 cm, are installed next to doors and windows to capture whether they are opened or closed for ventilation purposes. Auxiliary cases are connected to the kit through RJ45 cables (see Fig. 2). The Raspberry Pi board runs Raspberry Pi OS 11, the main development language is Python 3.9, and the local database engine is a SQLite 3. Regarding the back-end server, it runs a Linux, the web application is developed in PHP 7.2, and the database server is a MariaDB 10.1.

4 Preliminary Results

Although no data from classrooms have been collected yet, four kits have been deployed in our laboratory since March 2022. These kits have been placed at four different locations within the lab to imitate their deployment in classrooms. More specifically, KIT-1 is at the top of a cabinet in the middle of the room at 2 m height, KIT-2 is next to a window at 1.6 m height, KIT-3 is on a shelf of a cabinet in a corner of the room at 1.5 m height, and KIT-4 is on a desk, next to people. This experiment is not only useful to validate the robustness of the proposed platform, but it also allows sensing the environment at different points of the same room to study whether measurements are consistent. If differences are found, it will be important to know the exact kit's location in the classroom at the time of correlating these contextual variables with other variables.

Figure 3 illustrates the air temperature, relative humidity, CO_2 concentration and PM2.5 measurements of the four kits from 1 April 2022 to 15 May 2022. At

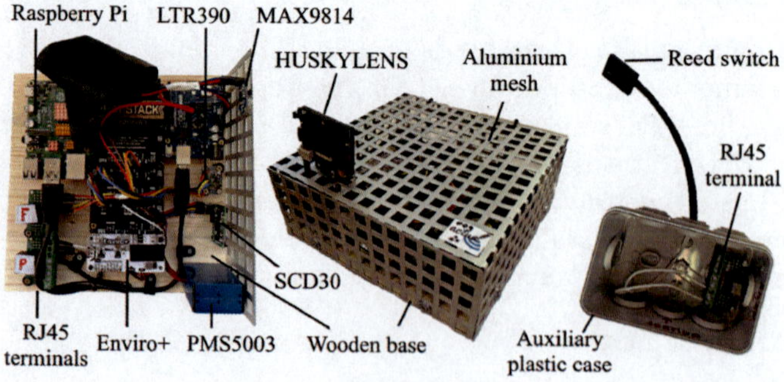

Fig. 2. Prototype of the sensing kit.

Table 1. Characteristics of the sensors used in our sensing kit.

Manufacturer	Sensor	Variable	Unit	Accuracy*	Output	Cost†
Sensirion	SCD30	Air temperature Relative humidity CO_2 concentration	°C % ppm	± 0.4°C + 0.023×(T[°C] − 25°C) ± 3% ± 30 ppm	Digital	60€
Adafruit	LTR390	UV Light	UVI lux	± 1 UVI when UVI < 5 25% across different light sources	Digital	5€
Pimoroni	Enviro+	Air temperature Relative humidity Barometric pressure Light	°C % hPa lux	± 1°C ± 3% ± 1% hPa Depends on light sources	Digital	45€
Adafruit	PMS5003	Particulate matter (PM1, PM2.5, PM10)	$\mu g/m^3$	± 10% in 100–500 $\mu g/m^3$	Digital	35€
RS Pro	Reed switch	Proximity (open/closed doors & windows)	Yes/No	Errors in distances ≈ 10 mm	Digital	3€
Adafruit	MAX9814	Noise level	dB	± 1%	Analog	7€
DFRobot	HUSKYLENS AI Vision	Movement level	Index	Poor performance in low light environments	Digital	45€

*Extracted from the official datasheets. †Approximate prices in June 2022.

first sight, results show a similar trend among kits' measurements, *e.g.,* all kits detect a temperature decrease during nights and a CO_2 concentration increase during work hours. However, measurements might slightly vary according to the kit's location. For instance, lower temperatures were recorded by KIT-2, which is likely influenced by the outdoor temperature due to its proximity to the window. These differences are more noticeable at nights. Humidity measurements are consistent as they are inversely proportional to the temperature values, *e.g.,* KIT-4 detected the lowest humidity but the highest temperatures. Regarding CO_2, peak values are clearly observed during work hours, while low and steady

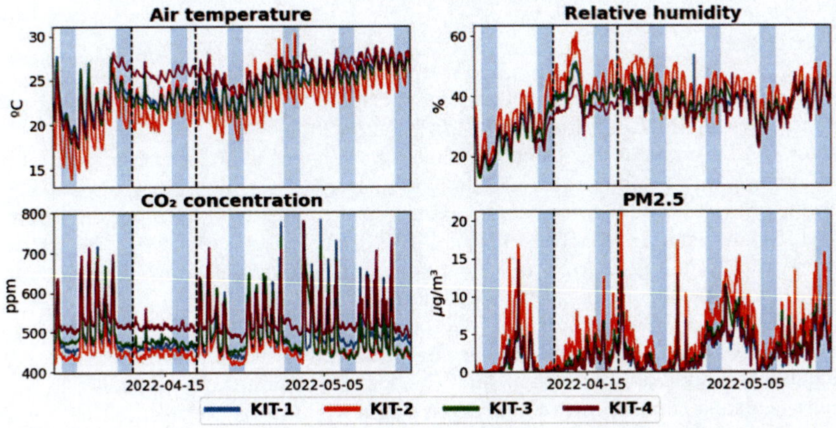

Fig. 3. Evolution of air temperature, relative humidity, CO_2 concentration and PM2.5 recorded by four kits within the same room during 45 days (weekends are painted in blue, and the start and ending of Easter are shown by dashed lines). (Color figure online)

values were obtained at weekends and during Easter, when no people were in the laboratory. Besides, since KIT-4 is closer to people's surroundings, recordings of CO_2 were higher in comparison to other kits. CO_2 stratification could be particularly interesting to study in this sense. Last, PM2.5 measurements present a similar behaviour among kits, even though highest values are mostly recorded by KIT-2, which may be influenced more by external factors like pollution. The analysis of data collected from many more kits installed in different schools will bring insightful knowledge on how to interpret these contextual variables for subsequent studies.

5 Conclusions

Being context-aware smart classrooms the natural evolution of today's classrooms, humongous volumes of data will be available through sensors. Environmental sensing opens the door to great research opportunities, such as the ability to create safer classrooms from an epidemiological perspective. This work has presented a low-cost kit able to sense over a dozen contextual variables supported by a scalable platform. With the imminent deployment of a large number of these kits in classrooms, further work will focus on cross-correlating these contextual variables with the epidemiological situation of the classrooms. Also, classrooms information, such as their dimensions, orientation and number of students, will be considered in these analysis too. Last but not least, evaluating the accuracy of our sensing kit in comparison to professional sensing devices is of interest.

Acknowledgments. This work is supported by the Government of Catalonia under grant 2020PANDE00103 (ACTUA), by URV's Institut de Ciències de l'Educació under grant 07GI2119, by MCIN/AEI/10.13039/501100011033 and "ERDF A way of making Europe" under grant RTI2018-095499-B-C32 (IoTrain), and by Agència de Gestió d'Ajuts Universitaris i de Recerca under grants 2017-SGR-896 and 2017-SGR-1516.

References

1. Saini, M.K., Goel, N.: How smart are smart classrooms? a review of smart classroom technologies. ACM Comput. Surv. **52**(6), 1–28 (2019)
2. Zhang, Y., Li, X., Zhu, L., Dong, X., Hao, Q.: What is a smart classroom? a literature review. In: Shaping Future Schools with Digital Technology, pp. 25–40 (2019)
3. Miraoui, M.: A context-aware smart classroom for enhanced learning environment. Int. J. Smart Sens. Intell. Syst. **11**(1), 1–8 (2018)
4. Huang, L.S., Su, J.Y., Pao, T.L.: A context aware smart classroom architecture for smart campuses. Appl. Sci. **9**(9), 1837 (2019)
5. ACTUA: Anàlisi Contextual dels factors de mitigació de la Transmissió de la COVID19 a l'aUlA. https://actua-urv.cat/ (In Catalan). Accessed 20 June 2022
6. Solanas, A., Patsakis, C., Conti, M., Vlachos, I.S., Ramos, V., Falcone, F., et al.: Smart health: a context-aware health paradigm within smart cities. IEEE Commun. Maga. **52**(8), 74–81 (2014)
7. Solanas, A., Casino, F., Batista, E., Rallo, R.: Trends and challenges in smart healthcare research: a journey from data to wisdom. In: 3rd International Forum on Research and Technologies for Society and Industry, pp. 1–6 (2017)

Modulation Recognition Based on BP Neural Network

Zhiwei Sun[1], Hua Wu[1,3](✉), Qinghe Zheng[2](✉), Yang Liu[1,3], Abdussalam Elhanashi[4], and Sergio Saponara[4]

[1] School of Information Science and Electrical Engineering (School of Artificial Intelligence), ShanDong JiaoTong University, Jinan 250357, China
wuhua@sdjtu.edu.cn

[2] School of Intelligence Engineering, Shandong Management University, Jinan 250357, China
15005414319@163.com

[3] CAAC Key Laboratory of General Aviation Operation, Civil Aviation Management Institute of China, Beijing, China

[4] Department of Information Engineering, University of Pisa, 56122 Pisa, Italy

Abstract. In the real wireless communication applications, there are always existing multiple signal modulation schemes. To complete the demodulation of modulated signals, it is necessary to understand the signal modulation method, so it is important for the signal receiver to have the ability to automatically identify the signal modulation schemes. In this paper, for six common digital modulation schemes, each modulation scheme modulates one hundred signals respectively, then extracts five feature parameters from each signal, and finally classifies them using back propagation (BP) neural network. The experimental results illustrate that the recognition accuracy can reach more than 90% when SNR > 10 dB. The model structure is simple and practical, and it can meet the requirements of automatic recognition of modulation schemes.

Keywords: Cognitive radio · Modulation recognition · Neural network

1 Introduction

Nowadays, as the signal analysis techniques [6, 7] have evolved faster, the automatic recognition of modulation scheme of communication signals has gradually become a popular research area. In the military communication field, electronic countermeasure has become an important part of modern war. In fact, identifying each other's modulation schemes [8] is an effective way to prevent the enemy from damaging our communication. In the civil fields, government departments can use the modulation recognition technique to combat the damage of illegal personnel to the wireless channel and improve the work efficiency of staff. Therefore, in the field of signal analysis, the automatic recognition of modulation mode is of great significance.

R. Berta and A. De Gloria (Eds.): ApplePies 2022, LNEE 1036, pp. 339–345, 2023.
https://doi.org/10.1007/978-3-031-30333-3_46

After years of development of modulation recognition technology, there are now two major research directions of mainstream modulation recognition technology at home and abroad, the first is the recognition method based on likelihood theory [1], and the second is the recognition method based on modulation signal feature extraction [2]. The main process of recognition algorithm based on likelihood theory is to first traverse all possibilities, calculate the likelihood function, and then select the modulation scheme corresponding to the maximum likelihood ratio as the result of recognition. The main process of classification algorithms based on modulated signal feature extraction is to first preprocess the signal, then to extract features from the signal, and finally to recognize it by a specific classifier.

With the continuous efforts of researchers, great progress has been made in modulation recognition techniques. Ali et al. [3] proposed a principal component analysis to further optimize the distinction of feature parameters of different types of modulated signals and used two supervised neural networks to maximize the approximation of the limits of the learning matrix to overcome the problem of susceptibility of signal feature parameters to interference under low signal-to-noise ratio (SNR) conditions. Satija et al. proposed to eliminate the interference of impulse noise using sparse signal decomposition on a hybrid dictionary for signals disturbed by impulse noise, and achieved high robustness for non-Gaussian impulse noises. Du et al. [4] proposed an expansive convolutional neural network for feature extraction and modulation recognition, which reduces the impact of overfitting on recognition accuracy to a certain extent. Alharbi et al. [5] proposed a robust automatic modulation classifier that can identify multiple signal modulations such as FSK, PSK, and OQPSK at high frequencies.

In this paper, six digital signal modulation schemes including 2ASK, 2FSK, 2PSK, 4ASK, 4FSK, and 4PSK are investigated. Firstly, 100 signals are modulated with different modulation schemes, then five features are extracted from these 600 signals, and finally the recognition process can be achieved by BP neural network.

2 Feature Learning and Recognition

2.1 Structure of Modulation Recognition

The proposed recognition method based on the extraction of the feature parameters of the modulated signal generally consists of three parts: the first part preprocesses the signal to provide suitable data for subsequent operations; the second part extracts the feature parameters; the third part is the classifier determining the modulation scheme. The modulation recognition process is shown in Fig. 1.

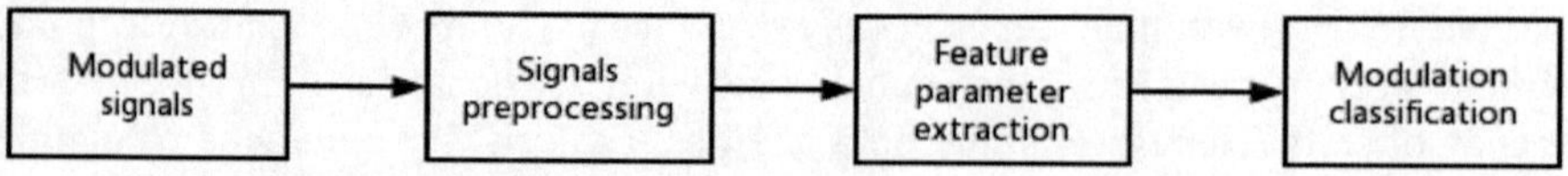

Fig. 1. Block diagram of modulation recognition.

2.2 Principle of Digital Signal Modulation

MASK Modulation. MASK uses M possible values to key the carrier amplitude and sends a carrier signal of one amplitude at each code element interval T_s. The time domain expression of MPSK can be expressed as

$$S_{MASK}(\mathrm{t}) = \sum a_n g(t - T_s)\cos(\omega_c \mathrm{t}) \tag{1}$$

where a_n is the amplitude corresponding to n^{th} symbol. For 2ASK, $a_n \in \{0, \ 1\}$. For MASK, $a_n \in \{2m - 1 - M|m = 1, 2...M\}$, $g(t)$ is the baseband pulse, and ω_c denotes the carrier angular frequency.

MFSK Modulation. MFSK modulation contains multiple digital information with several different carrier frequencies, and its time domain expression is

$$S_{MFSK}(\mathrm{t}) = \sum_n Ag(\mathrm{t} - \mathrm{T}_s)\cos(\omega_c t + 2\pi\Delta f_m) \tag{2}$$

where Δf_m $(m = 1, \ 2, \ ..., \ M)$ represents the offset value of the carrier frequency.

MPSK Modulation. The amplitude and frequency of MPSK modulation are both constant values, and different signal symbols modulate the phase of the carrier signal. The time domain expression of MPSK is

$$S(\mathrm{t}) = \sum_n a_n g(\mathrm{t} - n\mathrm{T}_s)\cos(\omega_c \mathrm{t} + \varphi_n) \tag{3}$$

where φ_n is the initial phase of n-th element, and M refers to the modulation order.

2.3 Feature Parameter Sets

The five feature parameters are selected in this paper.

The first feature is $\gamma_{\max}$, which can be calculated by

$$\gamma_{\max} = |DFT(\mathrm{A}_{cn})|^2/N \tag{4}$$

where $DFT(\cdot)$ represents the discrete Fourier transform, A_{cn} denoted the normalized central instantaneous amplitude, and N is the total number of signal samples.

The second feature is the absolute value standard deviation of the nonlinear component of the instantaneous phase σ_{ap}:

$$\sigma_{ap} = \sqrt{\frac{1}{N_c}(\sum_{A_n[n]>A_t} \phi_{NL}^2[n]) - (\frac{1}{N_c}\sum_{A_n[n]>A_t} |\phi_{NL}^2[n]|)^2} \tag{5}$$

where N_c denotes the number of samples satisfying $A_n[n] > A_t$, and A_t is a threshold to filter out low-amplitude signals with high sensitivity to noise. $\phi_{NL}[n]$ denotes the nonlinear component of the instantaneous phase of the nth signal sample.

The third feature is σ_{dp}:

$$\sigma_{dp} = \sqrt{\frac{1}{N_c}(\sum_{A_n[n]>A_t} \phi_{NL}^2[n]) - (\frac{1}{N_c}\sum_{A_n[n]>A_t} \phi_{NL}^2[n])^2} \tag{6}$$

The fourth feature is σ_{aa}:

$$\sigma_{aa} = \sqrt{\frac{1}{N}(\sum_{n=1}^{N} A_{cn}^2[n]) - (\frac{1}{N}\sum_{n=1}^{N} |A_{cn}[n]|)^2} \tag{7}$$

The fifth feature is σ_{af}:

$$\sigma_{af} = \sqrt{\frac{1}{N_c}(\sum_{A_n[n]>A_t} f_N^2[n]) - (\frac{1}{N_c}\sum_{A_n[n]>A_t} |f_N[n]|)^2} \tag{8}$$

2.4 BP Neural Network Classifier

In recent years, to solve classification problems, more and more people are using BP neural networks as their preferred method.

In the model, x_i is the input variable, ω_i is the weight of each propagation, and θ represents the threshold value of this neuron, and y is the output result of this neuron, which can be expressed as:

$$y = f(\sum_{i}^{n} \omega_i x_i - \theta) \tag{9}$$

where $f(x)$ is the activation function of this neural network. In the case of neural networks used for classification, it is customary to use the sigmoid function as the activation function. The functional expression of sigmoid can be expressed as:

$$f(x) = \frac{1}{1 + e^x} \tag{10}$$

By combining Eq. (9) and Eq. (10), the output of any neuron can be obtained by

$$y_j = \frac{1}{1 + e^{(\sum_{i}^{k} b_i \omega_{ij}^{'})}} \tag{11}$$

Since the parameters in the neural network are given randomly, there is a large error between the results obtained from the first calculation of the BP neural network and the real results.

3 Experimental Results and Analysis

3.1 Simulation Settings

Simulation experiments are performed by MATLAB 2021b for simulation, for six digital signal modulation methods, each modulation method modulates 100 signals, extracts five feature parameters, sets SNR = 10, 15, 20, 25, 30 dB, the five feature parameters are grouped as $[\gamma_{\max}; \sigma_{aa}; \sigma_{af}]$, $[\gamma_{\max}; \sigma_{aa}; \sigma_{ap}]$, $[\gamma_{\max}; \sigma_{aa}; \sigma_{dp}]$, $[\gamma_{\max}; \sigma_{ap}; \sigma_{af}]$, and $[\gamma_{\max}; \sigma_{ap}; \sigma_{dp}]$, which are called as group 1, group 2, group 3, group 4, and group 5, and the final recognition accuracy is obtained by using BP neural network to classify six digital signal modulation methods under different SNRs respectively.

3.2 Simulation Results

We first present the feature distributions of feature parameters at 20 dB in Fig. 2.

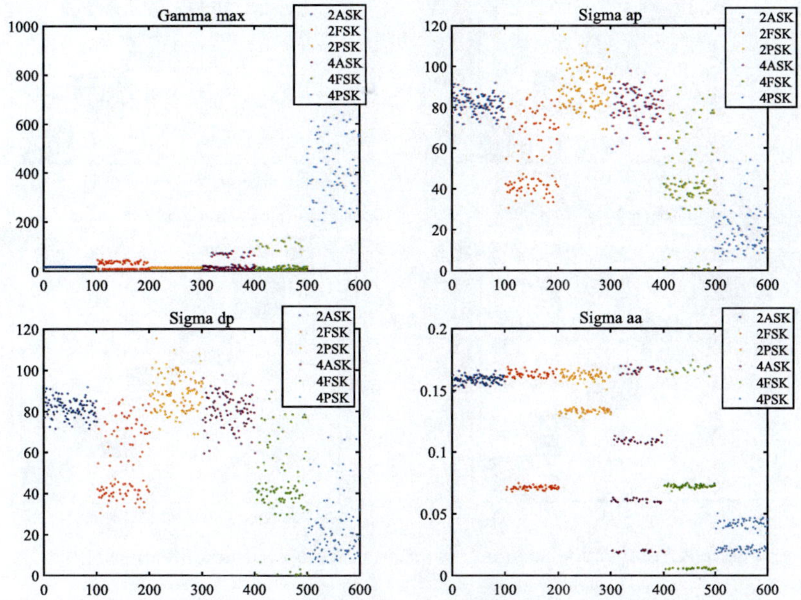

Fig. 2. Feature distribution map of six kinds of modulated signals.

A series of settings (SNR = 10, 15, 20, 25, and 30 dB) were compared to observe the classification accuracy of five features. The results are shown in Table 1. It can be analyzed that group 2 and group 3 achieve the highest recognition accuracy, and the results of these two groups are represented as confusion matrices, as shown in Fig. 3.

Table 1. Classification accuracy with different SNRs and different combos of parameters.

Group of feature parameters	Classification Accuracy (%)				
	10 dB	15 dB	20 dB	25 dB	30 dB
Group 1	60.33	73.67	61.83	91.50	80.67
Group 2	89.00	96.00	96.00	95.17	97.17
Group 3	83.33	95.67	95.83	95.00	98.33
Group 4	50.33	79.50	38.17	70.67	66.00
Group 5	62.33	69.00	87.33	77.17	87.17

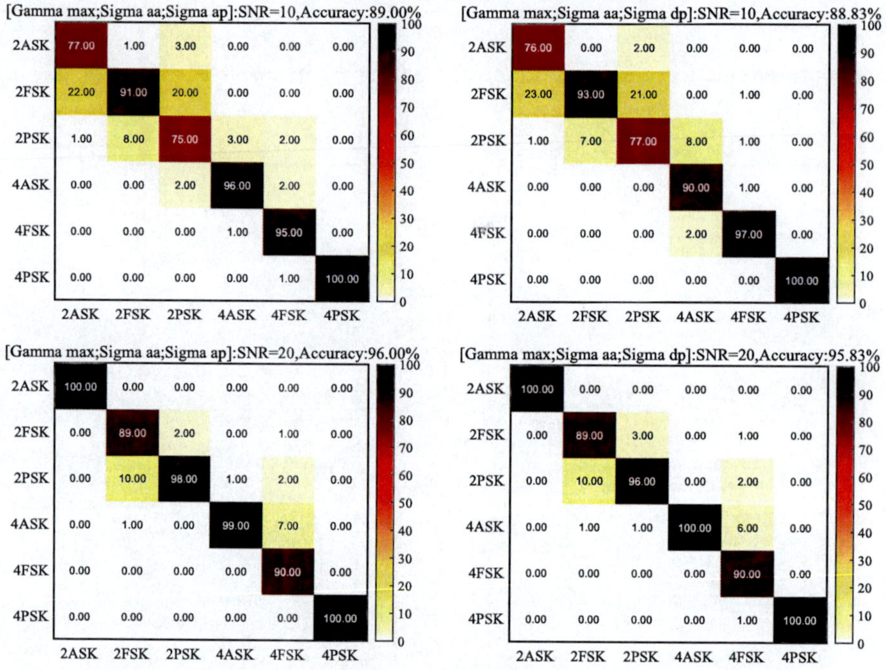

Fig. 3. Classification accuracy of different feature sets at different SNRs.

4 Conclusion

In this paper, six digital signal modulation schemes are classified using BP neural network. At first, 100 signals are modulated with different modulation schemes, then a series of features are extracted from these 600 signals, and finally the BP neural network is used for classification and recognition. The experimental results show that for the proper feature combinations achieve the recognition accuracy of more than 90% when SNR > 10 dB. The SNR of the experimental settings in this paper is relatively high, and the performance of the algorithm decreases rapidly under the condition of low SNRs, and the future research direction should consider the features of the signal with low

SNRs and extract features with high robustness and improve the recognition accuracy of the algorithm under the condition of low SNR.

Acknowledgements. This research was supported by Shandong Provincial Natural Science Foundation (Grant ZR2020MF151), CAAC Key Laboratory of General Aviation Operation (Civil Aviation Management Institute of China) (Grant CAMICKFJJ-2020-2), National Natural Science Foundation of China (Grant U1933130 and 71731001), and Research and Demonstration of Key Technologies for the Air-Ground Collaborative and Smart Operation of General Aviation (Grant 2022C01055).

References

1. Hong, L., Ho, K.C.: BPSK and QPSK modulation classification with unknown signal level. In: 21st Century Military Communications, Architectures and Technologies for Information Superiority, Los Angeles, CA, vol. 2, pp. 976–980 (2000)
2. Spooner, C., Brown, W., Yeung, G.: Automatic radio-frequency environment analysis. In: Conference Record of the Thirty-Fourth Asilomar Conference on Signals, Systems and Computers (Cat. No.00CH37154), Pacific Grove, CA, USA, vol. 2, pp. 1181–1186 (2000)
3. Ali, A., Erçelebi, E.: Modulation format identification using supervised learning and high-dimensional features. Arab. J. Sci. Eng. **48**, 1461–1486 (2022)
4. Du, R., Liu, F., Xu, J., et al.: D-GF-CNN algorithm for modulation recognition. Wireless Pers. Commun. **124**(2), 989–1010 (2022)
5. Alharbi, H., Mobien, S., Alshebeili, S., et al.: Automatic modulation classification of digital modulations in presence of HF noise. Wireless Pers. Commun. **124**(2), 989–1010 (2022)
6. Zheng, Q., Zhao, P., Li, Y., Wang, H., Yang, Y.: Spectrum interference-based two-level data augmentation method in deep learning for automatic modulation classification. Neural Comput. Appl. **33**(13), 7723–7745 (2020). https://doi.org/10.1007/s00521-020-05514-1
7. Zheng, Q., Zhao, P., Zhang, D., Wang, H.: MR-DCAE: manifold regularization-based deep convolutional autoencoder for unauthorized broadcasting identification. Int. J. Intell. Syst. **36**(12), 7204–7238 (2021)
8. Zheng, Q., Zhao, P., Wang, H., Elhanashi, A., Saponara, S.: Fine-grained modulation classification using multi-scale radio transformer with dual-channel representation. IEEE Commun. Lett. **26**(6), 1298–1302 (2022)

Towards Efficient Gateways and Servers for Biosensors

M. D. Grammatikakis(✉), S. Ninidakis, G. Kornaros, and D. Bakoyiannis

Department ECE, Hellenic Mediterranean University, 71410 Heraklion, Greece
{mdgramma,kornaros,d.bakoyiannis}@cs.hmu.gr, mtp178@edu.hmu.gr

Abstract. Smart E-health biosensors in wearable and mobile devices form an increasing technology trend. In most solutions, patient data is initially transmitted over Bluetooth to a gateway that in turn connects to a remote file server (or cloud). In this work, we focus on enhancing the performance of biosensor data flows across a Bluetooth-to-Ethernet gateway by examining lock-free concurrent queues and traditional lock-based circular FIFOs that embody single-producer single-consumer principles. Our results show that for large biosensor rates, the waiting time of the concurrent queue is smaller. Large rates, above 256 pulses/s, are not supported by the biosensor, but are examined through a self-similar, digital twin process. Finally, by considering a soft real-time analysis and animation application at the server, we explore interesting security vs QoS tradeoffs, leveraging the use of inexpensive crypto ICs.

1 Introduction

Miniaturized wearable sensors monitoring physiological parameters assist the proliferation of e-health services [1]. These sensors transmit biosignals to a remote server or cloud center, which stores, processes, and analyzes patient data, usually by invoking care personnel with expert knowledge. In a recent market study, we discovered that more than 50% of commercial pulse recorders, such as BodyGateway (BGW) [2], BodyGuardian [3, 4], Bittium Faros 360 [5], CardiBeat [6], D-Heart [7], and EC-12RM [8] support wireless Bluetooth which consumes less energy than Ethernet. Thus, it is necessary to support gateways that facilitate info exchange between sensor, gateway, and server.

Our study focuses on a gateway architecture managing a complex BT-to-Ethernet protocol stack. In particular, the gateway uses `rfcomm` to communicate over Bluetooth to a BGW pulse device, sending command packets to program the BGW device, enabling reliable data capture of raw BT packets, retrieving, and transmitting vital signals to a server via Ethernet.

Our gateway emulates a single-producer single-consumer bridge between Bluetooth and Wireless. Its implementation uses either a lock-based circular FIFO (called list) or concurrent queue (CQ) to store received BT packets. These data structures, although important performance-wise, have not been examined in prior work.

Existing state-of-the-art relates to efficient processing of ECG signals, i.e., filtering, smoothing, and arrhythmia analysis [9, 10]. Open and commercial Android platforms that provide a BT-to-Ethernet gateway have been described [11, 12]. Although our work is

R. Berta and A. De Gloria (Eds.): ApplePies 2022, LNEE 1036, pp. 346–352, 2023.
https://doi.org/10.1007/978-3-031-30333-3_47

similar in its foundation to [13] which deals with receiving, converting, and transmitting data packets between BT and Wireless, our focus is on efficient, Linux-based soft real-time processing and security, rather than interoperability.

Experiments examine the list and CQ waiting time delay at the gateway for different biosensor rates. While CQ behaves better for large biosensor rates, both implementations perform similarly for smaller rates. Finally, by leveraging crypto ICs, we examine interesting security tradeoffs in a soft real-time analysis & animation application, which consumes ECG data from the file server.

Next, in Sect. 2 we discuss the BT-Ethernet gateway. Section 3 focuses on the experimental framework and results. Finally, Sect. 4 discusses future work.

2 BGW Driver: Case-Study on BT-To-Ethernet Gateway

In our experiments, we use a single-lead, wearable patch (ST Micro BodyGateway, or BGW), supporting Bluetooth connectivity. The device is very similar to Boston Scientific's BodyGuardian mini and mini plus. Bodyguardian mini has been used to record data for almost 1 million patients per year in the USA.

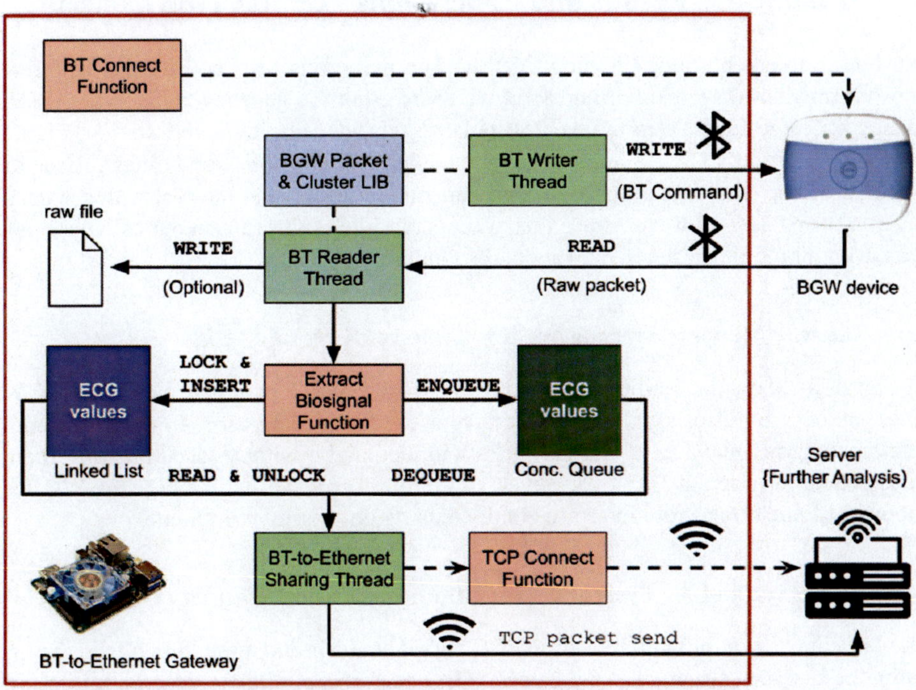

Fig. 1. The gateway architecture with a complex BT-to-Wireless protocol stack.

The gateway is an ARMv7 Odroid XU4 board. It manages a complex BT-to-Ethernet (IEEE 802.11) protocol stack using four software components, as shown in Fig. 1.

- The *BT Connect Function* uses `bluez` tools to pair with the BGW device.
- Then, the *BT Writer Thread* transmits command packets to configure and request periodic ECG data from the BGW device, i.e., up to 256 samples of ECG data per sec (with 12-bit precision); notice that the BGW device can also support the transmission of bioimpedance and accelerometer data.
- The *BT Reader Thread* operates in a loop. It first uses the `rfcomm` protocol to connect and periodically receive raw Bluetooth data at the specified rate [14]. Then, it extracts the ECG signal from the Bluetooth packet via bit operations, taking into account ARM's big-endian architecture. Finally, it inserts signal data either to a traditional producer-consumer lock-based "circular FIFO" (or a concurrent queue), as shown in Fig. 1.
- The *BT-to-Ethernet Sharing Thread* periodically dequeues all available biometric data in the concurrent queue (or resp. the "circular FIFO" of the list) and transmits them in real-time to a file server (an Odroid XU3 board) when their number exceeds the current rate; this optimization helps improve the waiting time. An isolated 2.1 Gbit/s TP-Link Archer C5400 router is used for the Ethernet connection between the gateway and the server.

3 Experimental Results and Conclusions - Digital Twin Concept

For both implementations (list and CQ), we run numerous tests considering different sensor transmission rates. Furthermore, we try to examine gateway and server performance for large ECG rates (above 256 pulses/s), by developing a self-similar, digital twin sensor process [15]. Accuracy is not excellent, since BGW sometimes misses its RTOS deadline, and data is subsequently transmitted in the next interval with a special incomplete packet. However, since this event is rare (less than 15% in our calculations), we have ignored this issue in our model development.

3.1 Gateway Metrics: Average Waiting Time vs BGW ECG Rate

In Fig. 2, we show the waiting time vs the BGW rate, for CQ and list. For small BGW rates (up to 256 pulses/s), CQ and List have a similar waiting time. In this case, both implementations manage to extract packets immediately, without accumulating them in the queue or circular FIFO. However, CQ outperforms the list for high BGW rates (above 512 pulses/s); these rates are obtained using our digital twin model.

3.2 File Server and Analysis and Animation Application: Security Overheads

Our last study examines soft real-time ECG signal analysis and animation on the server, using the List data structure at the gateway; CQ performs similarly, since both List and CQ implementations sustain average rates up to 4096 pulses/s (graphs omitted due to lack of space). ECG analysis and animation rely on extending open-source software libraries [16]: a) the Harvard Physionet WaveForm DataBase (WFDB) [17] for smoothing and standardizing ECG data to 200 samples/s according to ANSI/AAMI EC-13, and b) the EP Limited Open-Source ECG Analysis (OSEA) for low- and high-pass QRS filtering

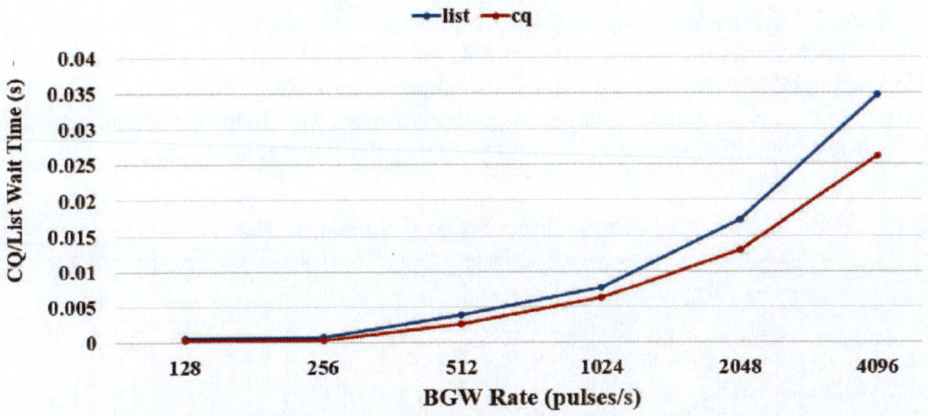

Fig. 2. Waiting time in the CQ or the list (circular FIFO) vs the BGW rate.

(via `easytest` scripts) to detect and classify normal or abnormal beats in soft real-time [18–20]. Our framework achieves a positive predictivity close to 99.8% with MIT/BIH and AHA arrhythmia databases. Finally, to visualize the annotated ECG waveform in soft real-time we use the Harvard Physionet `xview`-based WAVE software package [21].

Using this application, we explore tradeoffs between quality of service (QoS) and security & data privacy, two critical services in modern medical cyber-physical applications. QoS constraints require assurance of soft real-time when running the signal analysis and animation processes continuously. According to the Health Insurance Portability and Accountability Act (HIPAA) and European Data Protection Directives (e.g., EU Safe Harbor law), data privacy must be protected within every layer of an e-health platform. Hence, sensitive ECG data must be transmitted within a secure channel, from the BGW sensor to the server via the BT-wireless gateway [22].

To reduce security costs, we have used lightweight cryptography based on Microchip's ATECC608A, an inexpensive cryptographic IC [23, 24]; alternative solutions based on Zymbit Zymkey, and STSAFE100 modules are orders of magnitude more expensive (costing 20 to 50 USD vs 0.5 USD for ATECC608A).

ATECC608A supports tamper-proof data and key/certificate storage (in EEPROM) and a large variety of algorithms, such as AES-128, SHA-256 & HMAC HMAC, and ECDSA/ECDH (elliptic curves). The ATECC608A contains an EEPROM that can store keys, certificates, and user data securely. Storage regions are organized into 16 slots, with corresponding data and keys configured depending on the application. Locking the configuration and memory zones is required before any functionalities are exposed; locking once only also prevents malicious configuration.

In our platform, ATECC608A is connected to Odroid XU3/4 via an I2C interface using Microchip's SOIC/UDFN Socketed XPRO Board. The I2C baud rate is 400 kbps, while a level shifter circuit (by Hardkernel) converts power from the Odroid 1.8V operating voltage to 3.3V (ATECC operating voltage).

Security mechanisms are integrated into our gateway/server by porting: a) Microchip's `cryptoauthlib` library [24] and b) the `ATECC-util` package [25]. Thus, using HMAC-related commands from these libraries (e.g., `hmac-digest`), we have implemented authentication in our gateway/server. This helps avoid threats, such as man-in-the-middle, spoofing, and replay attacks (by integrating a nonce).

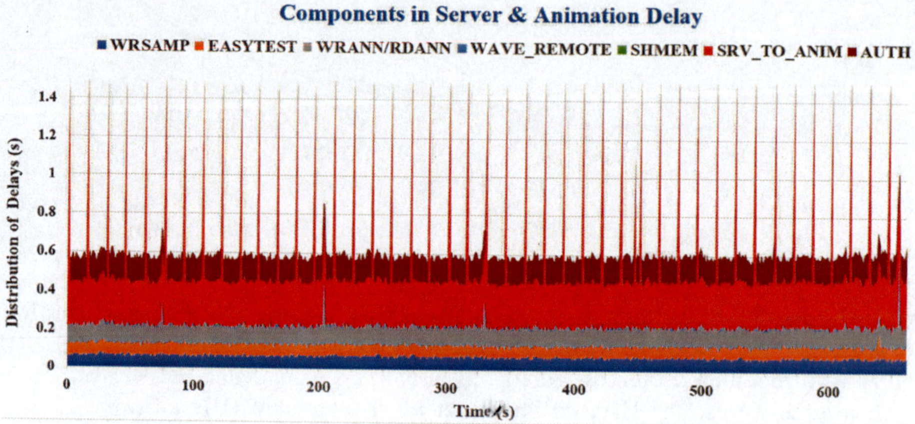

Fig. 3. Distribution of application delays, when BGW driver (*BT reader thread*) operates at 64 pulses/s, while BGW device operates at 128 pulses/s.

Concentrating on the server, Fig. 3 shows a typical delay distribution for more than 660 s; each second corresponds to an average of 128 samples since the BGW operates at 64 pulses/s and each pulse (21 bytes, including timing) is accompanied by an HMAC of equal size. We consider the authentication process, the file server process, and the analysis & animation process. The average delay for receiving data and appending it to the ECG file for the animator (`SRV_TO_ANIM`) takes ~0.298 s. This time represents the interval between reading consecutive ECG samples until the time these samples are processed by the analysis & animation process; rare peaks above 1s are due to synchronization, i.e., file locking between server and analysis & animation process. Despite using hardware-enabled cryptography, continuous authentication using 32-bytes HMAC (`AUTH`) has a comparable average delay of ~0.132 s. `WRSAMP` process spends 0.07s mainly for conversion to std EC-13, `EASYTEST` filtering takes ~0.06 s for heartbeat detection and classification, and `WRANN/RDANN` takes ~0.094 s for writing/reading to/from annotation files. Finally, asynchronous visualization (`WAVE-REMOTE`) takes ~0.005 s, and shared memory timing (`SHMEM`) contributes ~0.001 s.

Hence, the latency overhead for authentication using ATECC608A is manageable, since we can marginally sustain soft real-time when the sensor operates at the smallest rate (64 pulses/s). However, for higher biosensor rates (especially above 256 pulses/s) or when using complex security protocols, such as authenticated encryption, violation of end-to-end average rate (QoS) provisions occur, requiring system redesign.

4 Future Work

Future work focuses on performance and scalability of the gateway. In this direction, we hope to develop efficient and configurable security protocols using faster crypto peripherals, such as STMicro STSAFE100, and especially the NXP's nRF52840.

Acknowledgments. The authors acknowledge partial support from the EU Horizon 2020 project AVANGARD (Contract No. 869986) and travel support from the Public Investment Program (ΠΔΕ) of the Greek Ministry of Education & Religious Affairs."

References

1. Baker, S.B., Xiang, W., Atkinson, I.: Internet of Things for smart healthcare: technologies, challenges, and opportunities. IEEE Access **5**, 26521–26544 (2017)
2. BodyGateway, ST Microelectronics, 27 June 2022. https://usermanual.wiki/ST-Microelectronics-S-R-L/MHBGW1.Product-literature
3. BodyGuardian Mini, Preventice Solutions, 27 June 2022. https://www.preventicesolutions.com/patients/body-guardian-mini
4. BodyGuardian MiniPlus, Preventice Solutions, 27 June 2022. https://www.preventicesolutions.com/patients/body-guardian-mini-plus
5. Bittium Faros 360, 27 June 2022. https://www.bittium.com/medical/bittium-faros
6. CardiBeat, 27 June 2022. https://www.theheartcheck.com/cardibeat/index.html
7. D-Heart, 27 June 2022. https://www.d-heartcare.com
8. EC-12RM, 27 June 2022. https://www.omnia-health.com/product/ec-12rm-cardiospy-mobile-android-12-channel-resting-ecg-system
9. Pinto, J.R., Cardoso, J.S., Lourenço, A.: Evolution, current challenges, and future possibilities in ECG biometrics. IEEE Access **6**, 4746–4776 (2018)
10. Teplitzky, B.A., McRoberts, M., Ghanbari, H.: Deep learning for comprehensive ECG annotation. Heart Rhythm J. **17**(5), 881–888 (2020)
11. Elaarag, H., Bauschlicher, D., Bauschlicher, S.: System architecture of HatterHealthConnect. Int. J. Comp. Networks Commun. **5**(2), 1–22 (2013)
12. Personal Gateway, Pchalliance, 27 June 2022. http://www.pchalliance.org/personal-health-gateway-bluetooth-low-energy-manager
13. Shin, S-H., Suwon, S.: Apparatus and method for linking Bluetooth to Wireless LAN. US Patent 0071123A1, 15 April 2004
14. Corral-Acero, J., Margara, F., Marciniak, M., Rodero, C., et al.: The 'Digital Twin' to enable the vision of precision cardiology. European Heart J. **41**(48), 4556–4564 (2020)
15. Huang, A.S., Rudolph, L.: Bluetooth Essentials for Programmers. Cambridge Press, Cambridge (2007)
16. Grammatikakis, M.D., Koumarelis, A., Ntallaris, E.: Validation of soft real-time in remote ECG analysis. In: Proceedings International Conference Applications in Electronics Pervading Industry (ApplePies 2020), pp. 90–96 (2020)
17. WFDB, Physionet, 27 June 2022. https://archive.physionet.org/physiotools/wfdb.shtml
18. Hamilton, P.S., Patrick, S., Tompkins, W.J.: Quantitative investigation of QRS detection rules using the MIT/BIH arrhythmia database. IEEE Trans. Biomedical Engin. **12**, 1157–1165 (1986)
19. Tompkins, W.J.: A real-time QRS detection algorithm. IEEE Trans. Biomed. Eng. **3**, 230—236 (1985)

20. EP Limited, OSEA. https://www.eplimited.com/confirmation.htm
21. Physionet, WAVE, 27 June 2022. https://archive.physionet.org/physiotools/wug/wug.pdf
22. Hathaliya, J.J., Tanwar, S.: An exhaustive survey on security and privacy issues in Healthcare 4.0. Comput. Commun. **153**, 311–335 (2020)
23. ATECC608A, Microchip, 27 June 2022. https://www.microchip.com/en-us/product/ATECC608A
24. Cryptoauthlib, Microchip, 27 June 2022. https://github.com/MicrochipTech/cryptoauthlib
25. ATECC-util, 27 June 2022. https://github.com/wirenboard/atecc-util

Improvement of Sodium-Metal Halide Battery Electrical Equivalent Model Including Temperature Dependency

Gianluca Simonte(✉), Roberto Di Rienzo, Federico Baronti, Roberto Roncella, and Roberto Saletti

Dipartimento di Ingegneria dell'Informazione, University of Pisa, Via Caruso 16, 56122 Pisa, Italy
gianluca.simonte@phd.unipi.it,
{roberto.dirienzo,federico.baronti,roberto.roncella, roberto.saletti}@unipi.it

Abstract. Sodium-Metal Halide Batteries are a very promising alternative to the Lithium-ion ones for stationary applications, but their chemical complexity requires an accurate battery model to optimize their use. The electrical equivalent model of the battery is ordinarily used to this aim. The temperature dependency of the model parameters is studied in this work. Three characterization tests are carried-out at 270, 300, and 330 °C and analyzed to identify the model parameters. The parameters obtained are then compared with the literature showing that introducing in the model the temperature dependency can improve the accuracy of about six times.

Keywords: Electrical model · Sodium Metal Halide battery · Temperature · Battery Management System

1 Introduction

Sodium-Metal Halide Batteries (SMHBs) are an interesting alternative to Lithium-Ion ones for stationary applications, such as Uninterruptible Power Supply (UPS) and Smart Grid [1–3]. In fact, the SMHBs present a comparable energy density to the Lithium-Ion, but they are based on cheaper and safer materials [4,5]. The main drawback of this technology is the very high working temperature that goes from 250 to 350 °C. It imposes the use of a complex and bulky thermal management system.

An accurate battery model is key to maximize the battery performance. In fact, the model is used by the Battery Management System (BMS) to predict the battery behavior and extend the usable range of the battery State of Charge, still keeping the battery in the Safe Operating Area [6]. The electric equivalent model is the most promising one thanks to a very good trade-off between complexity and accuracy [7]. To the best of our knowledge, the most accurate SMHBs electric equivalent model is presented in [8] and reported in Fig. 1.

R. Berta and A. De Gloria (Eds.): ApplePies 2022, LNEE 1036, pp. 353–358, 2023.
https://doi.org/10.1007/978-3-031-30333-3_48

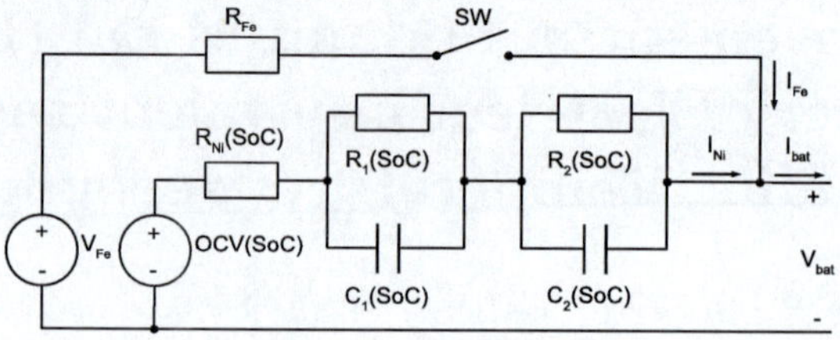

Fig. 1. Electric equivalent model of a Sodium-Metal Halide Battery [8].

This model is composed of two parallel connected branches that mimic the two main chemical reactions of the Sodium-Metal Halide Batteries: the nickel (Ni) and the iron (Fe) reactions, respectively. Each branch is composed of a voltage generator and a passive circuit. The former models the Open Circuit Voltage (OCV) of the reaction while the passive circuit models the voltage dynamic due to the current variations. As we can note, the iron branch contains a switch that excludes the branch from the model. The switch is open if the battery voltage is higher than V_{Fe}. V_{Fe} is obtained by multiplying the Fe reaction potential of 2.35 V by the number of series-connected cells composing the string. A deeper analysis of the model is reported in [8] in which the relationship between the electric model components and the physic-chemistry reactions is explained.

This model was improved in [9], in which a complete parameter identification procedure is proposed. However, the State of Charge (SoC) dependency of the model parameters is only considered in that work. It was demonstrated that the thermal dependency is not negligible in [10], so that the model should take into account the parameters variation due to the temperature. In particular, the authors carried-out three identical Pulse Current Tests (PCT) on a commercial SMHB with a temperature of 270, 300, and 330 °C. The PCT is a very common test used to identify the electric equivalent model [11] and it consists of a repetition of current pulses and rest time periods.

The aim of this work is to analyze the characterization test campaign presented in [10], by applying the model parameter identification method presented in [9]. The obtained parameters are compared to those presented in [9], in which the same commercial SMHB is modeled at 265 °C.

2 Characterization Test Campaign and Parameter Identification Procedure

The characterization test campaign was carried-out on the FZSonick 48TL200 commercial SMHB. This battery is composed of 5 strings of 20 series-connected cells. The capacity of each string is 40 Ah and its nominal voltage is 48 V. The BMS of the battery has been replaced with a custom electronic system that allows to completely control the working conditions of the strings [10].

The test campaign was carried-out on one string, but it can easily be extended to the others. The test campaign consists of three discharge PCTs starting with a fully charged string. Each PCT is composed of 9 current pulses followed by a rest time. Each current pulse discharge the battery of about 10 % of SoC with a constant current of 8 A for 30 min. The pulse is followed by a rest time of 2 h in which the string voltage can relax to the OCV value. The three PCTs are carried-out with a string temperature of 270, 300 and 330 °C, respectively. The string current and voltage during the three PCTs are reported in Fig. 2.

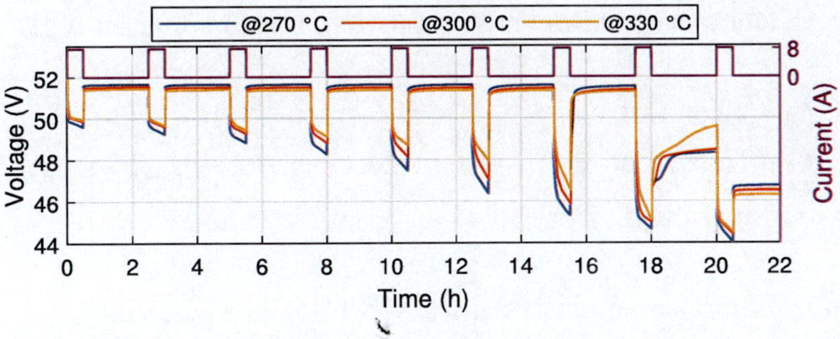

Fig. 2. The string current and voltage of the three Pulse Current Tests carried-out at 270, 300 and 330 °C, respectively.

The electric equivalent model parameters are identified using the procedure described in [9]. This procedure divides the PCT pulses in two groups: those in which only the nickel reaction occurs and those where both reactions take place. The first group leads to the identification of the Ni parameters only. R_{Ni} is identified using the string voltage drop produced by the current pulse. Instead, R_1, C_1, R_2, and C_2 are identified fitting with a double exponential function the string voltage acquired in the rest time. Finally, the OCV value is obtained as the final voltage value of the rest time when the string is completely relaxed. This procedure is applied to all the pulses belonging to the first group, obtaining the model parameters at different states of charge.

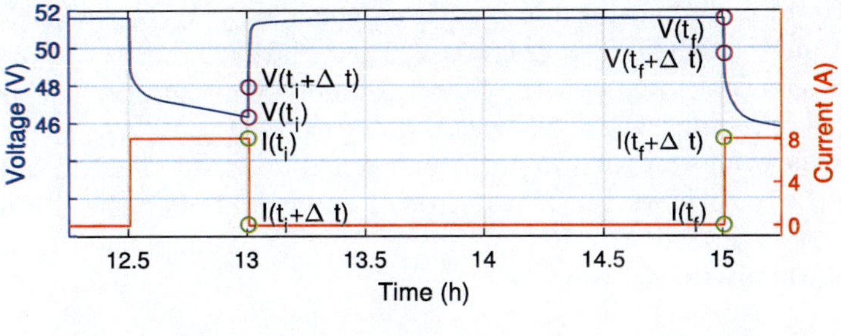

Fig. 3. The pulse used in the PCT @270 °C to identify R_{Fe}

The remaining pulses are used to identify also the iron parameters, which are SoC independent in the model. Since the string is composed of 20 cells, V_{Fe} is equal to 47 V. The identification of R_{Fe} is obtained by analyzing the response of the first pulse in which the string voltage becomes lower than V_{Fe}. Figure 3 reports the string voltage and current of this step for a PCT executed at 270 °C. We can note that the battery voltage is higher than V_{Fe} in t_f, and the Fe branch is open. Therefore, R_{Ni} can be identified as the ratio between the battery voltage difference and the current one between t_f and $t_f + \Delta t$. The identified R_{Ni} value is used in the first equation of (1) to determine the Ni current variation between t_i and $t_i + \Delta t$. This value is used in the second equation of (1) to obtain the Fe current variation that is finally used in the third equation to identify R_{Fe}.

$$\begin{cases} I_{Ni}(t_i) - I_{Ni}(t_i + \Delta t) = \frac{V(t_i) - V(t_i + \Delta t)}{R_{Ni}} \\ I_{Fe}(t_i) - I_{Fe}(t_i + \Delta t) = I(t_i) - I(t_i + \Delta t) - (I_{Ni}(t_i) - I_{Ni}(t_i + \Delta t)) \\ R_{Fe} = -\frac{V(t_i) - V(t_i + \Delta t)}{I_{Fe}(t_i) - I_{Fe}(t_i + \Delta t)} \end{cases} \quad (1)$$

3 Identification Results and Model Verification

The procedure described above is applied to the three tests, and the parameters obtained are reported in Fig. 4. τ_1, and τ_2 are the time constants of the RC groups. The obtained parameters are also compared with the parameters reported in [9] that are identified on the same commercial battery with a PCT carried-out at 265 °C. These literature parameters are very similar to those obtained from the test at 270 °C. Instead, the literature parameters become different when the battery temperature increases. For example, R_{Ni} is reduced of about 30 % when the battery temperature goes from 270 to 330 °C.

The parameter variation with temperature has a large impact on the model accuracy. For example, Fig. 5(a) compares the measured string voltage in a PCT at 330 °C with the string voltage obtained using the literature parameters and with the new one. As we could expect, the new model parameters fit the test voltage very well. As we can note, the literature model voltage differs to the measured one in both the pulses and rest time periods. This means that both the OCV and the passive circuit parameters are not well accurate and require to be improved including the temperature dependency. In fact, the new model parameters obtains a RMS error of only 110 mV that is about 0.22 % of the mean string voltage of 50 V. Instead, the literature model voltage has a RMS error of 651 mV that is about 1.3 % of the mean string voltage.

Finally, the same comparison has been carried-out with the PCTs at 270 and 300 °C. Figure 5(b) shows the RMS error of the model with the literature and new parameters. For all the PCTs the new set of parameters presents better results than the literature one.

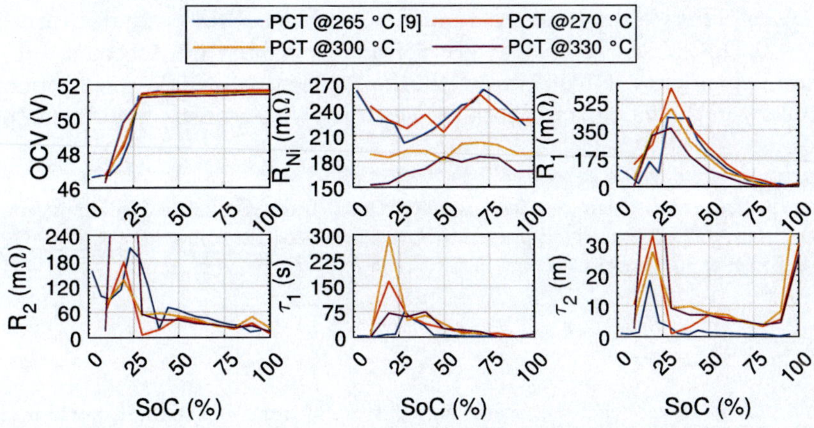

Fig. 4. Identified model parameters from the PCTs carried-out at 270, 300 and 330 °C, and comparison with the literature one (PCT @265 °C) presented in [9].

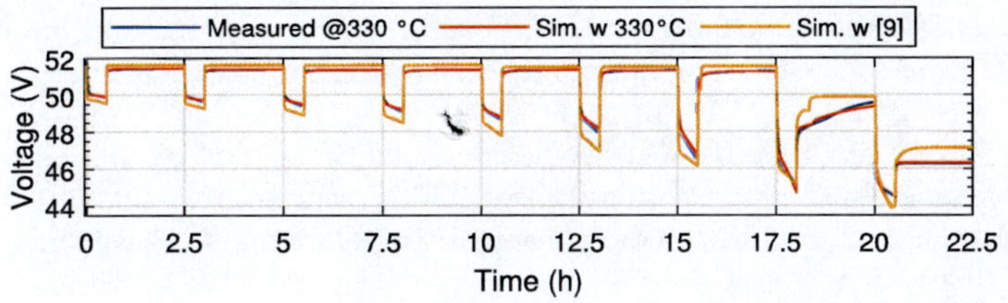

(a)

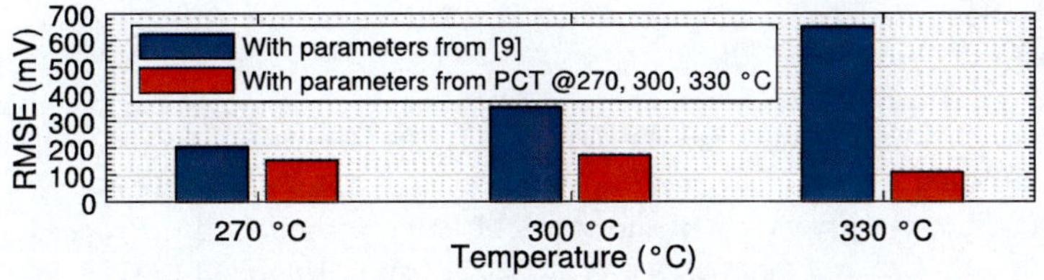

(b)

Fig. 5. Comparison between literature model parameters [9] and the new ones. (a) Comparison between the measured voltage in a PCT @330 °C with the model outcomes. The model voltages are obtained using the new identified parameters from the PCT at 330 °C and the literature ones, respectively. (b) RMS error of the model voltage obtained with the literature parameters and the new ones for the PCT carried-out at 270, 300 and 330 °C, respectively.

4 Conclusion

The comparison between the model parameters identified at 265 °C and presented in [9] with the new model parameters identified at 270, 300 and 330 °C shows the need to improve the Sodium-Metal Halide Battery models by including the temperature dependency. In fact, the model RMS error decreases from 1.3 % to 0.22 % of the mean battery voltage by taking into account the temperature dependency. Future works will be focused to include the temperature dependency in the model and check the accuracy with realistic application power profiles.

Acknowledgments. This research was partially funded by the University of Pisa Project PRA AUTENS, and supported by CrossLab project, funded by MIUR "Department of Excellence" program.

References

1. Telaretti, E., Dusonchet, L.: Stationary battery systems in the main world markets: part 1: overview of the state-of-the-art. In: 2017 IEEE International Conference on Environment and Electrical Engineering and 2017 IEEE Industrial and Commercial Power Systems Europe (EEEIC/I&CPS Europe). Institute of Electrical and Electronics Engineers Inc. (2017)
2. Sessa, S.D., Crugnola, G., Todeschini, M., Zin, S., Benato, R.: Sodium nickel chloride battery steady-state regime model for stationary electrical energy storage. J. Energy Storage **6**, 105–115 (2016)
3. Benato, R., et al.: Sodium nickel chloride battery technology for large-scale stationary storage in the high voltage network. J. Power Sources **293**, 127–136 (2015)
4. Dustmann, C.H.: ZEBRA battery meets USABC goals. J. Power Sources **72**(1), 27–31 (1998)
5. Benato, R., et al.: Sodium-nickel chloride (Na-$NiCl_2$) battery safety tests for stationary electrochemical energy storage. In: AEIT 2016 - International Annual Conference: Sustainable Development in the Mediterranean Area, Energy and ICT Networks of the Future (2016)
6. Boi, M., Battaglia, D., Salimbeni, A., Damiano, A.: A novel electrical model for iron doped-sodium metal halide batteries. IEEE Trans. Ind. Appl. **55**(6), 6247–6255 (2019)
7. Sun, K., Shu, Q.: Overview of the types of battery models. In: Proceedings of the 30th Chinese Control Conference, CCC 2011, pp. 3644–3648 (2011)
8. Boi, M., Battaglia, D., Salimbeni, A., Damiano, A.: A non-linear electrical model for iron doped sodium metal halides batteries. In: 2018 IEEE Energy Conversion Congress and Exposition, ECCE 2018, pp. 2039–2046 (2018)
9. Di Rienzo, R., Simonte, G., Biagioni, I., Baronti, F., Roncella, R., Saletti, R.: Experimental investigation of an electrical model for sodium-nickel chloride batteries. Energies **13**(10), 2652 (2020)
10. Simonte, G., Di Rienzo, R., Biagioni, I., Baronti, F., Roncella, R., Saletti, R.: Novel setup to extend the temperature characterization range of a sodium-metal halide battery. In: Saponara, S., De Gloria, A. (eds.) ApplePies 2021. LNEE, vol. 866, pp. 126–131. Springer, Cham (2022). https://doi.org/10.1007/978-3-030-95498-7_18
11. Morello, R., Di Rienzo, R., Roncella, R., Saletti, R., Baronti, F.: Hardware-in-the-loop platform for assessing battery state estimators in electric vehicles. IEEE Access **6**, 68210–68220 (2018)

Radio Frequency Drying of Wool Fabrics

Marco Cocci[1], Luca Pugi[1](✉), Enrico Boni[2], Massimo Delogu[1], Andrea Rocchetti[1], Luca Socci[1], and Nicola Andreini[1]

[1] Department of Industrial Engineering, University of Florence, Florence, Italy
luca.pugi@unifi.it
[2] Department of Information Engineering, University of Florence, Florence, Italy

Abstract. The drying processes of wool and other textile materials are energyintensive and not very efficient. In this work, the authors investigated an unconventional drying system able to produce a uniform drying preserving the quality of the treated fabric. The proposed process is also very efficient since the proposed radiofrequency technology can optimize the energy transfer to treated fabric minimizing undesired wasted heat fluxes to surrounding air or to other parts of the machine. For the investigation, a simulation model was developed to evaluate tissue heating.

1 Introduction

Wool needs in the various processing stages to be washed in water to eliminate the impurities that may be inside it such as powders, oleants etc... But also after the dyeing processes, where the fabrics are washed in water to eliminate dyes not fixed on the fabric. With the decarbonization process, companies will have to consume less and less fossil fuels and use renewable energy. Companies currently consume a lot of methane gas, for steam generation or thermal energy, so they will need to upgrade their drying systems to other, lower carbon renewables. There are various drying systems that are traditionally used:

- Hot air dryers: Hot air dryers can be direct gas or heat exchanger. In direct gas dryers, the air is heated directly by the combustion of methane gas which will have to dry the fabric. This system is much more efficient than dryers with exchangers, but it is more difficult to maintain a constant temperature and avoid the wool fabric as it can have color changes on the surface. Dryers with heat exchanger systems are less efficient, as a system is used, which is composed of a boiler generally powered by methane gas, this boiler heats a carrier fluid which can be steam, at a pressure that can vary between 7–20 bar, then the air inside the machine is heated by passing the carrier fluid through a heat exchanger. This drying method is not very efficient but allows to have an almost constant temperature on the heat exchanger, because this system uses the latent heat of vaporization at a constant temperature, therefore, it maintains a constant temperature throughout the exchanger. However, this system is very subject to clogging, further reducing efficiency.

R. Berta and A. De Gloria (Eds.): ApplePies 2022, LNEE 1036, pp. 359–365, 2023.
https://doi.org/10.1007/978-3-031-30333-3_49

- Infrared dryers: Infrared drying is used for very humid fabrics, this system can dry the fabric up to a maximum of 20% humidity, this system is very risky because it can create damage on the surface of the fabric. This drying system can be generated by electrical systems such as infrared lamps or with more efficient methane gas emitters.

1.1 Radiofrequency Heating

Material to be heated is placed between two metal plates working as an electric capacitor. The material works as a dielectric absorbing energy from the RF generator. The continuous inversion of polarity causes the heating of the material. The RF frequency bands used in dielectric heating are centered on 13.56 MHz, 27.12 MHz and 40.68 MHz. These frequencies are specifically reserved for use for industrial, scientific and medical purposes to avoid possible interference with other users of the radio spectrum. The RF energy is produced by an RF "generator" which basically comprises a power supply and an electronic oscillator. Generally, the triode amplifies the RF signal generated by the oscillator and behaves like an RF amplifier, the power can vary from 100 Watts to 150 kW or more, depending on the size of the triode and the supply voltage (Fig. 1).

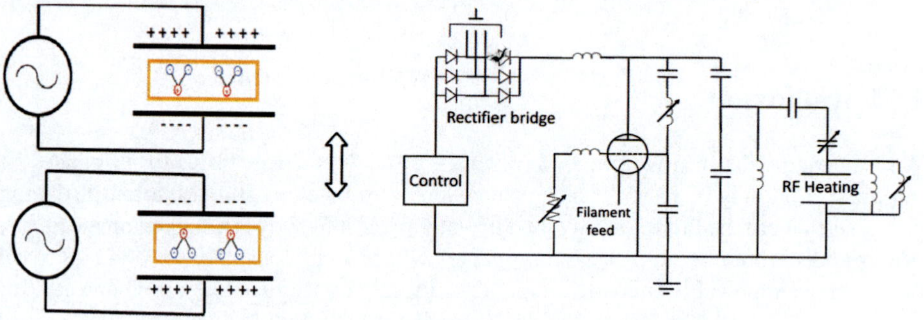

Fig. 1. RF Heating Principle of Operation (left) proposed machine layoout

The heating effect depends on the frequency used [1–3], the RF voltage range and the loss factor of the material to be heated. The transmitted power depends on Eq. (1) f is the frequency expressed in Hertz, E is the applied radiofrequency field in (Volt/meter), ε_0 is the permittivity of the space and is equal to 8,854 × 10^{-12} (Farad/meter), ε_r is the coefficient of the relative permittivity of the material to be heated and d is the loss angle. The loss factor is the product of the dielectric constant of the material ε_r and the loss tangent tan d.

$$\frac{power}{unit\ volume} = 2\pi f e_0 e_1 \tan dE^2 \frac{W}{m^3} \tag{1}$$

Equation (1) makes is used to determine the power required by the RF generator. It can be assumed that this value represents about 65–70% of the power absorbed by the power input. This calculation effectively determines the link between the RF voltage range and the required heating power. The maximum amount of water that can be removed is given by the latent heat of vaporization also considering a 10% of dispersed energy due to radiated heat losses.

2 Electro Magnetic Simulation Model

A simulation finite element simulation was performed to verify the heating of the wool fabric. It was supposed to make the fabric pass between two electrodes with motorized mechanical transmissions to make it pass several times through a single pair of electrodes as visible in Fig. 2, The fabric has a width of about 1.6 m, a thickness of about 2 mm. And it's modeled as a continuous tape inside the machine. Performed Electro-Magnetic simulation is a steady state one.

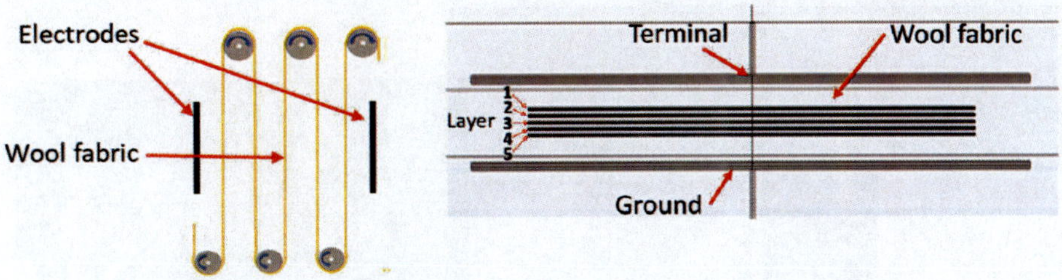

Fig. 2. Proposed Dryer layout (left) and FEM model right

Main parameters and e physical properties for each material, air and wet wool reported in Table 1.

Table 1. Main parameters of performed simulations[4, 5]

Parameter	Value	Parameter	Value
Electric conducibility (Wool)	1 [S/m]	Electric conducibility (air)	0 [S/m]
Relative permittivity (Wool)	18, 8–15	Relative permittivity(air)	1
Oscillating Voltage on Plates	10[kV]	Frequency	27.12 [MHz.]
Electrode Plate Size	2000 × 300[mm]	Humidity variation of Wool from 1st to 5th layer of wool*	20–19, 19–18, 18–17, 17–16, 16–15[%]
Wool Thermal Conductivity*	0.0349–0.0312 [W/mK]	Wool Density*	665–500 [kg/m^3]
Wool Specific Heat*	1908–1766 [J/kgK]	*Physic Properties of Wool are variable according to simulated humidity	

The physical properties of the wool considered are for a fabric that has an initial humidity of 20% which is reduced to 15% after 5 passages between dryer plates. In Fig. 3, the fabric has a greater distribution of power at the entrance and exit of the electrodes, this is due to the wavelength and size of the electrode. Distribution of heat power generated by RF heater seems to be quite unaffected by the level of humidity, but it's more sensitive by boundary effects due to concentration of electric field lines that should produce localized overheating on narrow boundary regions damaging the dried wool. This trouble can be largely solving by inserting lateral dielectric guides that as visible in Fig. 4: resulting distribution of power on wool boundary sections is much more homogeneous on boundaries.

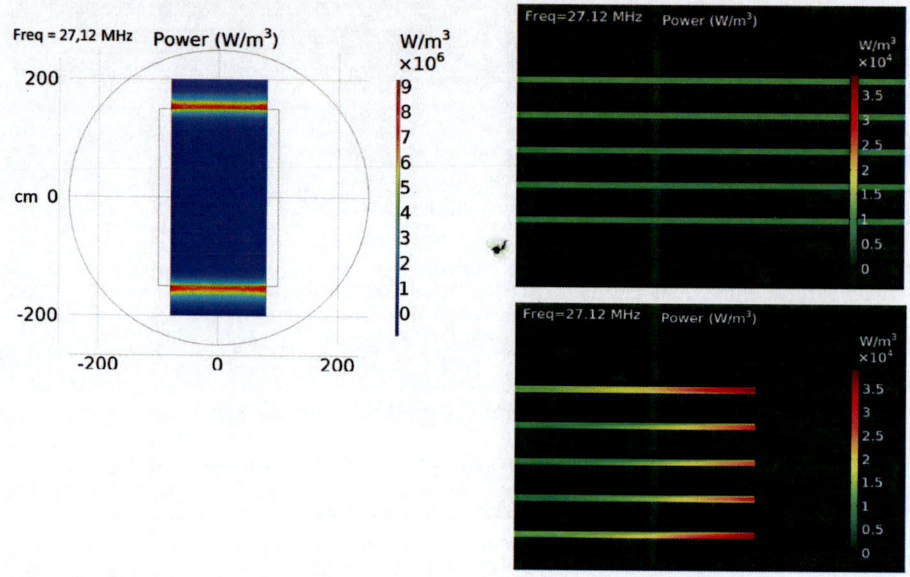

Fig. 3. Simulated distribution of generated heat power, top view (left), longitudinal and transversal sections (right)

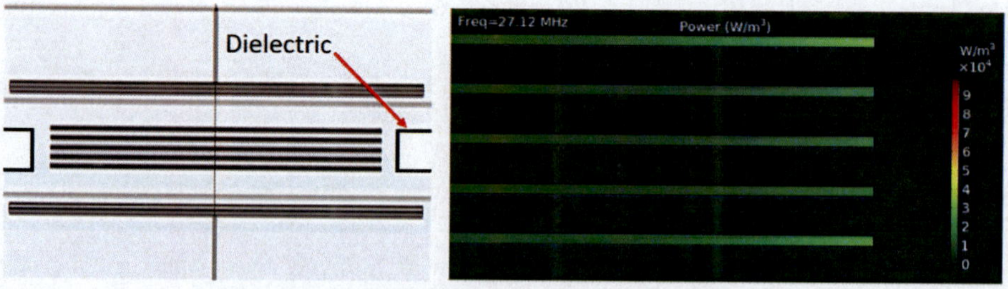

Fig. 4. Electrode and tissue and dielectric diagram

3 Thermal Simulation Model

Once distribution of RF heating power is known from electromagnetic simulations a bidimensional transient thermal model is assembled according to the scheme of Fig. 5: a mapped distribution of RF induced heat flux is applied to a transversal wool section where also thermal capacitive and thermal transfer phenomena are simulated. Wool is supposed to have an initial temp of 80 °C ad surrounding air has a temperature of 100 °C with a humidity of 28 g of steam per kg of dry air.

Forced ventilation is considered since air is insufflated with speed varying from 1 to 3.5 m/s in different sections of the machine.

The whole process has a duration of about 27 s and some results ae shown in Fig. 6: temperature inside the wool section is almost homogeneous from the beginning (time = 0[s]) to the end of the process (time = 27[s]). In Fig. 7 it is also shown the behavior of the temperature on single node, simulated temperature profile is quite smooth.

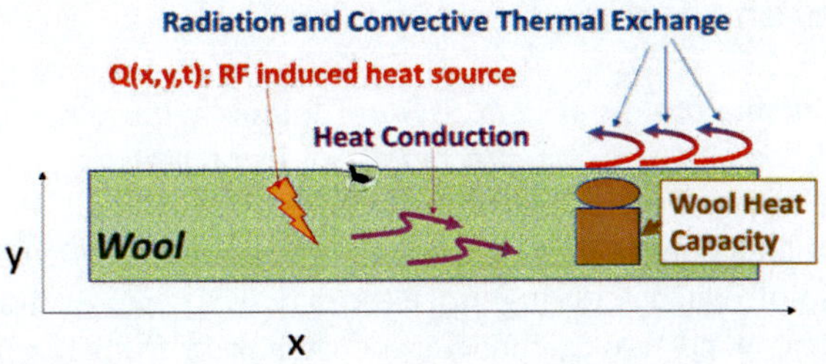

Fig. 5. Simplified Scheme of the Proposed Thermal Model

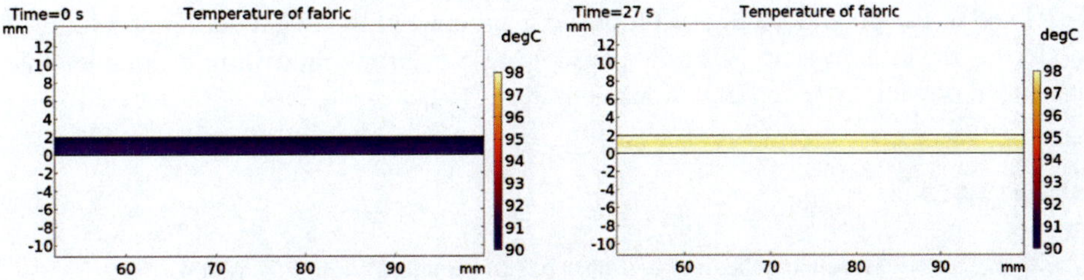

Fig. 6. Thermal Distribution at beginning (left) and the end right) of the RF heating process

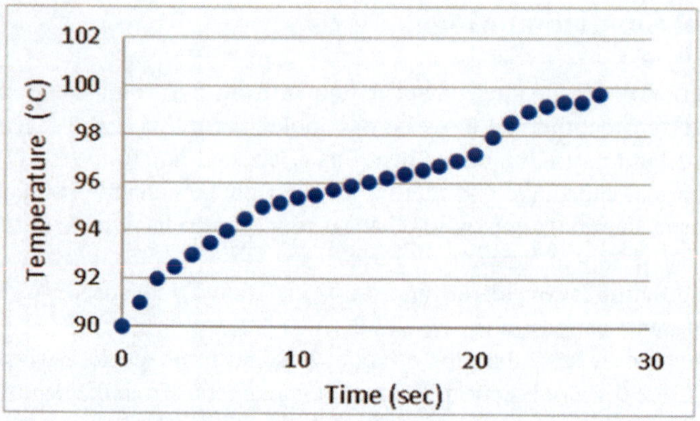

Fig. 7. Temperature at varying times of a single node

4 Conclusions and Future Developments

The optimization of the drying system of wool fabrics with radio frequency heating leads to improving the quality of the processed product but also a significant reduction in carbon oxides. The analyzed system facilitates the decarbonisation process of the textile industries, which currently consume large quantities of methane gas. The energy required for the radiofrequency drying process can derive from renewable sources such as photovoltaics and wind power, reducing consumption from fossil sources. In the future, the ventilation phase can be better studied also with the aid of CFD analysis. Moreover, it is possible to consider extending this technology to other very inefficient and very polluting textile industrial processes. Further development of proposed activities a erelated to design and testing of proposed machine and to the extraction of a lumped models of the system from FEM data exploiting bondgraph modelling techniques that have been previously tested in various activities [7–9].

References

1. Koral, T.: Radio frequency heating and post-baking. Bisquit World **7**(4) (2004)
2. Lesnikowsku, J.: Dielectric permittivity measurement methods of textile substrate of textile transmission lines. Politechnika Łódzka, Katedra Odzieżownictwa i Tekstroniki. ISSN 0033–2097, R. 88 NR 3a/2012 (2012)
3. Berzi, L., Cocci, M., Delogu, M., Pugi, L.: Sustainable revamping of wool carbonization systems. In: 21st IEEE International Conference on Environment and Electrical Engineering and 2021 5th IEEE Industrial and Commercial Power System Europe, EEEIC/I and CPS (2021)
4. Kim, H.A.: Moisture vapor permeability and thermal wear comfort of ecofriendly fiber-embedded woven fabrics for high-performance clothing. In: Materials MDPI (2021). https://www.mdpi.com/1996-1944/14/20/6205
5. Clarke, J.A., Yaneske, P.P, Pinney, A.A.: The harmonisation of Thermal Properties of building materials, Bepag (1991)
6. Europe 2021 - Proceedings, DOI: https://doi.org/10.1109/EEEIC/ICPSEurope51590.2021.958465

7. Pugi, L., Reatti, A., Corti, F.: Application of modal analysis methods to the design of wireless power transfer systems. Meccanica **54**(1–2), 321–331 (2019). https://doi.org/10.1007/s11012-018-00940-x
8. Pugi, L., Galardi, E., Carcasci, C., Lucchesi, N.: Hardware-in-the-loop testing of bypass valve actuation system: design and validation of a simplified real time model. Proc. Inst. Mech. Eng. Part E J. Process Mech. Eng. 231(2), 212-235 (2017). https://doi.org/10.1177/0954408915589513
9. Wu, Q., et al.: Freight train air brake models. Int. J. Rail Transp. (2021). https://doi.org/10.1080/23248378.2021.2006808

Preliminary Design and Simulation of a Transport and Winding System of an Innovative Radio Frequency Dryer

Marco Cocci, Massimo Delogu, Lorenzo Berzi, and Luca Pugi(✉)

Department of Industrial Engineering, University of Florence, Florence, Italy
luca.pugi@unifi.it

Abstract. Textile dryers are energy-consuming machines that are widely diffused in the textile industry. Authors are currently working on the complete redesign of a sustainable drier which is completely rethought to maximize productivity and sustainability. In this work, the authors focus their attention on modeling and design of winding and transport of the fabric across the machine comparing two different solutions: a conventional transmission and a direct drive version in which every pulley stage is actuated by a different electric motor.

1 Introduction

Application of Radio Frequency heating to the construction of drier for wool fabric is an innovative technology [1] that has been proposed to increase productivity and sustainability of different textile processes. As visible in the scheme of Fig. 1 the machine is composed by a drum on which the fabric is initially rolled, followed by a fabric loader/tensioner. Fabric is then convoyed to the section in which is dried using RF dielectric heating that is provided by an array of plates/antennas fed by a high voltage oscillator (10 kV) which is usually designed according to known schemes in literature [2]. Distance between RF heating plates is constrained by the maximum available voltage. To assure a high productivity speed of the fabric is quite high (about 0.5 m/s), so the route of the fabric tape inside the the RF heating stage must be prolonged to assure a permanency of the fabric between heating plates of about 30 s. To increase the heating time of fabric within RF heating plates and to optimize volume exploitation inside the heating stage, multiple double drawn pulleys are adopted according to the scheme of Fig. 1. Then the fabric after the RF heating stage is preloaded by a second loading/tensioner. Finally, the fabric is wounded on a motorized drum. Intermediate pulleys inside the RF heating stage must be also motorized to properly drive the fabric tape maintaining an optimal pull, tension: an excessive pull effort applied to wool should damage the fabric, while a too low tension should produce a directional instability and the loss of the correct trajectory of the textile tape. So, the way in which actuation and control of pulleys is performed is very important. In this work authors compare performance of two different actuation schemes: a first one in which many different pulleys are actuated by a single motor through a belt transmission system, a second investigated solution corresponds to the direct actuation of pulley with independent motor. Both schemes are investigated generating a digital

R. Berta and A. De Gloria (Eds.): ApplePies 2022, LNEE 1036, pp. 366–372, 2023.
https://doi.org/10.1007/978-3-031-30333-3_50

twin a complete mechatronic model of the system. Since a proper optimization of the actuation system is strongly constrained by friction and elastic properties of the treated fabric authors merge data from literature with experimental tests performed by authors.

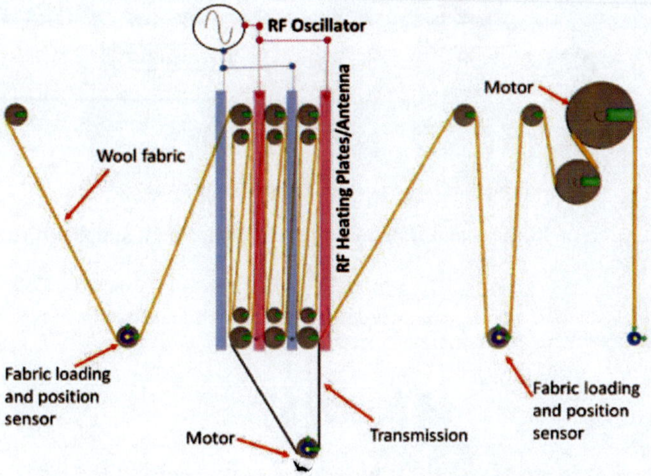

Fig. 1. Scheme of the investigated RF heating/drying system

2 Identification of Wool Fabric Properties

To properly model the properties of wool fabric tape that is drawn between machine rollers/pulleys, authors need to introduce some identification of manipulated materials in particular for what concern density, friction and elastic properties. There is a tech literature [3–6] in which these properties are described but it is incomplete respect to proposed applications: data available in literature suffer of a high statistical variability due to different kind of treated fabrics and application; For proposed application wool fabric specimens are available so authors preferred to identify this mechanical properties through the simple procedure described in Fig. 2: wool is loaded with increasing load on a fixed roller surface measuring fabric elongation, values of applied loads, values of applied loads that should cause the sliding of the wool on the surface.

3 Modelling of the Transmission System

As visible in Fig. 3 two different transmission systems are simulated using a Siemens Simcenter™ model in which mechanical behavior of wool fabric is modelled as an equivalent flat extensible belt, interacting through friction which pulleys that have the same physical properties of machine rollers in terms of geometric and inertial properties. Since fabric is dried along the machine properties of the belt in terms of inertia, stiffness and friction change along the machine.

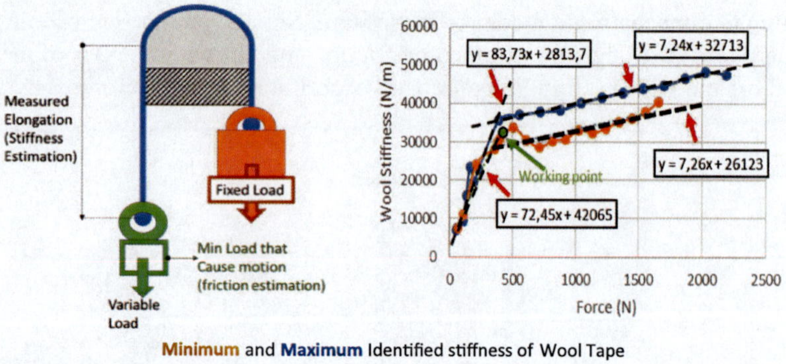

Fig. 2. Identification of stiffness and friction properties of the wall, simplified static procedure

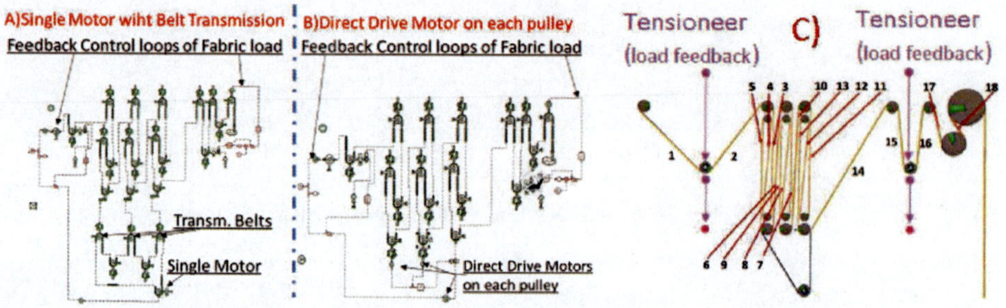

Fig. 3. Investigated configuration of the machine (a/b) and common disposition of sensing tensioners in the first (1&2) and final (15&16) branches of the fabric tape

Both investigated configurations regulate tractive load applied of dried wool measuring the displacement of two tensioners respectively applied in the entering and ending sections of the machine as visible in Fig. 3/C. In both layouts the last tensioner (between fabric branches 15&16) is used to evaluate tractive effort on the wool to regulate the torque applied by a motor on the last pulley. This feedback assure that wool is drawn with a constant tractive effort.

The same loop is performed by the entering tensioner (fabric branches between 1&2) where measured tractive efforts are regulated by increasing the tractive efforts of the intermediate motor/motors. Here is the difference between the two layouts: in the conventional one (Fig. 3/a) a single motor move every intermediate pulley trough a transmission system; in the other layout (Fig. 3/b) each roller is actuated by a direct drive motor. In this second configuration several motors have to be controlled at the same time. All the direct drive motors aiming to control intermediate rollers are regulated to follow in parallel the same torque reference. This control architecture is also described in literature in literature [7].

Main properties of simulated working conditions are described in table 1.

Table 1. Parameter adopted for performed simulations

Parameter	Value	Parameter	Value
Weight of 1st tensioner	60[kg]	Fabric stiffness	32000[N/m]
Weight of last tensioner	80[kg]	Fabric weight for meter	0,5–6 [kg/m]*
Bearing friction coefficient	0,018[m]	Fabric Speed	0,52 [m/s]
Roller Inertia	0.04[kgm^2]	Fabric-Roller Contact	0,016–0,026*
*Variable Properties since fabric i	Friction's dried alogn the machine		

4 Simulation Results

Starting from the model described in Fig. 3, authors performed some simulation to compare the behavior of different control configurations. As example in Fig. 4/a/b, it is compared the behavior of traction forces on fabric in different sections/branches of the machine during the activation of the machine and the gradual acceleration of the fabric from standstill condition to the final traveling speed of about 0.5[m/s]: in both machine layouts the duration of transients is limited to few seconds however speed oscillations on each textile branch have a quite different shape: in the single motor machine it's evident the discontinuity in speed oscillations of textile branches within the heating sections respect to the other ones. In the machine with independent motors oscillations are still visible but there is less difference along the fabric tape so this behavior is preferable. This smoother and more precise control of tractive efforts applied on fabric can be also recognized analyzing the steady state distribution of tractive efforts for both solutions as described in Table 2 and in Fig. 5: differences between two machine configurations are not very big, however distribution of tractive efforts in the machine with direct drive actuation of roller is clearly better.

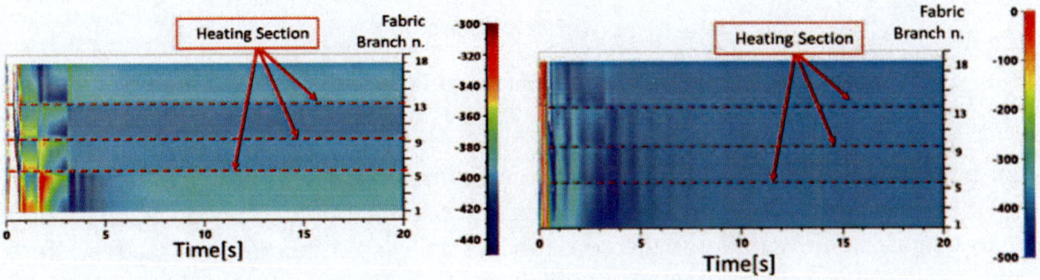

a) b)

Fig. 4. Simulated acceleration transients for both machine configurations a) single motor, b) direct drive motors on rollers

A direct drive actuation is also better respect to a single motor in case of a typical failure of this kind of machines, the blockage of wool fabric: fabric is not correctly guided by rollers this should happen for different causes ranging from troubles on rollers

Table 2. Steady State distribution of traction forces along wool fabrics, a)single motor, b)direct drive motors on rollers

Single Motor on Intermediate Rollers				Direct Drive Motors on Rollers			
Branch	T. Force	Branch	T. Force	Branch	T. Force	Branch	T. Force
1	393[N]	10	411[N]	1	393[N]	10	394[N]
2	394[N]	11	412[N	2	394[N]	11	394[N]
3	393[N]	12	412[N]	3	393[N]	12	394[N]
4	393[N]	13	412[N]	4	393[N]	13	395[N]
5	393[N]	14	394[N]	5	393[N]	14	394[N]
6	428[N]	15	396[N]	6	393[N]	15	396[N]
7	429[N]	16	396[N]	7	394[N]	16	396[N]
8	429[N]	17	397[N]	8	394[N]	17	397[N]
9	429[N]	18	397[N]	9	394[N]	19	397[N]
***Max differencence 36N Max difference between two adiacent cells is 35N**				***Max differencence 4N Max difference between two adiacent cells is less than 1N**			

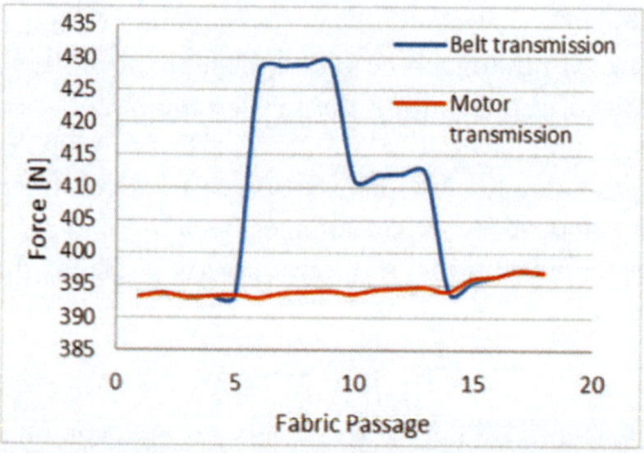

Fig. 5. Steady state behavior of tractive effort behavior of compared solutions along the machine

(bearings blocked, deformed rollers) to defects on manipulated fabric. In this conditions wool is blocked around roller and as visible in Fig. 6, if the failure is not detected for both the investigated configurations, the consequence is a rapid increase of tractive efforts which can cause various kind of troubles ranging from the blockage of the machine due excessive torques on motors to the deformation and consequently to the destruction of rollers. Respect to this failure both configurations produce in theory the same effects as visible in Fig. 6. However, it should be considered a clear advantage of the direct drive solution respect to the conventional one: since each roller is motorized in an independent way and each motor drive can provide speed and torque measurements, the whole process is much more monitored and the failure can be prevented in an easier way.

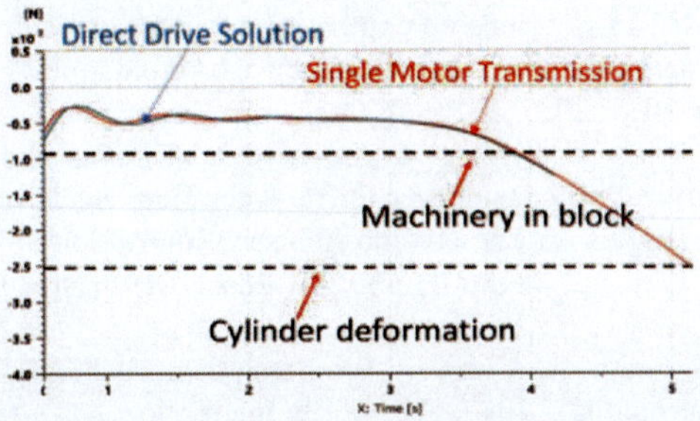

Fig. 6. Simulation of machine blockage (blockage on rollers between branch 7&8), behavior of maximum traction forces of wool fabric

5 Conclusions and Future Developments

In this work authors have studied the actuation of a RF drier comparing different solutions. Respect to investigated solutions a direct drive system seems to be largely preferable. Also, in terms of costs this solution is affordable since installed motors are smaller and avoid the installation of many other roller belts and components, simplifying the overall design of the machine. This work is a part of a more complex activity (the description of the RF drier system is the object of another research paper presented in the same congress) that will produce a complete redesign of the machine and the manufacturing of a much more efficient and performing generation of driers within one or two years. Currently Authors are focusing their attention of executive design of the new machine and in particular to the introduction of some improvements in the design of the RF system through the usage of Finite Elements techniques that have been recently exploited for the simulation of various kind of wireless power transmission equipment[8, 9] including also capacitive ones[10].

References

1. Berzi, L., Cocci, M., Delogu, M., Pugi, L.: Sustainable revamping of wool carbonization systems. In: 21st IEEE International Conference on Environment and Electrical Engineering and 2021 5th IEEE Industrial and Commercial Power System Europe, EEEIC / I and CPS Europe 2021 - Proceedings (2021). https://doi.org/10.1109/EEEIC/ICPSEurope51590.2021.9584659
2. Cottee, C.J., Duncan, S.R.: Design of matching circuit controllers for radio-frequency heating. IEEE Trans. Control Syst. Technol. **11**(1), 91–100 (2003)
3. Kothari, V., Gangal, M.: Assessment of frictional properties of some woven fabrocs. Indian J. Fibre Text. Res. **19**, 151–155 (1994)
4. Lindberg, J., Gralen, N.: Measurement of friction between single fibers. frictional properties of wool fibers measured by the fiber-twist method. Swedish institute for Textile Research, Githenburg, Sweden, pp. 284–301 (1948)
5. Speakman, J.: The rigidity of wool and its change with adsorption of water vapour. Trans. Faraday Soc., 92–103 (1928)
6. Feughelman, M., Collins, J.: The torsional rigidity of wool fibers extended in water. The University of New South Wales, School of Textile Technology, Kensington, pp. 627– 629 (1774)

7. Wen, P., Stapleton, C., Li, Y.: Tension control of a winding machine for rectangular coils. In: 2008 10th International Conference on Control, Automation, Robotics and Vision, 17 December 2008, pp. 2031–2037. IEEE (2008)
8. Corti, F., Grasso, F., Paolucci, L., Pugi, L., Luchetti, L.: Circular coil for EV wireless charging design and optimization considering ferrite saturation. In: 5th International Forum on Research and Technologies for Society and Industry: Innovation to Shape the Future, RTSI 2019 - Proceedings, art. no. 8895601, pp. 279–284 (2019). https://doi.org/10.1109/RTSI.2019.8895601
9. Allotta, B., Pugi, L., Reatti, A., Corti, F:. Wireless power recharge for underwater robotics. In: Conference Proceedings - 2017 17th IEEE International Conference on Environment and Electrical Engineering and 2017 1st IEEE Industrial and Commercial Power Systems Europe, EEEIC/I and CPS Europe 2017, art. no. 7977478 (2017). https://doi.org/10.1109/EEEIC.2017.7977478
10. Reatti, A., Pugi, L., Corti, F., Grasso, F.: Effect of misalignment in a four plates capacitive wireless power transfer system. In: Proceedings - 2020 IEEE International Conference on Environment and Electrical Engineering and 2020 IEEE Industrial and Commercial Power Systems Europe, EEEIC/I and CPS Europe 2020, art. no. 9160627 (2020). https://doi.org/10.1109/EEEIC/ICPSEurope49358.2020.9160627

Design and Test of an LSTM-Based Algorithm for Li-Ion Batteries Remaining Useful Life Estimation

Andrea Begni, Pierpaolo Dini(✉), and Sergio Saponara

Department of Information Engineering, University of Pisa, Pisa, Italy
a.begni@studenti.unipi.it, pierpaolo.dini@phd.unipi.it, sergio.saponara@unipi.it

Abstract. The article describes how to extrapolate a useful time-series of features from a raw dataset of complete charging/discharging cycles of Li-ion batteries. The extrapolation of such time-series based on features is helpful to reduce the size of the LSTM (Long Short-Term Memory) as much as possible, differently from classical approaches with LSTM applied to raw data time-series. After a data pre-processing step, this work implements a features-extraction process that allows selecting the best features to describe the performance degradation of the batteries during the time and to estimate the RUL (Remaining Useful Life) during the battery life.

1 Introduction

Nowadays Li-ion batteries are becoming widely applied in very large range of application fields, such as smartphones [1], home automation [2], robotics and industry 4.0 [3], energy storage systems [4], vehicles [5] and power electronics [6]. Anyway, one of the most important issues with Li-ion batteries is their fast degradation process. This is the reason why for safety-critical applications is important to design efficient algorithms able to estimate the remaining useful life of the batteries. In this work we propose the design of an Artificial Intelligence methodology based on a data-driven learning process to exploit real-data availability, obtaining an algorithm able to estimate the RUL of Li-ion batteries. After the analysis of the work motivations, hereafter we review the state-of-art of batteries RUL estimation. In EBA (Experience Based Approach) [8–11] the RUL is estimated through a stochastic distribution analysis describing the degradation itself. Such an approach has in general very low theoretical complexity but is often characterized by high computation cost. In MBA (Model-Based Approach) [12–15], the real behaviour is estimated with the «a priori» knowledge through a mathematical model of the battery, exploiting state observer theory. In DDA (Data-Driven Approach) [16–19] the model learns by real acquired data the variable's relationships and extrapolates a model for the battery's degradation estimation. If complete battery life measurements are available, DDA can be defined as the most effective methodology.

R. Berta and A. De Gloria (Eds.): ApplePies 2022, LNEE 1036, pp. 373–379, 2023.
https://doi.org/10.1007/978-3-031-30333-3_51

2 Dataset Description

The dataset considered in this work is composed of data from 4 lithium-ion batteries, see Table 1, taken from the repository of the NASA Ames Prognostics Centre of Excellence [20]. Each dataset records information about the charging, discharging and impedance cycles of the batteries. The charge and discharge process of lithium-ion batteries is organized as follows. Each observation is represented by timeseries in terms of direct measurements. Both in Charge and Discharge cycles, observations are in terms of battery terminal voltage, battery output current, battery temperature, current measured at charger, voltage measured at charger.

Table 1. Dataset Organization

Battery ID	N. Charge Cycles	N. Discharging Cycles
B0005	170	168
B0006	170	168
B0007	170	168
B0018	134	132

3 Preliminary Dataset Analysis and Manipulation

Figure 1 shows the characteristic of lithium batterie. Therefore, EOL is defined as the nominal capacity multiplied by the % residual capacity. They represent the charge-discharge cycle where the Capacity falls below the EOL threshold.

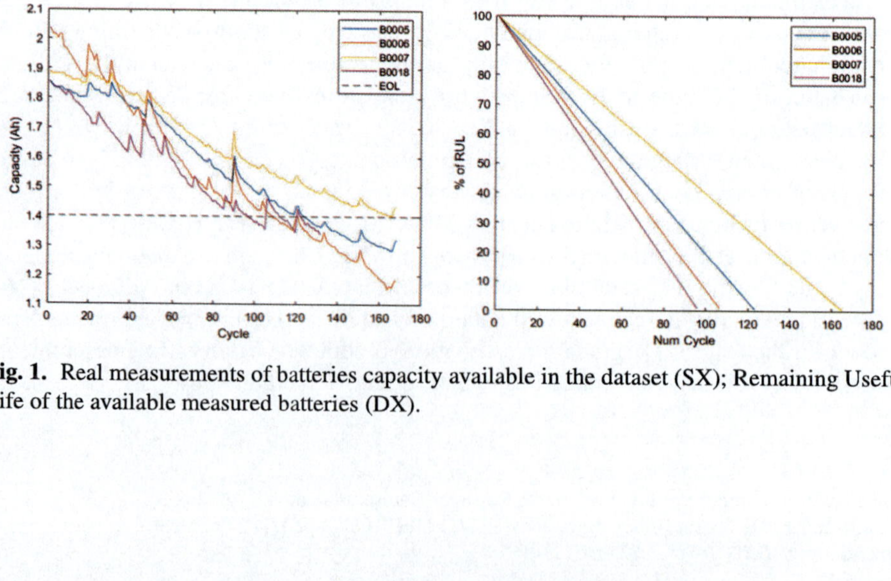

Fig. 1. Real measurements of batteries capacity available in the dataset (SX); Remaining Useful Life of the available measured batteries (DX).

We exploit the MATLAB toolbox Signal Features, which provides general signal-based statistical metrics for time-series features extraction. From the Statistical Features, it can be expected to change as a deteriorating fault signature intrudes upon the nominal signal. In Table 2 are reported the Statistical chosen features. In Table 3 are reported the Impulsive Metrics which are related to signal peaks. Signal Processing Metrics, shown in Table 4, From the 10 measured time series are extracted 13 features, resulting in a new 130 features-based dataset. From the dataset of features, the Capacity is not considered. To evaluate extracted features, we propose the Monotonicity Analysis.

Table 2. Impulsive Metrics

Feature	Formula
Peak Value	$x_p = \max_{i}\lvert x_i\rvert$
Impulse Factor	$x_{IF} = \frac{x_p}{\frac{1}{N}\sum_i \lvert x_i\rvert}$
Crest Factor	$x_{cr} = \frac{x_p}{x_{rms}}$
Clearance Factor	$x_{clear} = \frac{x_p}{\left(\frac{1}{N}\sum_i \sqrt{\lvert x_i\rvert}\right)^2}$

Table 3. Selected Statistical Features.

Feature	Formula
Mean Value	$x_m = \frac{1}{N}\sum_i x_i$
Standard Deviation	$x_\sigma = \sqrt{\frac{\sum_i (x_i - x_m)^2}{N-1}}$
Root Mean Square	$x_{rms} = \sqrt{\frac{\sum_i x_i^2}{N}}$
Skewness	$x_{skew} = \frac{\frac{1}{N}\sum_i (x_i - x_m)^3}{\left[\frac{1}{N}\sum_i (x_i - x_m)^2\right]^{3/2}}$
Kurtosis	$x_{kurt} = \frac{\frac{1}{N}\sum_i (x_i - x_m)^4}{\left[\frac{1}{N}\sum_i (x_i - x_m)^2\right]^2}$
Shape Factor	$x_{SF} = \frac{x_{rms}}{\frac{1}{N}\sum_i \lvert x_i\rvert}$

From the dataset of features, the Capacity is not considered. It cannot be directly measured through voltage and current sensors and must be calculated through a specific method offline. To evaluate extracted features, we propose the Monotonicity Analysis. The monotonicity quantifies the monotonic trend in condition indicators as the system evolves toward failure. The values of Y range from 0 to 1, where Y is 1 if X is perfectly

Table 4. Signal Processing Metrics

Feature	Formula
Signal-Noise Ratio (SNR)	$SNR = \frac{P_{signal}}{P_{noise}}$
Total Harmonic Distortion (THD)	$THD = \frac{\sqrt{\sum_{n\geq 2} P_{n\_rms}^2}}{P_{fund\_rms}}$
Signal-Noise and Distortion Ratio (SINAD)	$SINAD = \frac{P_{signal}+P_{noise}+P_{distortion}}{P_{noise}+P_{distortion}}$

monotonic and 0 if X is non-monotonic. After several tests, the best monotonic threshold to select the features for training the neural network resulted to be 0,4. So the features that do not achieve monotonicity equal to or higher than the threshold will be discarded. A high monotonicity trend describes better the trend of degradation of the characteristic of the battery. Using all features might lead the neural network to not understand the RUL trend over time and give wrong estimates.

$$x_{istand} = \frac{x_i - \mu}{\sigma} \tag{1}$$

4 LSTM Design and Validation

LSTM is a special kind of RNN that can process not only single data points, but also entire sequences of data (see Fig. 2). After several tests, we propose an LSTM constituted by (see Fig. 3): I) an Input layer for 13 input features array; II) an LSTM layer with 300 hidden units (LSTM neurons); III) a Fully Connected layer with 150 hidden units; IV) a Dropout layer configured with probability equal to 0.5; V) a Fully connected layer with one neuron; vi) an Output layer with one regression output. Given the small size of the dataset, several combinations are tested. Considering the low amount of usable data, 4

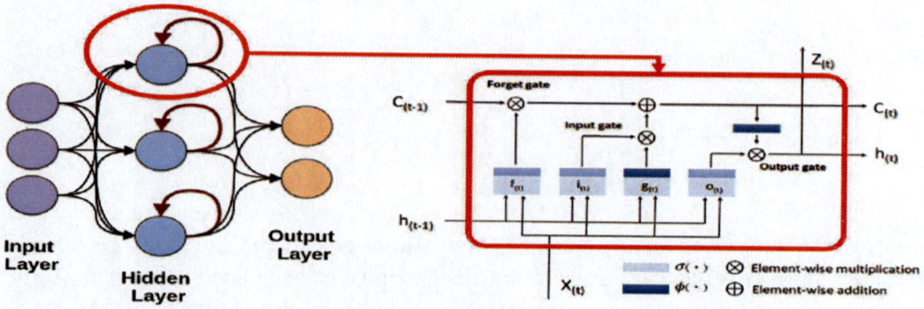

Fig. 2. Schematic representation of LSTM neural network and characterisitc neuron of hidden layers.

tests were conducted to verify the learning capacity of the network and the effectiveness of the proposed method even with little available dataset (Fig. 4).

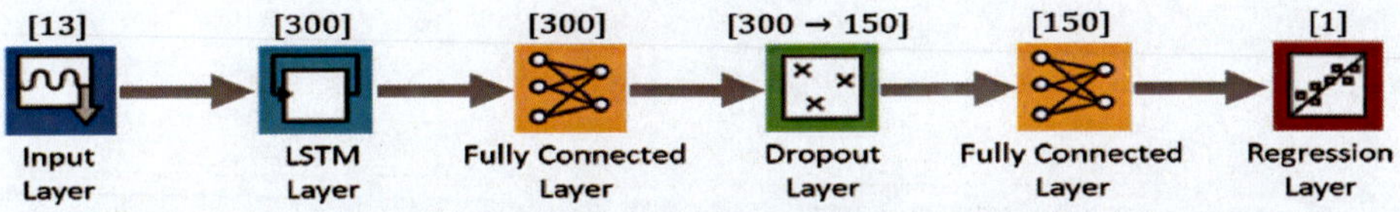

Fig. 3. Architecture of the applied LSTM model in the Mathworks environment.

Since the dataset is constituted of timeseries about only four monitored batteries, to avoid (as much as possible) dependency by selection choice, for training and validation phase, we propose 4 different Test configurations. Test 1 use (B0007, B0006, B0005) for the LSTM training & validation, while use B0018 for the final performance evaluation. Similarly, Test 2 use (B0018, B0006, B0005) in training and B0007 for evaluation; Test 3 use (B0018, B0007, B0006) in training and B0005 for evaluation; and Test 4 use (B0018, B0007, B0005) for training and B0006 for evaluation. In Tests 1 and 3, the RUL predictions follow the line of the real RUL. In several points the estimation and the real RUL are overlapped. For the Tests 1, 2 and 3 the LSTM output follows better the real RUL after having accumulated some sample of the time series. It means the network can recognize the characteristic slope of the curve after several cycles and achieves a good estimation of the EOL. The Test n. 4 follows the downward trend of the RUL, also if the local estimation is not so precise, could be still used for qualitative understanding the remaining useful life. Results are shown in Fig. 5.

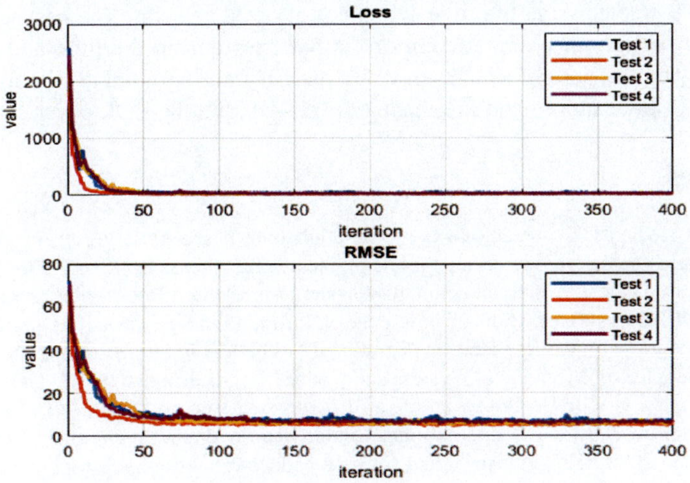

Fig. 4. RMSE and Loss of Training & Validation phase.

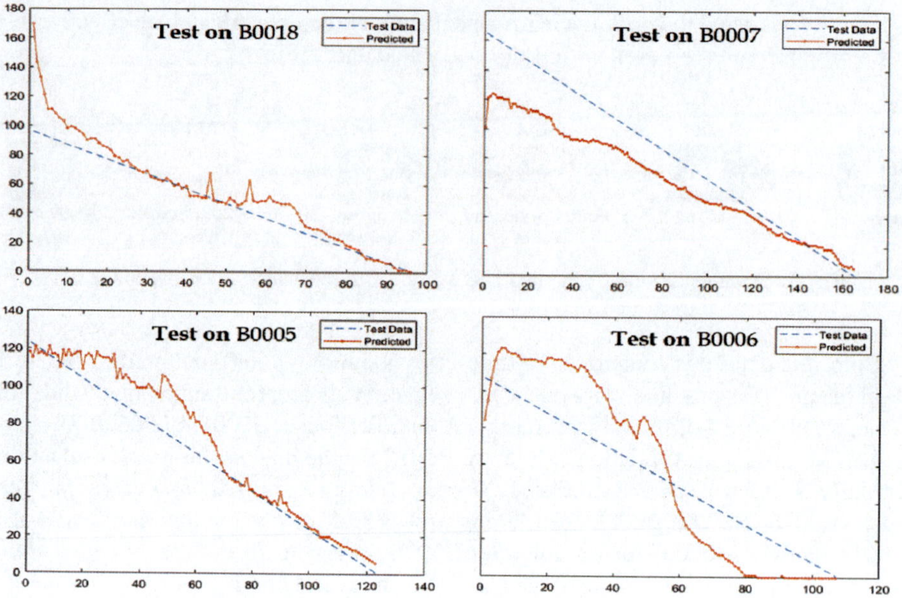

Fig. 5. Test result in the possible combinations.

5 Conclusion and Future Works

The article has described how to manage the dataset of complete charging/discharging cycles of Li-ion batteries, with the goal to design an LSTM neural network to estimate the remaining useful life from directly collected measurements at NASA. The study of the features extracted from the time series has allowed selecting the best features that describe the performance degradation of the batteries during the time and estimating the RUL during the battery life. Even if the dataset training is relatively small, it was possible to estimate quite good the shape of real RUL and the EOL point.

References

1. Ali, H., Khan, H.A., Pecht, M.G.: Evaluation of li-based battery current, voltage, and temperature profiles for in-service mobile phones. IEEE Access **8**, 73665–73676 (2020)
2. Yang, L., et al.: Design of home automation system based on ZigBee wireless sensor network. In: 2009 IEEE First International Conference on Information Science and Engineering (2009)
3. Li, Y., et al.: The impact of radiation degraded li-ion battery to mobile robots. In: 2017 IEEE Int. Conference on Mechanical, System and Control Engineering (ICMSC) (2017)
4. Benedetti, D., et al.: Design of an off-grid photovoltaic carport for a full electric vehicle recharging. In: 2020 IEEE International Conference on Environment and Electrical Engineering and 2020 IEEE Industrial and Commercial Power Systems Europe (EEEIC/I&CPS Europe) (2020)
5. Benedetti, D., et al.: Design of a digital dashboard on low-cost embedded platform in a fully electric vehicle. In: 2020 IEEE International Conference on Environment and Electrical Engineering and 2020 IEEE Industrial and Commercial Power Systems Europe (EEEIC/I&CPS Europe) (2020)

6. Dini, P., et al.: Electro-thermal model-based design of bidirectional on-board chargers in hybrid and full electric vehicles. Electronics **11**(1), 112 (2021)
7. Dini, P., et al.: Analysis, design, and comparison of machine-learning techniques for networking intrusion detection. Designs **5**(1), 9 (2021)
8. Khelif, R., et al.: Experience based approach for Li-ion batteries RUL prediction. IFAC-Papers Online **48**(3), 761–766 (2015)
9. Lyu, J., et al.: Remaining useful life estimation with multiple local similarities. Eng. Appli. Artifi. Intell. **95,** 103849 (2020)
10. Barré, A., et al.: Statistical analysis for understanding and predicting battery degradations in real-life electric vehicle use. J. Power Sources **245**, 846–856 (2014)
11. Shi, G., et al.: Determination of optimal indicators based on statistical analysis for the state of health estimation of a Lithium-ion battery. Front. Energy Res. **9**, 262 (2021)
12. Duan, B., et al.: Remaining useful life prediction of lithium-ion battery based on extended Kalman particle filter. Int. J. Energy Res. **44**(3), 1724–1734 (2020)
13. Wang, Y., et al.: A comprehensive review of battery modelling and state estimation approaches for advanced battery management systems. Renew. Sustainable Energy Rev. **131**, 110015 (2020)
14. Xue, Z., et al.: Remaining useful life prediction of lithium-ion batteries with adaptive unscented Kalman filter and optimized support vector regression. Neurocomputing 376, 95–102 (2020)
15. Tang, X., et al.: Battery incremental capacity curve extraction by a two-dimensional Luenberger–Gaussian-moving-average filter. Appli. Energy **280**, 115895 (2020)
16. Ren, L., et al.: Remaining useful life prediction for lithium-ion battery: A deep learning approach. IEEE Access **6**, 50587–50598 (2018)
17. Liu, K., et al.: A data-driven approach with uncertainty quantification for predicting future capacities and remaining useful life of lithium-ion battery. IEEE Trans. Indust. Electron. **68**(4), 3170–3180 (2020)
18. Ge, M.-F., et al.: A review on state of health estimations and remaining useful life prognostics of lithium-ion batteries. Measurement **174**, 109057 (2021)
19. Gao, D., et al.: Prediction of remaining useful life of lithium-ion battery based on multi-kernel support vector machine with particle swarm optimization. J. Power Electron. 17(5), 1288–1297 (2017)
20. Saha, B., Goebel, K.: Battery Data Set, NASA Ames Prognostics Data Repository (http://ti.arc. NASA. gov/project/prognostic-data-repository). NASA Ames Research Centre, Moffett Field, CA. NASA AMES Prognostics Data Repository (2007)

A Clinical Tool for Prognosis and Speech Rehabilitation in Dysarthric Patients: The DESIRE Project

Massimiliano Donati[1](✉), Alessio Bechini[1], Clelia D'Anna[2], Bruno Fattori[3], Marco Marini[1], Martina Olivelli[1], Susanna Pelagatti[4], Giulia Ricci[3], Erika Schirinzi[3], Gabriele Siciliano[3], Mirko Tavosanis[5], Francesca Torri[3], Nicola Vanello[1], and Luca Fanucci[1]

[1] Department of Information Engineering, University of Pisa, Pisa, Italy
massimiliano.donati@unipi.it
[2] Department of Surgical, Medical, Molecular and Critical Area Pathology, University of Pisa, Pisa, Italy
[3] Department of Clinical and Experimental Medicine, University of Pisa, Pisa, Italy
[4] Department of Computer Science, University of Pisa, Pisa, Italy
[5] Department of Philology, Literature and Linguistics, University of Pisa, Pisa, Italy

Abstract. Dysarthria is a motor disorder of speech characterized by alteration of articulation and intelligibility of speech. The goal of dysarthria management is to optimize communication effectiveness for as long as possible. To help clinicians in monitoring disease progression and rehabilitation outcomes, the DESIRE tool analyzes several reading sessions in which the patients pronounce predetermined selected words aloud, elaborating a measure of how much the patient's pronunciation deviates from those of previous sessions and the expected performance. In addition, the electronical record offers a comprehensive view of patient's status, and the web access allows the care team to remotely monitor progresses, so that they can tailor rehabilitation programs over time. Through the possibility to understand the patient difficulty about specific phonemes, word length, consonant clusters, this innovative tool offers a method to assess and monitoring dysarthria, to address therapeutic strategies, and to provide useful requirements for clinical trials readiness.

Keywords: dysarthria · speech rehabilitation · remote monitoring · ICT

1 Introduction

Speech is a very complex behavior requiring the synchronous activity of muscle groups associated with respiration, laryngeal function, airflow transit, orofacial movements and prosody. In several neurological disorders an alteration of these complex motor processes induces dysarthria, a dysfunction in the execution of speech conditioning its intelligibility. It has been estimated that at least 40% of patients suffering from different neurological disorders involving central or

R. Berta and A. De Gloria (Eds.): ApplePies 2022, LNEE 1036, pp. 380–385, 2023.
https://doi.org/10.1007/978-3-031-30333-3_52

peripheral nervous system (e.g. stroke, amyotrophic lateral sclerosis, myotonic dystrophy) present an associated speech disorder [1].

An accurate evaluation of the pathophysiological aspects of the different sources of the motor speech disorder and the analysis of their interaction are mandatory to better define clinical characteristics of the impairment. In doing that, diagnostic tools used up to now in clinical practice are not always able to promptly detect small changes which impair speech function and can be quite invasive [2]. To define follow-up programs and, when possible, to trace a prognostic trajectory of the speech impairment, should be desirable, also to address tailored rehabilitation treatment plans.

Obtaining valid clinical outcomes measures needs the application of rigorous and repeatable assessment protocols able to quantify, confirm and refine highly subjective perceptual ratings often unreliable within and across evaluators. In fact, as observed in patients affected by amyotrophic lateral sclerosis [3], acoustic, kinematic and strength variables of the speech are to be considered and the use of tools able to catch and analyze speech signals and recognize specific dysarthric pattern should be recommended [4].

The aim of this work is to provide neurologists, phoniatrists and speech language therapists (SLT) with an ICT-based aid tool to be exploited both in the diagnostic field and in the evaluation and rehabilitation field of dysarthria for a targeted classification of the patient's language difficulty. Such a tool offers the possibility of creating a specific electronic health record, containing the patient's clinical information deriving from neurological and phoniatric visits and from the speech therapy assessments carried out. Additionally, the results of questionnaires, examinations and tests carried out can be entered, in particular those that evaluate the articulatory and perceptive aspects of the patient's dysarthria (e.g. speech and respiration, resonance, articulation and prosody, etc.). Moreover, it is possible to take notes on the communication strategies adopted and the subjective analysis of the SLT allowing to monitor patient performance over time and also to have a reference baseline shared among all clinicians involved in the patient management. Indeed, it is possible to record the patient's voice during each treatment session proposing a set of standard words, starting from the first medical-speech therapy interview. By obtaining automatic feedback on the quality, quantity and type of words articulated by the patient, the SLT is supported in the functional classification of the patient's dysarthria and in monitoring the patient's performance over time.

The evaluation of dysarthria includes both an objective and subjective analysis of the speech subsystems, as well as their interactions [5]. So, the acoustic parameters evaluated with the DESIRE tool will permit to have a numerical and objective quantification of severity and progression of speech disorders [6].

In order to have a precise view of the patient clinical status, the tool must propose an appropriate set of words to be monitored and the construction of the database is an important activity that, especially in some early stages, must include a review of the linguistic features. First of all, it is important that the recorded speech includes words that can actually be pronounced by patients in real circumstances. Furthermore, the words must be selected to cover the repertoire of Italian phonemes, taking into account factors such as dialectal variation.

The set of words has been defined in the context of the realization of the IDEA database [7]. IDEA is the first Italian dysarthric database, whose aim is to improve the performance of automatic speech recognition (ASR) tools. These systems, in fact, are difficult to use for people affected by dysarthria, because they are trained using data not representing the peculiar characteristics of the dysarthric speech. IDEA is composed by the recordings of the selected words said aloud by different patients, during multiple sessions spread along time. The tool used to perform and manage such recordings is called RECORDIA and it is the starting point of our running project, called DESIRE, which aims at expanding its feature in order to meet doctors' need in treating dysarthria.

2 Materials and Method

The DESIRE tool is based on RECORDIA [7], a Java application originally developed to build the IDEA database and to provide patient characterization and voice recording features. RECORDIA provides a Graphical User Interface (GUI) that helps medical professionals to manage recording sessions, presenting the set of words to be said out aloud, allowing the user to stop, resume and repeat the session and to later listen to the recordings. Patient characterization includes a basic registry person and clinical information, such as pathology and its assessment scale, impairment level [8] and dysarthria classification [9], a measure of the quality of life [10] and measure scale of the therapy outcome [11]. This information and all the recordings (.wav files) are stored in the user's local PC, and the user can send them to a server, placed at the University of Pisa, using a File Transfer Protocol (FTP) client external to the RECORDIA software.

In order to meet doctors needs in treating dysarthria, while increasing the usability of the software, RECORDIA has been enhanced with some functionalities and integrated with a Spring-based server application for user interactions and data consultation, REST web services for bi-directional data exchange and a centralized MySQL relational database for data storing. Figure 1 shows the DESIRE tool architecture, in which the RECORDIA and IDEA modules are highlighted.

In this new configuration, the RECORDIA client is used not only to enroll new patients in the system and to perform the recording sessions, as already described, but also to actively support clinicians during therapy sessions. In addition to local storage in the file system, data is also sent to the centralized database using a dedicated REST service, so that they can be later requested for consultation using the web application or the client itself.

Indeed, the web application gives to authenticated users, using any internet browser, the access to patient's registry, medical records and recording sessions as they are working on RECORDIA client. The main difference is that an internet connection is required to use the web application with respect to RECORDIA software. Two different professional user profiles are provided: medical doctor

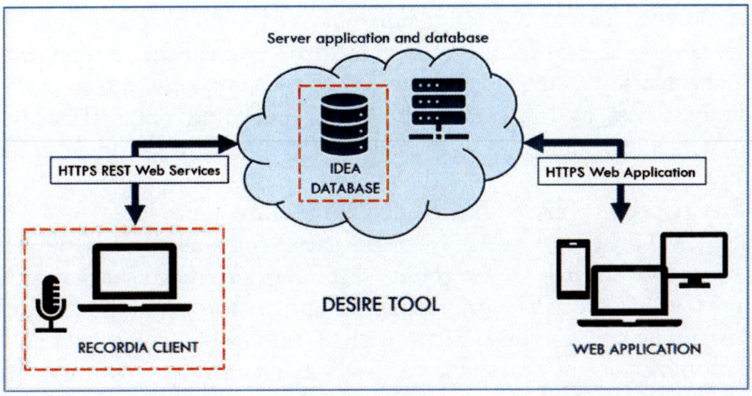

Fig. 1. The DESIRE tool architecture

and researcher/scientist. In the former case, the user has full access to the information of his/her followed patients, whereas, in the latter, only anonymized recordings and clinical data (with patient characterization) are provided. Additionally, also the patient (possibly with the help of a care giver) can access the DESIRE web application to perform personalized exercise sessions at home.

In order to facilitate the work of SLT, the application has been enriched with the real-time computation of features on the single recording both in the time domain (i.e. duration of the sound and syllables, via threshold-based algorithms) and in the frequency domain (i.e. spectral representation and average frequency, using Fast Fourier transform). Since these are indicators of dysarthria progression, for each patient, it is useful to compare them among different recordings of the same word and with respect to expected values. For this reason, both the RECORDIA client and the web application allow the user to compare and visualize their trends over time. Such indicators are first computed when recording the voice, and further sent to the REST service when uploading the entire session results. With this approach it is possible to keep RECORDIA temporarily independent from the presence of the internet connection, while providing full functionalities. Moreover, this architecture facilitates the sharing of patient's data among all the members of patient's care team.

Overall, for each recorded word the information stored comprehends the .wav file (44.1 KHz sample rate, 8 bit sample size), the indication of the target word and the statistics computed over it. As a result, the gathered data can expand the IDEA database, hence enabling the scientific community to further exploit such a data for example to improve ASRs systems accuracy for dysarthric people.

3 Discussion

In general, clinical practice is based on the definition and analysis of outcome measures derived from subjective different sources (patients self-reported, observer-reported (caregiver), clinician-reported) and objective performance-based parameters derived by use of standardized clinical scales or specific tools.

The identification of digital biomarkers offers the advantage to continuously and carefully assess at the same time in a smart and not intrusive manner, functional motor parameters also in real-life settings. These can represent more reliable and repeatable measure indexes, reducing errors derived from methodological bias related to more classic clinical measurements [12].

In fact, the DESIRE tool also has potential for speech therapy rehabilitation: it is possible to propose exercises, tests and specific tasks, tailored to the patient, on the basis of the performance found in the evaluation. Considering that the management of the patient with dysarthria is complex and requires careful customization, it is possible to insert some activities to work on specific tasks such as slowing down speech or reading balanced phoneme sentences with gradually progressing target phonemes according to the objectives, in favor of a better intelligibility, also reading of words and phrases with different intonation for a work on prosody, exercises on the voice (tonal pitch, intensity). The SLT can also monitor the patient remotely, favoring work in tele-medical mode (avoiding movements of complex patients) with real-time feedback to the patient who can thus try to find useful strategies to complete the exercise and be motivated, and the SLT can also maintain a history of the treatments carried out over time. It is useful to monitor the voice even remotely, where not in a therapeutic environment but at home, as the voice can vary many times throughout the day and it is possible for us to have important indications for the treatment (e.g. fatigue, times of the day where the patient is most tired, help from the caregiver, communication strategies adopted and so on).

Another important aspect will be the possibility to provide effective requirements for clinical trial readiness, which firstly need a deep understanding of disease mechanisms and evolution over time, allowing a correct identification and stratification of study populations to set targeted interventional clinical trials.

At present the tool have been successfully used by five medical doctors and STL at Pisa University Hospital with about thirty patients. In addition to the medical treatment benefits already mentioned, the tool was highly rated by both therapists and patients for ease of use, which is very important for daily examination in the hospital visit and compliance to at home exercise.

4 Conclusion

There is still little experience in identifying useful digital biomarkers for clinical practice to trace the evolution of speech disorders. DESIRE may challenge evidence in a gray area in which the phenotypic heterogeneity and still lack of unconfutable biomarkers makes difficult to choose what type, time interval, sequence and combination of diagnostic and therapeutic interventions can be appropriate. The additional advantage is to have an objective tool able to give clinicians objective measures of language production to be added to the subjective evaluations of the speech therapist, which may limit diagnostic and clinical errors. When used appropriately in combination, perceptual and objective measures allow for the correct identification of speech features that most severely affect naturalness and intelligibility, dysarthria type and severity, and therapeutic targets. They also enable clinicians to track change over time [13].

Our current limit is to evaluate only the articulation of words, but we set ourselves as a future goal the evaluation and monitoring of articulation in connected language (i.e. sentences and passage) in order to evaluate other parameters that can interfere in the intelligibility of speech (e.g. respiratory support, breath pauses, articulation speed rate, prosody, fatigue, and so on).

Acknowledgments. This work is supported by University of Pisa Research Project Program (PRA).

References

1. Enderby, P.: Disorders of communication: dysarthria. Neurological. Rehabilitation **110**, 273–281 (2013)
2. Finch, E., Ivanek, M., Wenke, R.: The who, why, when, where, what and how of using outcome measures in dysarthria: a qualitative exploration of speech-language pathologists' perspectives. Int. J. Speech-Lang. Pathol. **24**, 12–21 (2022)
3. Lee, J., Madhavan, A., Krajewski, E., Lingenfelter, S.: Assessment of dysarthria and dysphagia in patients with amyotrophic lateral sclerosis: review of the current evidence. Musc. Nerve **64**, 520–531 (2021)
4. Ishikawa, K., MacAuslan, J., Boyce, S.: Toward clinical application of landmark-based speech analysis: landmark expression in normal adult speech. J. Acoust. Soc. Am. **142**, 441–447 (2017)
5. Yorkston, K.: Management of Motor Speech Disorders in Children and Adults (2010)
6. Chiaramonte, R., Pavone, P., Vecchio, M.: Speech rehabilitation in dysarthria after stroke, a systematic review of the studies. Eur. J. Phys. Rehabil. Med. **56**, 547–562 (2020)
7. Marini, M., et al.: IDEA: an Italian dysarthric speech database. In: 2021 IEEE Spoken Language Technology Workshop (SLT), pp. 1086-1093 (2021)
8. Defazio, G., Guerrieri, M., Liuzzi, D., Gigante, A., Nicola, V.: Assessment of voice and speech symptoms in early Parkinson's disease by the Robertson dysarthria profile. Neurol. Sci. Off. J. Ital. Neurol. Soc. Ital. Soc. Clin. Neurophysiol. **37**, 443–449 (2016)
9. Duffy, J.: Motor Speech Disorders E-Book: Substrates, Differential Diagnosis, and Management. Elsevier Health Sciences (2019)
10. Piacentini, V., Zuin, A., Cattaneo, D., Schindler, A.: Reliability and validity of an instrument to measure quality of life in the dysarthric speaker. Folia Phoniatrica Et Logopaedica **63**, 289–295 (2011)
11. Enderby, P., John, A., Petheram, B.: Therapy Outcome Measures for Rehabilitation Professionals: Speech and Language Therapy, Physiotherapy. Occupational Therapy. Wiley, Hoboken (2006)
12. Jaddoh, A., Loizides, F., Rana, O.: Interaction between people with dysarthria and speech recognition systems: a review. Assist. Technol. **18**, 1–9 (2022)
13. Sevitz, J., Kiefer, B., Huber, J., Michelle, S.: Troche: obtaining objective clinical measures during telehealth evaluations of dysarthria. Am. J. Speech-Lang. Pathol. **30**, 503–516 (2021)

Author Index

R. Berta and A. De Gloria (Eds.): ApplePies 2022, LNEE 1036, pp. 387–389, 2023.
https://doi.org/10.1007/978-3-031-30333-3